Water

Sheng Meng · Enge Wang

Water

Basic Science

Sheng Meng
Institute of Physics
Chinese Academy of Sciences
Beijing, China

Enge Wang
Peking University
Beijing, China

ISBN 978-981-99-1540-8 ISBN 978-981-99-1541-5 (eBook)
https://doi.org/10.1007/978-981-99-1541-5

Jointly published with Peking University Press
The print edition is not for sale in China (Mainland). Customers from China (Mainland) please order the print book from: Peking University Press.
ISBN of the Co-Publisher's edition: 978-730-12-5145-4

This Springer imprint is published by the registered company Springer Nature Singapore Pte Ltd.
The registered company address is: 152 Beach Road, #21-01/04 Gateway East, Singapore 189721, Singapore

Contents

Chapter 1
Forewords

Water is the most abundant and important substance in nature, yet it is also the most studied but barely understood. Clean water resource is one of the greatest challenges for humanity in the twenty-first century, especially for the vast majority of developing countries, and it is particularly evident in China. On the other hand, water molecules are composed of hydrogen and oxygen, which can be decomposed into hydrogen and oxygen by absorbing enough energy. These gases in turn produce water by combustion, thus water can be used as a carrier and working medium of renewable and clean energy source. If this process can be easily implemented, it will undoubtedly make sustainable development of the energy industry possible. Basic research in water sciences plays a key role, both in water purification, where basic livelihoods are at stake, and in the acquisition and utilization of renewable energy sources as high-tech developments. The bottleneck that currently restricts these solutions from being put into large-scale applications is the lack of inexpensive and highly efficient materials and devices, and there is an urgent need to seek breakthroughs at the level of basic science, especially in the study of the interactions between water and material interfaces.

In 2005, Nobel laureate Richard E. Smalley listed the top ten challenges for mankind in the next 50 years, with "water" ranking second, just after "energy". Considering that energy generation in photosynthesis in nature is through the decomposition of water, and that water/semiconductor is an important system for artificially simulating the production of H_2 energy, water research is also an important direction for solving the energy problem. In addition, both food and the environment (third and fourth items respectively) are inseparable from the remediation of water pollution. Therefore, it is no exaggeration to say that the research on water (water decomposition mechanism and pollution treatment) is the core of solving the most important first four problems in this list, and is the key to building a harmonious society and running towards a better tomorrow for human beings.

The progressing of human society is greatly dependent on the forms of energy used and the exploitation of those energy sources. In recent years, energy consumption and demand for energy generated by human production and living activities

S. Meng and E. Wang, *Water*,
https://doi.org/10.1007/978-981-99-1541-5_1

have been increasing exponentially. In the traditional energy industry, the proportion of fossil energy sources (oil, natural gas and coal) used is as high as about 85%, which on the one hand poses increasingly serious environmental problems, such as the global greenhouse effect caused by carbon dioxide emissions and acid rain caused by sulphur dioxide emissions; on the other hand, the massive exploitation of fossil energy sources is constrained by their limited reserves, which are now in a trend of depletion. According to estimates of proven reserves and expected demand, the Earth's coal resources can sustain supply for another 75 years, oil resources for about 50 years, and natural gas resources for only 25 years. The energy crisis, and the closely related environmental issues, are the most serious challenges facing humanity today. Finding new and renewable clean energy sources, while continuously improving living conditions, is the biggest proposition that basic science urgently needs to address. Nature decomposes water and CO_2 through photosynthesis, which provides food and energy for organisms on Earth. Water can be used as a carrier and working medium for a renewable clean energy source, because water molecules absorb energy to decompose into hydrogen and oxygen, and hydrogen combustion generates water. This makes the sustainable development of the energy industry possible on the one hand and the protection of the environment on the other. Research has shown that the use of solar energy and with the help of suitable catalysts and device structures, hydrogen can be produced by electrolysis of water with high efficiency, which provides an inexhaustible new resource for the energy industry and is also the cleanest, most environmentally friendly and safest way to consume energy, thus promising as a coping strategy to solve the energy crisis.

The Earth's clean water resources are in danger of being depleted. The Earth's environment and water resources are under enormous pressure due to the rapid growth of the global population. Water pollution is one of the most important factors affecting environmental security, and it is an important component of national security and stability, along with military and economic security. Drinking water safety has become a major issue affecting human survival and health. A survey by the World Health Organization shows that 80% of human diseases are related to water pollution. Nature magazine reported in 2008 that the current global population of 2.6 billion people's drinking water lack adequate purification treatment; 1.2 billion people lack safe and clean water sources; more than 2.4 million people die from water pollution each year; an average of 3,900 children die from infectious diseases caused by drinking water every day. It is estimated that nearly half of the world's population will be highly water-stressed by 2030. Based on these facts, on 28 July 2010, the United Nations listed access to clean water and domestic water sanitation as a basic human right.

These problems are most prominent in developing countries. The water pollution caused by the direct discharge of domestic sewage and industrial pollution is becoming increasingly serious every year. With the increasing amount of industrial wastewater, urban and rural domestic sewage, pesticides and fertilizers, many drinking water sources are contaminated and the content of pollutants in the water seriously exceeds the standard. Due to the deterioration of water quality, it is difficult to guarantee the quality and sanitary conditions of drinking water for urban and rural residents who directly drink surface water and shallow groundwater.

As a large developing country on the rise, the problem of water pollution in China has seriously affected the development of industry and agriculture, the coordination of the ecological environment and people's quality of life and health. In China, the drinking water of about 360 million people in rural areas alone is unsafe, and about 190 million people have problems such as excessive fluoride and arsenic content, excessive iron and manganese content and brackish water in their drinking water. This has led to disease epidemics and, in some places, outbreaks of typhoid, paratyphoid and major infectious diseases such as cholera, and a high incidence of cancer in individual areas. According to surveys, 90% of the rivers flowing through cities in China are polluted and 75% of lakes are eutrophic. Among the 46 key cities, only 28.3% of urban drinking water sources have excellent water quality, 26.1% have good water quality and up to 45.6% have poor water quality. The direct consequence is that the drinking water of about 100 million people in China's cities can not fully meet the sanitary standards of drinking water. Water pollution accidents occur frequently, and the more serious ones in recent times include: tap water pollution in Yancheng, Jiangsu Province, on 20 February 2009; the hepatitis A incident caused by barrel-filled water in Guiyang on 31 March 2008; the cyanobacteria outbreak caused by contaminated water in Taihu Lake, Wuxi, on 29 May 2007; the mega pollution accident in Tuojiang River in March 2004, etc.

Along with China's accession to World Trade Organization and the globalization of trade, barriers and obstacles to the import and export trade of related products caused by new types of pollutants such as highly toxic and non-degradable organic pollutants are becoming more and more obvious. China's environmental protection has experienced the treatment of conventional air pollutants (such as SO_2, dust, etc.), conventional pollutants in water bodies and heavy metal pollution control, and is now developing towards the control and abatement of new highly toxic and non-degradable organic pollutants in water. Such new pollutants include persistent organic pollutants, environmental endocrine disruptors, personal care products, antibiotics and other drug residues, (multi) halogenated substances, etc., and conventional pollutants co-exist in water, when the formed composite pollutants present a new great challenge on conventional water treatment technology. The environmental hazards caused far exceed those of acid gases, heavy metals and general organic pollutants, which are not only carcinogenic, teratogenic and mutagenic, but also have environmental hormonal effects, directly threatening the survival and reproduction of wildlife and even humans. The elimination of these water pollutants requires the study of the microscopic mechanism of the surface interface between water and pollutants and various new catalytic materials from the basic water science problems, and then the development of various unit treatment methods based on different principles, and the creation and development of new water pollution treatment technologies. These studies will help to solve the increasingly serious water pollution problems in China and to achieve the safe recycling and utilization of water resources.

Currently, water cleaning and supply have formed a huge industry: it costs about $250 billion/year and is expected to cost $660 billion/year by 2020. Convenient and efficient water cleaning technologies for industrial and domestic use in large quantities are imminent.

On 19 May 2011, the G8 and the National Academies of Science of 13 countries, including India, Brazil and Mexico, issued a joint statement noting that water and water-related health problems, such as aquatic bacteria (cholera, diarrhea, etc.), the spread of water-related diseases (malaria, etc.), increasing organic contaminants in water caused by industry, agriculture and groundwater treatment, significantly affect people's economic activity and social development (reducing working life by 4%), as well as education (500 million school days lost each year due to aquatic diseases) and public health. The statement strongly calls on governments to strengthen the construction of basic water sanitation treatment systems, improve education levels, and fund research into low-cost, effective water treatment technologies and related disease prevention methods. The "No. 1 document" issued by the Chinese government on 29 January 2011, with the theme of "decision on accelerating water conservancy reform and development", was the first time that water conservancy reform systematically and comprehensively deployed in the new China, highlighting the current concern for water issues at the decision-making level. In 2006, China issued the National Medium- and Long-term Scientific and Technological Development Plan (2006–2020), which proposed to meet the challenges within 15 years and over-deploy major special projects, cutting-edge technologies and basic research, intending to build China into an innovative country by 2020. The 12th Five-Year Plan, which has just been formulated, also emphasizes attention to people's livelihood and leapfrogging development.

The currently known issues of water resources use and management cover macro areas of socio-economic-political-technical aspects, but have not yet penetrated into the basic issues of water science. This book focuses on basic research in water science. Basic research in water science is simply defined as: using fundamental knowledge of physics and chemistry, to develop new theoretical methods and advanced experimental equipment, to study the structure of water and the processes and laws of its interaction with other material surfaces at the atomic/molecular level, to reveal its microscopic mechanisms and to propose solutions to key industrial technologies, to trigger an industrial revolution in the rational use and effective management of water resources.

Therefore, actively layout and accelerate the development of basic research in water science in China is of great strategic significance to promote the construction of people's livelihood, guarantee national security, achieve the goals of the national medium- and long-term scientific and technological development plan, and improve the development level of China's economy, cutting-edge science and major projects. At present, China has accumulated a certain amount of basic research in water science and gathered many high-level research scholars, but a large number of studies are still limited to the observation and description of macroscopic phenomena, or limited to the publication of low-level papers, and still lacks systematic disciplinary planning and stable support, and lacks scientific objectives to guide the solution of practical problems, especially in the study of microscopic physicochemical mechanisms of contact between water and material surface interfaces.

Basic water science is a matter of national importance and faces enormous challenges. The current status of and challenges to the development of basic water science in China are summarized as follows.

1. Basic research in water science is of great significance, but there is a lack of sufficient attention at the national strategic level, research is still in a spontaneous state, the disciplinary system is incomplete, the systematic organization, guidance and support for basic water research activities are lacking, and there is a lack of awareness of the strategic status, depth and breadth of basic water science.

While people are saluting our economic miracle, the "soft constraint" of water scarcity is becoming increasingly evident. Our per capita water resources are only 28% of the world average. At present, the average annual water shortage in China is 50 billion cubic metres, and about 2/3 of the 600 cities have different forms of water shortage, with groundwater overdraft covering an area of 190,000 km^2. While China's economic construction continues to develop, it is important to do a good job in environmental protection to prevent water pollution, develop advanced and feasible drinking water source management technologies and improve the quality of drinking water to protect people's health and develop the economy.

To address the technological bottlenecks in the safe use of water resources (environmental and energy), there is an urgent need for basic research in water science, particularly on the fundamental forms and laws of interaction between water and surface interfaces. For example, can safe, non-toxic and efficient durable water purification materials be found to deal with the water quality crisis? Can inexpensive and efficient catalysts be designed to produce hydrogen or hydrogen compounds from light and water? The bottlenecks that are currently limiting the large-scale application of this solution are the materials and cost issues. Specifically, to identify suitable low-cost water cleaning materials, the optical energy conversion efficiency of semiconductor materials should be further improved and the cost of water cleaning and energy conversion devices should be reduced. This requires research including the interaction between the material surface and water, the microstructure of surface water, the infiltration, chemical activity, and stability of new materials, the mechanism of photoelectric decomposition of surface water, the absorption efficiency of photoelectric conversion materials to sunlight, the stability and efficiency of catalysts, the interaction of groups in water with the material surface, and the process of their adsorption and desorption on the surface. This series of problems involves many disciplines such as surface, chemistry, and materials, and breakthroughs need to be sought at the level of fundamental research.

Research on water issues (e.g., water decomposition and water cleanup) is a very complex project, with basic and applied basic research in water science accounting for about 10% of the overall water problem, but its impact radiates across all aspects of the water problem. We need to have a clear understanding of the current status and trends of domestic and international research, to conceptualize an overall framework for comprehensive research and utilization of water issues, and to establish a preliminary roadmap for the relationship between basic water science research and water development/utilization in new energy and environmental protection.

There is currently a lack of awareness of the strategic position of basic water science in China and a lack of understanding of the depth and breadth of the dimensions involved in basic water science. Water is the most important component of the environment, clean water is the basic guarantee of life and production, and water may also be the main way to solve energy problems in the future. Due to the limited availability of clean water resources and the sudden outbreak of nasty accidents such as water pollution, it has been said that (clean) water is more of a problem for developing countries. However, the limited attention and investment in basic water science issues in our country compared to developed Western countries is not consistent with the strategic role that water issues play in the dynamic development of our society and the position we occupy in the world. We must be conscious of this.

2. There is a lack of unified planning for the development of basic theories and experimental technologies in water science at the national strategic level, and there is even a one-sided substitution of water research development planning for water basic science development planning.

At present, there is no unified planning and guidance for basic theoretical research and experimental technology development in water science in China, and scientific research in a strategic position, such as basic water science, is still in a laissez-faire state. This situation may not be harmful to general basic science problems and scientific development for the time being, but it will have a negative impact and cause considerable damage to the solution of basic science problems such as water, which is related to the people's livelihood of the country and is in an urgent state.

In fact, through the survey, we found that compared with the United States, Europe, Japan, Korea and other western developed countries and some important neighboring countries, our researchers are obviously not concerned enough about the basic science of water issues, and the research strength is very weak. The survey shows that from 2001 to 2010, the number of international papers on basic water science research showed an overall growth trend, with an average annual growth rate of 4.3%. The top countries in terms of the number of papers published on basic water science research from 2001 to 2010 were the United States, Germany, Japan, the United Kingdom and China. Among them, the United States in the field of basic water science research results accounted for about 38.9% of the total number of research results, equivalent to 3.7 times the No. 2 Germany. China ranks 6th in the number of papers, accounting for only 7% of the total. The National Science Foundation (NSF) has been continuously funding basic research on water, funding about 50 researchers each year to explore the fundamental properties of water. 2004, Science magazine selected the research results in water structure and water chemistry as the top ten breakthrough research progress, of which all but one is the research results of Stockholm University in Sweden, the rest are in the United States completed.

One example is the scanning tunneling microscopy (STM) technique invented in the early 1980s, which is being used extensively to investigate microstructures of water layers at the molecular level on surfaces, especially on metal surfaces; almost all of the breakthroughs in interfacial water structure in the last two years (completed in Japan, the US and the UK respectively) are the result of research using this technique. In contrast, although China has purchased a large number of commercial scanning tunneling microscopes and even built its own scanning tunneling microscope equipment with international leading resolution capabilities and special features, these devices are often used to explore more traditional surface physics, basic chemical research, but rarely used to explore water, which seems to be a commonplace, unobtrusive, simple but important, practical system.

On the other hand, due to (1) the destructive nature of experimental probing techniques, (2) the lack of surface probing techniques at solid–liquid-gas interfaces, and (3) the special "fragility" (the strength of hydrogen bonds connecting water is 1/10 to 1/5 that of normal chemical bonds) and complexity of the water molecular system (more than 15 bulk phases and an uncounted number of "nanophase", all these require theoretical studies and simulations of the behavior of water at the molecular scale, especially using modern powerful computational simulation techniques and knowledge of quantum mechanics, which together with related experimental studies can solve major scientific problems for mankind, such as the production of hydrogen by photodegradation of water and the treatment of water pollution. There are several problems in the current theoretical and simulation studies of water: (1) lack of simple and effective models of water; (2) lack of first-principles approaches for large-scale simulations; (3) lack of accurate description of the quantum behavior of H in water; (4) lack of description of the properties and kinetic processes of the excited states of water. These fundamental aspects of important research problems have not been planned at the national level in our country, much less carried out practically.

What is more worrying is the current tendency to simply replace basic water research and development planning with "water resources" research and development planning, which can cause great confusion and greater harm. Water resources research and development is a macro-scale water science research, biased towards engineering regulation and engineering applications, often does not involve the structure and nature of the water molecule itself, does not involve the mechanism of water and material interaction; or water resources research is the use of human scientific knowledge about water itself to observe the distribution, change and deployment of water from a macroscopic perspective, to provide services for people's production and life activities. But where does the basic knowledge of water science needed for water resources research come from? From basic water science research, of course. Basic water science research is different from water resources development research, which is a science that provides an important and necessary scientific basis for the latter.

The development of water resources research often requires breakthroughs at the level of basic water science. For example, the use and conservation of water resources require water purification treatment, and existing methods are often inadequate to meet the challenges posed by current large-scale industrial production and water depletion. To develop new means and methods of water purification, it is necessary to carry out basic water science research, to understand the mechanism and change rules of interaction between water and external substances, the surrounding environment at the molecular scale, and to design and develop new materials for the decomposition and purification of water. On the other hand, only when there is a breakthrough in basic water science research, it is possible to innovate the existing water resources water engineering treatment technology and invent new methods of polluted water treatment to meet the challenges posed by water resources depletion and water pollution. It should be recognized that engineering research on water, in general, does not lead to these fundamental breakthroughs and technological revolutions.

As far as we know, a number of research centres related to water issues have been established in various universities and research institutes in China, such as the Water Resources Centre of Peking University, the Water Science Research Centre of Zhengzhou University and the Hangzhou Water Treatment Technology Research and Development Centre. The main strength of these research centers is still placed on the development, utilization and protection of water resources. At present, specialized research institutions or platforms engaged in basic water science research in China need to be strengthened. We call on the state and relevant departments to pay attention to the research on basic water science issues, unify planning and guidance, strengthen the professional research force, and develop a development strategy for basic water science research.

3. In basic water science research, there is a lack of long-term goals for the development of new materials for water pollution treatment and water decomposition.

Most of the basic research activities in water sciences that have been carried out in China are in a spontaneous state, lacking a clear direction and application goals. In general, the goal of basic science development lies either in understanding the mysteries of nature or in laying the foundation for the development of potential new applications and technologies. We believe that the development of basic research in water science in our country is aimed more at the latter, especially in the development of new materials for water treatment based on the understanding of the interaction of water and matter, to make important contributions to water pollution control, clean water treatment, anti-icing, etc., which are important livelihood issues.

Materials are the cornerstone of human social progress. The invention and widespread use of a key new material is a sign that mankind has entered a new phase of social development: such as the Stone Age, the Bronze Age, the Iron Age, and the Electronic Age. Because of the important role and widespread impact of water on daily life, production, and environmental protection, water treatment materials are also one of the key symbols of human development and have a critical impact on the development of human society. For example, it has been suggested that the use of copper pipes, which can produce toxins from long-term contact with water, as drinking water pipes, thus causing chronic poisoning and a gradual decline in health, may have been an important reason for the decline and fall of the ancient Roman Empire. In the current era, the increasing depletion of clean water resources, the massive use of large industrial production and electronic devices causing increasing environmental pollution, fossil energy sources leading to possible climate change, these problems are more serious in the majority of developing countries, including China, all these put forward higher requirements and more urgent challenges to the development of new water treatment materials to combat environmental pollution and develop new energy sources.

At present, there is no unified planning and long-term objectives for carrying out basic water science research and developing new materials for water treatment, and the existing sporadic research lacks clear objectives to guide it. The ultimate goal of carrying out basic water science research is to serve the development of our society and science, and to solve the environmental and energy challenges facing our country in the development process. The development of new materials for water treatment for these major applications and the formulation of long-term goal planning and development tasks for each step are the current imperative.

4. There is a lack of experimental apparatus and means to study in-depth the physical mechanisms of water/material interface reactions at the molecular or even atomic level.

One of the most important aspects of basic water science research is to study the interaction between water molecules and other substances, and the interaction between water and the outside world is realized through the interface, which requires the study of some interfacial water properties and the interfacial properties of water itself (collectively referred to as "interfacial water" properties). A deeper understanding of these interactions requires microscopic, especially at the atomic-molecular level, to explore the microstructure and charge distribution, transfer, and other laws of interfacial water. This is a complex and multidisciplinary frontier science. In recent years, due to the success and wide application of density function theory (DFT), some progress has been made in the theoretical treatment of interfacial water microscopic mechanisms and interactions relatively; however, more accurate treatment of weak interactions such as hydrogen bonding and van der Waals forces, the establishment of models describing the structure of real water layers (liquid, several to tens of nanometers thick) at (solid, liquid) interfaces, and the full quantum treatment are still to be solved. Experimentally, interfacial water is difficult to study, and there is a great lack of simple methods to probe and characterize the structure and properties of interfacial water at the atomic-molecular level, and because the interactions between water and surfaces and water molecules are weaker than normal chemical bonds (a fraction of the strength of normal chemical bonds), the interfacial structure is easily destroyed in experimental probes. There is a need to vigorously develop interface-sensitive, non-destructive experimental methods and tools, such as nonlinear optical methods such as sum-frequency vibrational spectroscopy, and new scanning probe techniques.

5. There is a lack of integrated planning for research and industry at the national level. On the one hand, there is a one-sided pursuit of low-level applications and a lack of breakthroughs in existing ideas, systems, and methods from the perspective of basic research; on the other hand, there is a tendency for some basic research to pursue the publication of low-level papers, which is not yet organically integrated with practical objectives.

Due to the current lack of awareness of the strategic position of basic water science in China, the lack of planning for basic research activities, and the lack of target guidance and key technologies for the development of new materials for water treatment, there is no unified planning and management at the national level for a series of research activities and related industrial development from basic water science research, to applied research development, to macro water resources governance, etc. There is a serious disconnect between the various links, rather than forming a smooth and effective communication and mutual promotion, affecting the healthy development of the whole water science. This is reflected in the following.

(1) Industrial applications, the one-sided pursuit of low-level macro applications, the lack of breakthroughs in existing ideas, systems, and methods from the perspective of basic research. Although we have a large number of water resources enterprises and water pollution control materials, products and methods, the sources are often based on backward methods and concepts, operating with old knowledge and macro experience, the microstructure of the material, and the mechanism of action is not well understood; Individual enterprises tend to import the so-called "advanced" equipment and technology from abroad, but lack the determination and strength of independent research and development and independent transformation. These low-level macro applications can not meet the current environmental and energy challenges, macroscopic engineering development and utilization can never bring a breakthrough in new methods of water treatment, nor can benefit from the current rapid technological development. In order to change this unfavorable situation, we must lay out the basic water science, study the mechanism of the interaction between water, impurities in water, and various natural and artificial materials, study the structure and dynamics of the interface, so as to obtain the breakthrough of new materials and new methods of water treatment (water pollution control, water decomposition), so that the existing practical engineering application technology continues to develop substantially forward.
(2) Some existing basic water science research still tends to pursue the publication of low-level papers, and has not yet formed a virtuous circle with practical applications. Due to the lack of appropriate planning and guidance, the current scattered basic water research is in a spontaneous state, many researchers just repeat some experimental results and theoretical models at a low level, in order to publish some low-level papers, rather than to solve one or some scientific and practical problems in the application. Scientists are often isolated from practical application problems in life and production, making such research lack the necessary roots and vitality, resulting in a waste of human and material resources. And in fact, it is not that there are no problems in real production applications, but that there are a large number of unresolved and difficult problems and issues. There is no suitable channel for these problems to be transmitted from the actual production activities to the scientists engaged in basic research at the front line. This lack of connection between basic research and practical problems, difficult to communicate is more serious in our country. This is partly due to the lack

of tradition and strength of basic research in our traditional water treatment enterprises, on the other hand, mainly because we have not yet established a complete system from basic research in water science to the practical application in industry, thus failing to form a virtuous cycle from basic research to practical application.

6. There is a serious shortage of cross-disciplinary talents, and the training and reserve of future talents engaged in basic research in water sciences are insufficient.

Water science is a comprehensive discipline that involves many disciplines such as physics, chemistry, materials science, biology, and engineering. Due to the progress of science and technology and the complexity of water environment pollution, researchers engaged in water science research need to have basic knowledge of physics, chemistry, biology, materials science, engineering, and other aspects in order to carry out water science-related research work well, which has put forward high requirements for the training of personnel engaged in water science research.

At present, most of the personnel engaged in water science research at home and abroad have a background in water supply and drainage, environmental engineering, or other related engineering disciplines. The training of talents in these disciplines is mostly oriented to solving practical problems in water pollution control engineering, with a preference for practical knowledge learning and engineering applications, but there is a large lack of basic knowledge in the disciplines required for water science research, and a lack of knowledge background in understanding, analyzing and solving basic problems in water science, especially a relative lack of knowledge in basic disciplines such as physics and chemistry. As a result, it is difficult to conduct high-level research in water science by relying on the talents trained in the existing models and disciplines.

Professor Werner Stumm, former head of the Department of Chemistry at Harvard University, recognized this problem in the 1970s and acted on it, relying on his profound knowledge of colloid and coordination chemistry to carry out many years of basic research in water chemistry, initially establishing the theoretical system of water chemistry, and the Swiss Federal Institute of Aquatic Science and Technology (EAWAG) lead by him, has been built into a leading international research institute in basic water science. His book "Aqueous Chemistry" has become a classic work in the field of water science and has made outstanding contributions to the research and development of water science in the world. Therefore, considering the demand for talents in water science research and the current situation of disciplinary settings in China, we should learn from the advanced experience of Professor Stumm and EAWAG, vigorously strengthen the water basic cross-science education system, and cultivate high-level talents who will be engaged in water science research in the future.

In the light of the above, there is an urgent need to develop a national development strategy for basic water science research in China. The currently known problems of water resource utilization and management involve macro areas of social-economic-political-technological aspects, but have not yet penetrated into the basic problems of water science. In order to accelerate the development of basic water science research in China, it is recommended to vigorously develop theoretical and experimental methods for basic water science research, and establish a national platform for basic water science research centered on exploring the basic physical and chemical properties of water and its applications in the environment and energy, cultivate cross-cutting talents, and actively plan and organize the development of basic water science in China. The specific proposals are as follows.

1. Plan the overall development strategy and objectives of basic water sciences in China and formulate medium- and long-term development plans according to the overall development situation and needs of the country.

Water pollution, water shortage, and other water resources problems are alarming, and their solution requires a concerted socio-economic-political-scientific effort. Because water problems and the environmental and energy problems brought by them are global and are not limited to one place at a time, and are prominent in developing countries and populous countries like China, we suggest that we should fully recognize the significant role of basic research in water science in solving global water resources and environmental problems and the strategic position of basic water science, and set up major national research projects as soon as possible, and establish a national research platform with basic research in water science and applied basic research in environment and energy as the core.

The platform should have the central task of organizing and coordinating the national forces of basic water science research and unifying the planning, organizing, and leading the development of basic water science issues in China. It focus on promoting the attention of all sectors of society, especially the environmental resources sector, industry and the scientific community, to basic research in water science; establishing channels of communication and mutual stimulation in a series of links from basic research, materials development and research, to industrial production applications, and the national water resources and water environment management and improvement; strengthening the training of interdisciplinary talents with a certain depth and breadth of knowledge to meet the needs of basic water science research in the future; planning the overall development strategy and objectives of China's basic water science, and formulating medium- and long-term development plans according to the overall development situation and needs of the country.

2. Select several key issues in basic water science research (e.g., water/interface interactions and structure at the molecular level, elimination of pollutants in water, interfacial nano-water film research, etc.), identify major directions, and strive for major breakthroughs.

Select some key basic water science problems and concentrate on tackling them, to make major breakthroughs in basic science problems that are closely linked to major applications and key technologies. For example, research should be strengthened on the following key issues.

(i) The microstructure and dynamic behavior of water on material surface/interface. Both photodegradation and purification of water are achieved and controlled through surface interfaces. An important aspect of understanding water is to understand the interaction patterns and relationships between water and external materials represented by solid surfaces. Solid surfaces are also a model system for studying the interaction of water and other complex systems such as water and proteins. We need to study the adsorption configuration of water molecules and water clusters on solid surfaces, the interactions between water molecules and with solid surfaces in confined systems, the molecular structure and molecular orientation of interfacial (liquid) water layers of tens of layer thicknesses and specific photoelectric properties of the water layer, the distribution of ions at interfaces and the effect on the water structure, the dissolution of solid substances in water, the solidification and melting of water, etc. The exploration can give a clear picture of various physical and chemical processes of water at the surface/interface of the material. This is the basis for studying the structure of water and its interaction with external matter.

(ii) Photo-elimination of contaminants in water. After various traditional pollutants and new highly toxic and hard-to-degrade pollutants comprise composite pollutants enter natural water bodies, some of them will degrade naturally in the presence of light, oxygen, solid particulate matter, microorganisms, etc. Whether chemical, photo- or bio-oxidation is catalytically enhanced by micro-interfacial processes between the composite pollutant and the particulate matter and natural organic matter in the water bodies. Micro-interfaces tend to accelerate electron transfer processes and generate more oxidation radicals, leading to different interfacial chemistry and reactions. Natural Fe–Mn oxide particles and sediments play a large role in the redox degradation of organic matter. Light is one of the important factors controlling the form of iron species present and light-initiated electron transfer can directly lead to the reduction of trivalent iron (Fe(III)) to divalent iron (Fe(II)), while oxidizing species in the environment such as Cr(VI), O_2/H_2O_2, $O_2^{\cdot -}$, ·OH, etc. can convert Fe(II) to Fe(III). Through this cyclic process, Fe species in various states contribute to the valence changes of other elements in the environment, the degradation of organic pollutants, and the production and consumption of reactive oxygen species. Only a systematic and in-depth study of the mechanism of the interaction process between complex pollutants and natural or artificial material micro-interfaces is expected to effectively transform the complex pollutants in natural water bodies. The new highly toxic and hard-to-degrade organic pollutants have three distinctive features compared with the traditional pollutants in water: low concentration, generally 3–9 orders of magnitude lower than the concentration of natural dissolved substances or nutrients in the water, namely the so-called background substances; high toxicity, with unique lipid solubility or bioconcentration, which can lead to acute and persistent toxicity; hard to degrade, with half-lives of decades or even centuries, and some pollutants such as dioxins have almost no biological enzymes corresponding to decomposition in nature. The three characteristics determine the environmental effects and the safe elimination of such new pollutants has become the most urgent hot research topics need to address in international water treatment today. At the same time, these new pollutants are difficult to be removed by traditional physical and chemical methods (such as sedimentation, adsorption, reverse osmosis, wet oxidation, etc.), so it is urgent to propose or establish new principles and methods for efficient and selective elimination. One of the most concerned and promising methods in the world to eliminate these new pollutants is to use environmentally friendly oxidants (O_2 or H_2O_2) in the presence of catalysts (e.g. TiO_2, transition metals, etc.) and advanced oxidation technology under light irradiation, which has the characteristics of mild conditions, complete oxidation, no secondary pollution, and environmental friendliness, etc. How to effectively activate O_2/H_2O_2 to achieve efficient selective elimination of such pollutants is an important basic research direction in the field of water science.

(iii) Interfacial nano-water film studies. A nanoscale thick water film covering a solid surface in the atmospheric environment possesses unique physical/chemical properties due to specific interactions between the solid surface and water molecules and between water molecules. For example, Hu Jun and Salmeron et al. discovered a layered water- "room temperature ice" structure on the surface of mica. Recent studies have shown that the self-assembly process of peptides in such solid surface nano-water films proceeds differently from the self-assembly at the macroscopic water/solid interface, reflecting the uniqueness of nano-water films. In the air, when the scanning probe is in contact with the solid surface, water condenses between the two forming a nano- "water bridge". This nano "water bridge" can be used to transport chemical and biological molecules adhering to the probe to the surface, and because of the precise and controlled position of the probe on the surface, nanopatterns with specific chemical and biological functions can be formed. Such nano "water bridges" can also be used for nanomodification of surfaces by electrochemical reactions. The characterization and manipulation of interfacial water present on a solid surface will provide fundamental information about the nano-water film or water droplet and is a very important approach to our understanding of the fundamental nature of water.

(iv) Energy conversion processes and new materials for photocatalytic decomposition of water. The exploration of new catalyst materials for photodissociation of water and the accurate understanding of the microscopic mechanism of water molecule decomposition are among the key issues in basic water science. We need to compare two methods of photocatalytic direct water decomposition and photocatalytic water electrolysis to elucidate the light-energy-chemical energy conversion mechanism in each process respectively. We also use time-resolved optical absorption spectroscopy, electronic spectroscopy, and precise electrical measurements, combined with theoretical analysis and simulation, to study the mechanism of light-energy-chemical energy conversion and explore ways to improve the energy conversion efficiency. To elucidate the influence of the electronic energy band structure of catalyst materials on the decomposition process of water molecules, and the influence of different materials on the processes of water decomposition, hydrogen/oxygen generation, and desorption, to find efficient and stable catalyst and electrode materials.

3. Significantly strengthen the development of important challenging core technologies and develop new technologies for the study of the basic physicochemical properties of water.

For three decades there has been an increasing understanding of water—especially at the interfacial water molecular scale in detail—the most common and complex fundamental material system and its physicochemical properties through experimental and theoretical studies. However, due to the special characteristics of water such as transparency, softness, and weakness, the existing experimental means often have some inherent drawbacks: (1) experimental detection techniques (ion beam, electron beam, current, etc.) are destructive to the water structure being probed, (2) lack of detection techniques sensitive only to the interface, (3) experimental means have limited spatial separability and temporal resolution, etc. In particular, due to the special "fragility" of the water molecule system (the strength of hydrogen bonds connecting water is 1/10 to 1/5 of normal chemical bonds) and complexity (water has more than 15 bulk phases and an uncountable number of "nanophases"), the fundamental issues of water and surface interaction remain very challenging using existing experimental techniques. This reality has led to, as the well-known German scientist D. Menzel put it: any discovery or insight into water is highly controversial and ambiguous.

It is urgent to study the fundamental problems of water science and to vigorously develop core and new technologies with important challenges. We believe that existing techniques for interfacial water sum-frequency vibrational spectroscopy, high-resolution surface energy spectroscopy, and advanced accurate first-principles methods for large-scale calculations should be further developed; and that emphasis should be placed on the development of a new generation of scanning probe techniques and ultrafast photoexcitation probes in complex environments (room temperature, non-vacuum, etc.) to study the physicochemical details of interfacial water structure and decomposition dynamics.

In recent years, sum-frequency vibration spectroscopy (SFVS) has been widely used to study the interface structure of various kinds of water. SFVS is a nonlinear laser spectroscopy technique with high surface resolution developed by Yuan-Yang Shen's group at the University of California, Berkeley, more than two decades ago. So far, it is the only experimental technique that can provide a realistic surface (interfacial) vibrational spectra of water. SFVS vibrational spectroscopy allows one to obtain structural information of the surface (interface) at the molecular level. In particular, it can be applied to complex in situ environments, including interfaces between various gases, liquids, solids, and water, without demanding requirements on temperature, pressure, etc. In combination with other complementary techniques and theoretical calculations, SFVS can be used to study many important fundamental problems in water interactions. Synchrotron radiation is an excellent performance light source, and advanced synchrotron radiation techniques (e.g., angle-resolved photoelectron spectroscopy (ARPES)) are the preferred experimental tools of choice for studying a variety of materials and systems as well as novel quantum phenomena. X-ray absorption spectroscopy, X-ray emission spectroscopy, and X-ray imaging are the primary techniques for studying the microstructure of water (including surface, interface, and bulk phases) from the atomic and electronic levels, and are the primary techniques for investigating the normal state and extreme conditions of the hydrogen bonding structure of water and its dynamics, and are beginning to play an irreplaceable and important role in the field of microscopic fundamental research in water science such as confined systems, interfacial imaging, nanomaterials, and nano biological composite systems. Meanwhile, the development of soft X-ray nanoprobe CT methods, water windowing techniques, and other related technology platforms will all provide the most innovative approaches to the study of water science. Scanning probe microscopy (SPM) has its irreplaceable application in basic water science research and is the "eyes" and "hands" of scientific researchers to observe and modify the microscopic world. However, traditional SPM technology is difficult to achieve reliable nano/sub-nano resolution in the field of water interface research, and its resolution capability and application scope are greatly limited. To address the major scientific needs in water interface research, there is a need to develop a high-resolution non-contact frequency modulation (FM) atomic force microscope (AFM) system and a cryogenic scanning tunneling microscope with high stability, high resolution, and stable inelastic tunneling spectroscopy. Femtosecond laser spectroscopy with high temporal resolution is a powerful tool for studying light-matter interactions and the various physicochemical processes induced by them, while SPM with high spatial resolution can study the surface electronic states of samples at the single-atom scale. The combination of these two advanced techniques allows an in-depth study of the mechanism of the adsorption conformation, light absorption, photocatalytic chemical reactions, and light-induced processes such as surface desorption and electronic state jumping of individual water molecules and H^+/OH^-/H and O atoms/H_2 and O_2 molecules, etc.

On the theory side, the introduction of the universal empirical potential into density functional theory could be tried out, along with the development of methods for a more precise discussion of van der Waals interactions and hydrogen-bonding interactions based on first principles, as well as the development of full-quantum Monte Carlo simulation calculations. These new methods will allow a more precise treatment of hydrogen-bonding interactions between water molecules and interactions between water and solid surfaces, and will also be applied to treat quantum vibrational effects of light elements (especially H), leading to detailed studies of the quantum properties and laws of hydrogen in the interaction of water and solid materials. In addition, we need to vigorously develop a first-principles approach to study the electronic structure and energy states of interfacial water, especially the electronic structure and dynamics of surface excited states. So far, the theoretical study of surface excited states is very difficult. Research in this area is very necessary and urgent, since it is essential for understanding the mechanism of photochemical dynamics of surfaces.

It should be noted that experimental detection techniques for studying the microstructure and properties of water at the solid–liquid–gas interface are still very lacking. In addition to the vigorous development of the above-mentioned techniques for their application to the study of interfacial water, there is still a great need for the vigorous development of new technical tools such as ambient pressure SPM and optically coupled SPM, and more accurate and effective theoretical models, to contribute to the understanding of the structure and properties of actual interfacial water systems at the microscopic scale.

4. Vigorously promote the integration of basic water basic research with clean water and clean energy industrial application projects, strengthen collaboration with other related disciplines (e.g. climate, geography, energy, nanotechnology, etc.), and unite various research groups in basic and applied water sciences. Gradually establish a unified system from basic research to practical application to promote innovation and sustainable development in water science and technology in China.

Nowadays, the traditional engineering application methods of water pollution control and clean water treatment, which are limited to macro-management and macro regulation of water resources, no longer meet the needs of production and life in the new century. Current methods tend to focus on the use and deployment of large systems, but require a correspondingly large amount of money, engineering, and facilities, the cost is too high; while the use of chemical methods and other simpler methods, is particularly time-consuming and ineffective. Therefore, we need to vigorously promote the integration of basic water research with clean water and clean energy industrial applications projects, and to strengthen multidisciplinary collaboration on water research, for example, with climate, geography, energy, nanotechnology, and other related disciplines.

Current major application problems such as water pollution treatment and water catalytic decomposition face new challenges and opportunities: that is new nanotechnology based on water interface interactions. Bai Chunli of the Chinese Academy of Sciences proposed that "the potential of nanotechnology to solve environmental pollution problems should be fully exploited". For example, many systems of photocatalytic materials already exist, many semiconductors with efficient absorption of visible light are easily corroded under photocatalytic conditions, and there are very few stable photocatalytic materials with high catalytic efficiency under solar illumination. It is imperative to discover and produce a new generation of efficient and stable visible light nano catalytic materials by designing the energy band structure of nanomaterials and controlling the electron transport on the surface of nanomaterials.

Water issues involve all aspects of human life and social production, with different requirements at different levels and different perspectives of consideration. We should acknowledge that as the level of understanding evolves, there are now research teams and production groups in a range of areas, from basic to applied research, from industrial engineering production to integrated water resources management, involving various disciplines such as physics, chemistry, biology, geography, health, environment, and medicine. We propose to integrate all aspects of these water issues, unite research groups from basic to applied and across multiple disciplines, and gradually establish a unified system from basic to practical applications and even industrial production, to promote mutual exchange and a virtuous cycle among basic research, industrial applications, and water resources governance. To promote innovation and sustainable development in water science and technology in China, and to provide solid professional strength to effectively address the challenges of water issues in the twenty-first century.

5. Vigorously cultivate multidisciplinary and cross-field talents in basic water sciences.

Water sciences research is a long-term and arduous task that must be comprehensively considered and deployed at the national level. Plans for the training of future talent in water sciences research need to be developed and implemented immediately. The main tasks in this area are the cultivation of talents in various fields such as theoretical model building and simulation calculation, surface physical analysis, material preparation and characterization, optical/electronic measurement and technology, chemistry, and environmental science, including the introduction of high-level talents, the construction of talent echelons at all levels, and the condensation of teams to undertake large-scale research tasks. The aim is to cultivate a multidisciplinary and multi-level talent team including top leaders, to promote highly oriented basic research in water science, and to promote the development of water environment engineering and the new energy industry in China.

Considering the current situation of China's water science research talent needs and discipline settings, it is recommended that a few universities with strong science and engineering disciplines be selected to set up pilot projects to train high-level talents engaged in water science research in the future. These schools have a strong research base and scientific research strength in the fields of physics, chemistry, and other disciplines. From the students entering these universities majoring in physics, chemistry, biology, and other basic disciplines select outstanding students who are interested in water science research, face the international frontier of water science, scientifically set up a high standard of undergraduate—master—doctorate through the training program of excellence, strengthen the teaching and research of frontier knowledge in water science, strengthen the training of independent innovation and scientific research ability in water science.

(i) Theoretical teaching of broad physical and chemical foundation cultivation. The training of students in the pilot class will face the frontier of water science development, aiming at the international first-class level, emphasizing foresight, advancement, and practicability. Theoretical teaching link will be through the introduction of small class courses, on the basis of water science core courses teaching, strengthen the core courses in physics, chemistry, and mathematics disciplines, and employ masters and scholars in water science research at home and abroad to teach cutting-edge courses. On the basis of strengthening curriculum experiments, the practical teaching link will complete a series of open experimental projects and undergraduate research projects, focusing on strengthening the practical skills cultivating and research quality training, so that students basically have the basic quality of independent water science research after the completion of a complete water science experimental projects.

(ii) Practical teaching for high scientific research quality and innovative talent development. The pilot class implements the student mentor system and internship system, through the basic research in water science in relevant institutes of the Chinese Academy of Sciences and practical research in water science in the research institutions of famous water companies (such as Veolia Water, Suez Water, Thames Water, Dow Group, etc.), early exposure to scientific research, exposure to excellent research and development teams and personnel, this will help students to understand as early as possible the significance and role of basic and applied water science research in the development of new technologies and principles of water pollution control, strengthen their understanding of basic and applied water science research, help them to understand the importance of applying their learning to their studies as early as possible, and give them a stronger motivation for academic growth during their study phase, in order to cultivate innovative talents in the field of water science.

(iii) Undergraduate education-based undergraduate-master-doctoral water science research excellence training system. Through the education of the undergraduate pilot class, students have been equipped with good quality to engage in water science research. In the postgraduate training stage, emphasis is placed on the improvement of students' basic scientific research ability and the cultivation of scientific research elite potential. Students are guided and encouraged to target the frontiers of major scientific problems in water science, select their topics and carry out scientific research in water science independently; among the various departments of universities and the different institutes of the Chinese Academy of Sciences, we try to adopt a multi-mentor system, so that students' minds are guided by different tutors and their ideas are subject to the collision of different academic ideas and disciplinary knowledge; we encourage postgraduates to go to international laboratories that are excellent in water science research for short-term visiting studies to expand their academic horizon; during the study period, students are guided to apply for, organize and hold small international symposiums on water science research, which can not only build a relevant academic network among young students engaged in water science research, but also exercise students' organizational skills and leadership ability in undertaking water science research projects.

In conclusion, basic research in water science is to use the basic knowledge of physics and chemistry, develop new theoretical methods and advanced experimental equipment, study the processes and laws of water and its interaction with materials at the atomic/molecular level, and reveal its microscopic mechanisms and propose solutions to key industrial technology problems, to trigger an industrial revolution in the rational use and effective management of water resources. The active layout and accelerated development of basic research in water science. Actively distributing and accelerating the development of basic water science research in China is of great strategic significance in promoting the construction of people's livelihood, safeguarding national security, achieving the goals of the national medium- and long-term scientific and technological development plan, and improving the development level of our economy, cutting-edge science and major engineering.

Chapter 2
Understanding the Structure and Function of Water at the Molecular Scale

2.1 Magical Watery World

If you had to choose "what is the most important substance in nature", I'm afraid most people's answer would be: water! This is probably one of the beliefs that have remained constant throughout the history of mankind. We have to ask: What is it about water and why is it so magical?

Water is the most abundant substance in nature. Our planet, on its surface, is actually a big "water" ball. Rivers, lakes, and oceans, and all around it, there is water everywhere. It is therefore also associated with the most common types of natural phenomena. For example, water in the soil is associated with rock weathering and permafrost; water in living things, where all kinds of life processes, like ion transport and protein folding, take place; water in life, where iron rusts and corrodes; water in the atmosphere, where it forms clouds and contributes to lightning and rainfall; and even water in outer space, where it is a major part of the nucleus of comets and some stars. Because of its abundance, water has become a readily available, inexhaustible, and inexhaustible resource.

Water is also one of the most important substances in nature. It is the medium through which life exists. DNA replication, protein synthesis and folding, cell membrane formation and fractionation, and nerve signal transmission are all dependent on the action of water molecules. The Earth's temperature regulation, acid–base balance, and environmental pollution control can not be achieved without the credit of water. Not only that, it is also a major object of research in the fields of catalysis, solutions, electrochemistry, sensors, meteorology, etc. In fact, catalysis was discovered in 1823 when Döbereiner observed the formation of water from oxygen and hydrogen on a platinum surface at room temperature [1]; later, Berzelius named this phenomenon "catalysis" [2].

That is why since ancient times people have been filled with a keen interest in and rich affection for water. The birth of human civilization in ancient times, and even the emergence of great figures in our time, could not have been possible without the

S. Meng and E. Wang, *Water*,
https://doi.org/10.1007/978-981-99-1541-5_2

nurturing and nourishment of the great waters. Deep-minded philosophers contemplate its spirit. Water, wood, gold, fire, and earth were considered by the ancient Chinese to be the five most basic elements that make up the world, and were called the “five elements”. The ancient Greek philosopher Thales claimed that water is the origin of the world. In the Tao Te Ching, Lao Tzu praised the spirit of water: “The best kindness is like water. Water is good at benefiting all things but has stillness, and dwells in the place hated by all people, so it is almost like the Tao.” Confucius looked at water and lamented that “those who pass away are like that”, and said that “water has five virtues”, so “when a gentleman meets water, he must look at it”. Artists have spared no effort to express the charm of water. Musicians have high mountains and flowing water, the painter Leonardo da Vinci sketched the gesture of water, literary scholars have more, “the sunset and lone rustling flying together, autumn water together with a long sky”, “draw the knife to break the water more flow”, “the moon shines between the pine, the clear water flowing on the stone”, and so on. The image of water is integrated into people’s culture and life, and even the three-dotted water has become one of the most commonly used radicals of Chinese characters.

Is water so common, so ordinary, that we don’t understand enough about it? Is there anything we don’t know about it? Yes, although the deepening of the knowledge and use of water has been with mankind throughout history, we still cannot say that we understand all the wonders of water until today, and we can even say that our current understanding is only a very small and superficial part.

Water is actually a very peculiar and complex substance. Water possesses many wonderful properties that are different from other substances. A few simple examples are as follows.

- Water has a very high melting and boiling point, much higher than that obtained by extrapolation to congeners (see Fig. 2.1)!
- Different from the normal thermal expansion and contraction behavoirs, water is most dense at 4 °C!

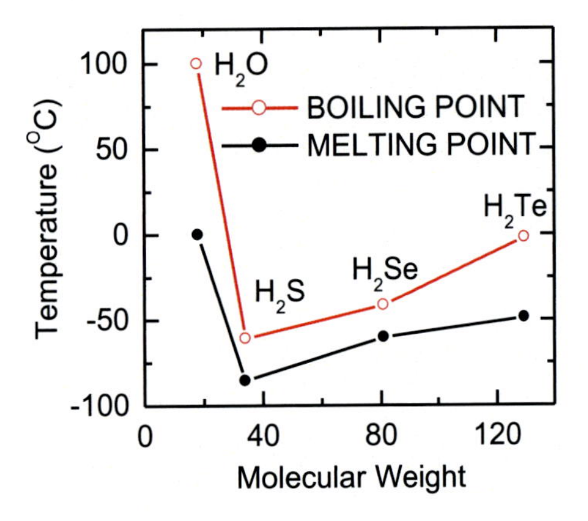

Fig. 2.1 Boiling and melting points of hydrides. The boiling and melting points of water are unusually higher than their extrapolation by molecular weight

- Under certain conditions, however, hot water freezes before cold water (i.e. the Mpemba effect)!
- Under certain pressure, instead of water solidifying into ice, the ice melts into water!
- The viscosity of water is also strange. While the viscosity of a non-polar, oily molecular liquid can increase infinitely in a thin layer of a restricted few molecular layers, the viscosity of water increases only by a factor of 2 to 3 over the bulk state [3]!
- Also, the heat capacity of water is uncharacteristically high!

The exotic nature of water is related to its structure. The flexibility of water molecules and the hydrogen bonds that connect them cause the complex and varied structure of water. Water in its solid-state, also known as ice, has been found in as many as eleven types of crystals [4] (numbered Ice I and Ice XI, see Fig. 2.2 Phase diagram of water). At temperatures around 100 K, there is also a difference between high-density amorphous ice and low-density amorphous ice; whether the transition between them is a primary or secondary phase transition is still under intense debate [5, 6]. Recently, there has also been a great deal of debate about whether there is a phase transition between high-density supercooled water and low-density supercooled water and the critical point for the phase transition. Why do warm-blooded animals have body temperatures around 40 °C? This is thought to be related to the detailed structure and dynamic properties of water at this temperature. The temperature difference between the boiling and freezing points of water at 100 °C is also thought to be related to the two different strengths of hydrogen bonding structures and vibrations that may exist in water [7].

How can these unusual properties and complex structures of water be understood? How to further explore the states and roles of other ions (e.g. OH^-, H^+, K^+, Na^+)

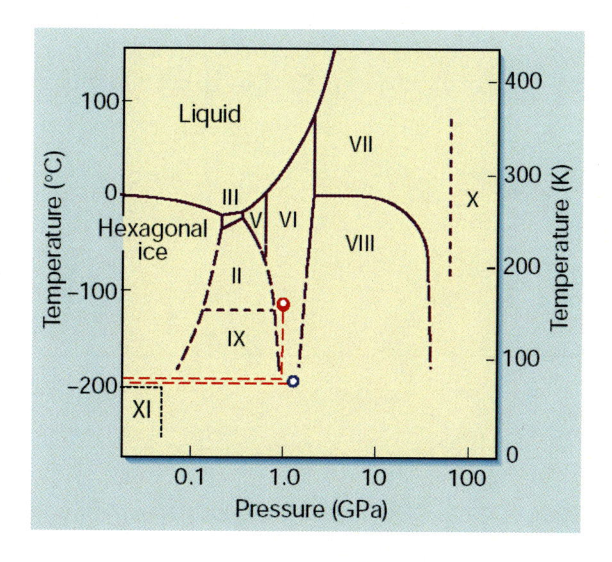

Fig. 2.2 Schematic diagram of the phase diagram of water (ice) [4]. Shows that water or ice has a rich structure

and molecules (e.g. DNA, proteins, phospholipids) in water in this regard? More importantly, how can water be controlled and manipulated at the microscopic level, in addition to macroscopic water engineering? To find the answer, we must use the scientific method to explore and study the secrets of water. In a sense, to understand the mystery of water is to understand the mystery of the whole world.

But what does water really look like to scientists working in basic science? It starts with the microscopic structure of water (molecular and electronic structure).

2.2 Water at the Molecular Scale

2.2.1 Water Molecule

It is well known that water consists of two elements, hydrogen (H) and oxygen (O). Two hydrogen atoms and one oxygen atom form a V-shaped structure (Fig. 2.3), which constitutes the water molecule (H_2O). The experimental determination of the OH bond length of the gaseous free water molecule at room temperature and pressure is 0.9572 Å (1 Å = 10^{-10} m), and the HOH bond angle is 104.52°. This is very different from the linear structure of CO_2 (bond angle O–C–O = 180°). The bond lengths and bond angles of water can all vary over a wide range with the surrounding environment (bonding conditions, temperature, pressure), and together with its V-shaped structure, these make the water molecule appear very flexible. This peculiar V-shaped structure is related to the electronic structure of water. In any case, from now on we have to remember that the V-shaped structure of the water molecule is its basic characteristic.

The V-shaped structure of the water molecule is due to sp-electron hybridization. An O atom has two 2s electrons and four 2p electrons in its outermost layer, and an H atom has only one 1s electron. To form an 8-electron full-shell structure, an O atom needs to bond with two H atoms by sp hybridization of electrons outside the nucleus,

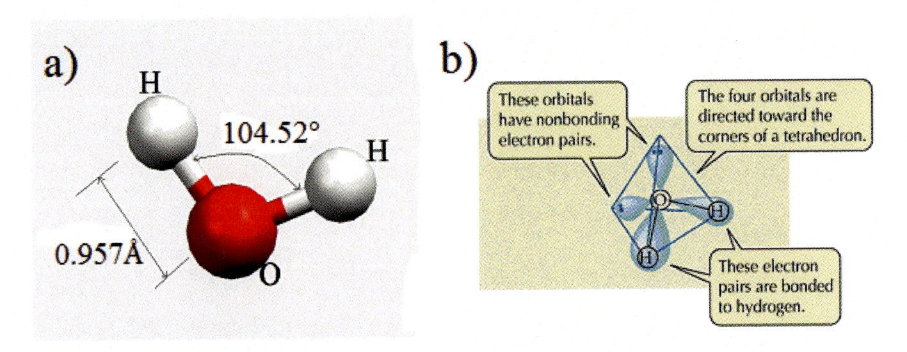

Fig. 2.3 The **a)** geometric structure and **b)** electron distribution of the water molecule

forming two OH bonds. The remaining two pairs of unbonded electrons are called lone pairs. They often play a decisive role in the interaction of water molecules with the outside world (especially metals or cations). So the electron distribution around a water molecule is an approximately tetrahedral structure (Fig. 2.3), corresponding to sp^3 electron hybridization. It is said to be an approximate tetrahedron because the center of the Wanier function of the OH bond orbital and the lone pair of electron orbitals, namely the center of orbital localization, are 0.52 Å and 0.30 Å, respectively. The lone pair orbital angle is 114°, and together with the HOH bond angle (104.52°), both deviate from the ideal angle of 109.5° in the tetrahedron.

Due to its V-shaped structure, the water molecule is very polar. Due to bonding, the H atom loses electrons and is positively charged; the lone pair electron region gathers excess electrons (negative charge). The dipole moment is threaded from the O-terminus to the H-terminus along the HOH angle bisector. The dipole moment for a single water molecule is 1.855 D (1 D = 10^{-18} esu·cm). For comparison, the dipole moments of CO and NO, the other two most common polar molecules, dipole moments are 0.1 D and 0.16 D, respectively. The larger polarity of the water molecule makes it easy to participate in interactions with other polar molecules or ions, thus allowing many reactions to occur in aqueous solutions.

The geometry of a free water molecule has a C_{2v} symmetry. Since a water molecule is composed of three atoms, it has nine degrees of freedom: three translational degrees of freedom, three rotational degrees of freedom, and three vibrational degrees of freedom. The first two are directly related to the external environment, while the latter are internal molecular degrees of freedom, but are subject to frequency shifts by the external environment. The corresponding 3 internal degrees of freedom are OH symmetric stretch, OH asymmetric stretch, and HOH scissor or bending motion. Their corresponding atomic dynamics are shown in Fig. 2.4. In the free case, the frequencies (ν:0 → 1) of these three vibrational modes are 3657 cm^{-1}, 3756 cm^{-1}, and 1595 cm^{-1}, respectively.

In energy space, the electronic orbitals of water molecules can be divided into five energy levels: $1a_1$ (539.7 eV, core electronic state), $2a_1$ (32.2 eV), $1b_2$ (18.5 eV), $3a_1$ (14.7 eV), and $1b_1$ (12.6 eV). Typically, if corresponding to localized orbitals, $1b_2$ is considered to be the bonding energy level, $1b_1$ corresponds to an unbonded lone pair electrons, and $2a_1$, $3a_1$ are a mixture of bonded and unbonded orbitals. Taking away an electron from the strongly bonded $1b_2$ orbital causes the decomposition

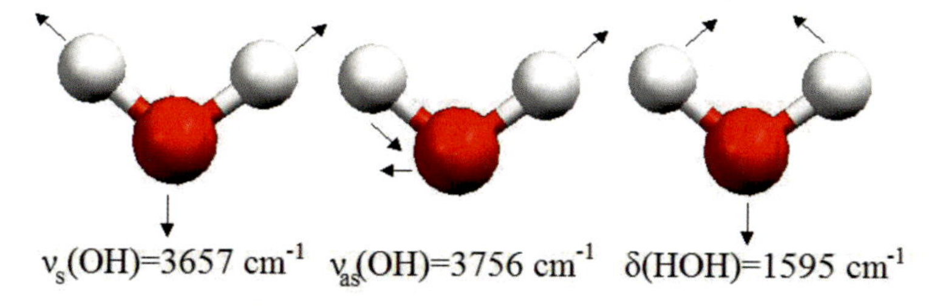

Fig. 2.4 The three modes of vibration of the water molecule

of the water molecule H_2O into H^+ and OH^-. The $3a_1$ orbital plays a key role in maintaining the V-shaped geometry of the water molecule, and removing an electron from this orbital causes the V-shaped structure of the water molecule to be converted into a linear structure similar to that of CO_2. The distribution of these orbitals in three dimensions is shown in Fig. 2.5.

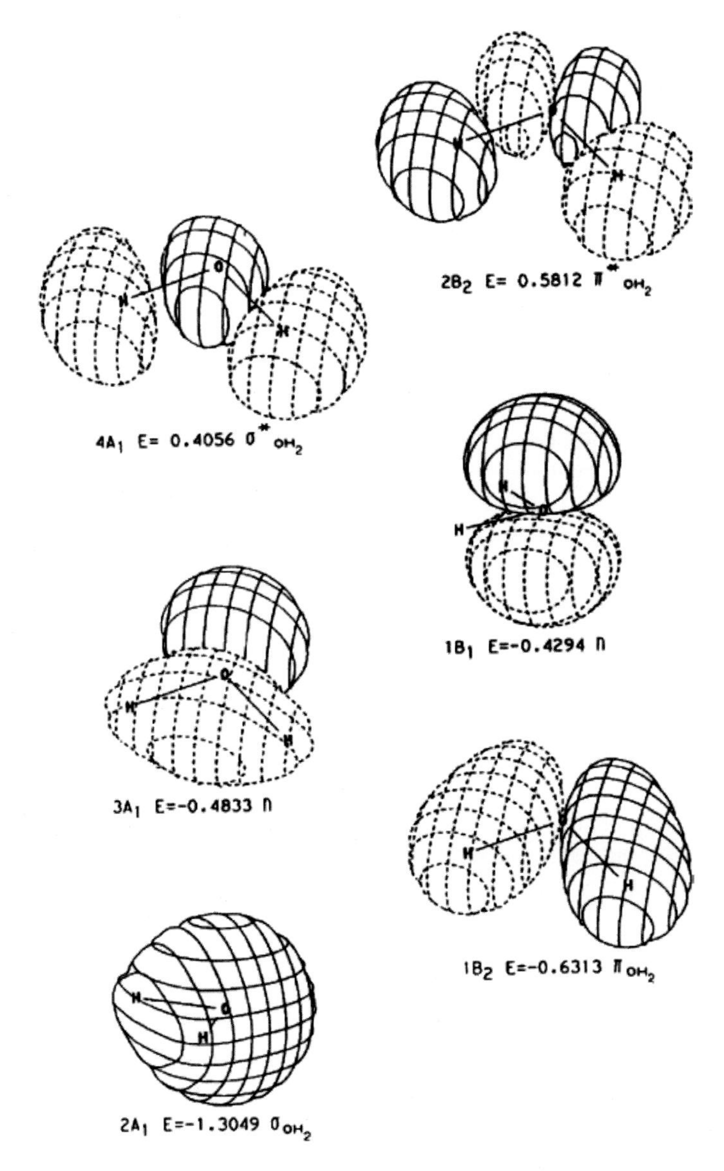

Fig. 2.5 Non-localized electronic orbitals of the water molecule [8]. The unit of energy level is 1Ry = 13.6057 eV

Table 2.1 Basic physical parameters of water molecules

Physical parameters	Numerical value
OH bond length	0.9572 Å
HOH bond angle	104.52°
Van der waals radius	1.45 Å
Moment of inertia	1.0220, 1.1.9187, 2.9376 × 10^{-40} g^{-1} cm^{-2}
Dipole moment	1.855 D
Average polarization	1.444 × 10^{-24} cm^3
OH bond dissociation energy	5.18 eV

Some of the basic physical parameters of the water molecule are listed in Table 2.1.

2.2.2 Hydrogen Bond

Water molecules are bonded to each other by hydrogen bonds. Hydrogen bond refers to the interaction between a highly electronegative atom (such as oxygen or nitrogen) and a hydrogen atom that is already covalently bonded to oxygen or nitrogen, for example, by mutual attraction. Hydrogen bonds mostly occur between H and atoms such as N, O, and S. It is relatively weak (a few tens to two or three hundred meV) and lies between the usual chemical bond and the much weaker Van der Waals bond. Hydrogen bonds exist widely in nature, and it is not only one of the most common intermolecular forces, but also exists inside molecules. The most famous example is the double helix structure of DNA which is maintained by a set of hydrogen bonds, and the three-dimensional structure of proteins is often determined by hydrogen bonds between different groups. It is generally accepted that hydrogen bonds arise mainly from electrostatic Coulomb interactions between H atoms, which are positively charged due to covalent bonding, and atoms such as O and N, which are strongly electronegative. But a small part also originates from induced polarization and intermolecular dispersion forces. Curiously, although based on coulomb attraction and weakly, hydrogen bonds are still saturated and directional as chemical bonds are.

The interaction between water molecules that we usually refer to is hydrogen bonding. It is formed by the mutual attraction of the element H in water with the lone pair of electrons of O in another nearest neighboring water molecule. In the free state, i.e., the gaseous double water molecule (dimmer, $(H_2O)_2$) structure, the O–H–O bond length is about 2.976 Å, the bond angle is close to 180° (linear), and the strength is about 23 kJ/mol. The party providing H is called the proton donor and the party receiving H is called the proton acceptor. About 0.02 electrons are transferred from the proton acceptor to the proton donor. Since a water molecule has only two lone electron pairs, it accepts up to two H's at the same time to form hydrogen bonds, plus its own two H's can also form hydrogen bonds with other molecules, and an H_2O molecule can form up to four hydrogen bonds to form a localized tetrahedral structure. This local tetrahedral network structure is a fundamental feature of water, both in ice and in liquid form (Fig. 2.6).

Most of the peculiar properties of water are caused by the hydrogen bonds between water molecules. Because hydrogen bonds are much stronger than other types of intermolecular interaction in general, the water molecules in water are very strongly bound, resulting in extremely high melting and boiling points and great heat capacity. This is quite inconsistent with extrapolation by molecular weight, since water has a very small molecular weight of 18, and if intermolecular bonding by Van der Waals interaction, water should vaporize at – 75 °C (Fig. 2.1); yet the high melting and boiling points and large heat capacity of water are the very necessary conditions for the Earth's environment to be able to produce life and be suitable for its development. You can see how big a role the hydrogen bonds between water molecules play!

The strange property of water below 4 °C to shrink when exposed to heat is also caused by hydrogen bonds. Because of the directional and saturated nature of hydrogen bonds, to maintain the strongest hydrogen bonding in water, the water molecules must be arranged in a certain geometric shape (tetrahedral structure) and each water molecule must occupy a certain volume. When the temperature increases, the kinetic energy of the water molecules increases so that some of the hydrogen bonds break and the constraints on the water molecules to remain in a larger spatial configuration are reduced, so that the water molecules are arranged more densely and the whole volume is smaller. In turn, the forced reduction of the space occupied by water molecules causes some of the hydrogen bonds to break due to the disruption of the perfect spatial structure, resulting in a phase transition from ice to water, which is the cause of pressure-induced water melting.

Some of the peculiar properties of water and their hydrogen bonds explanations, as well as practical examples, are summarized in Table 2.2. Note that these examples are decisive for life to be able to exist.

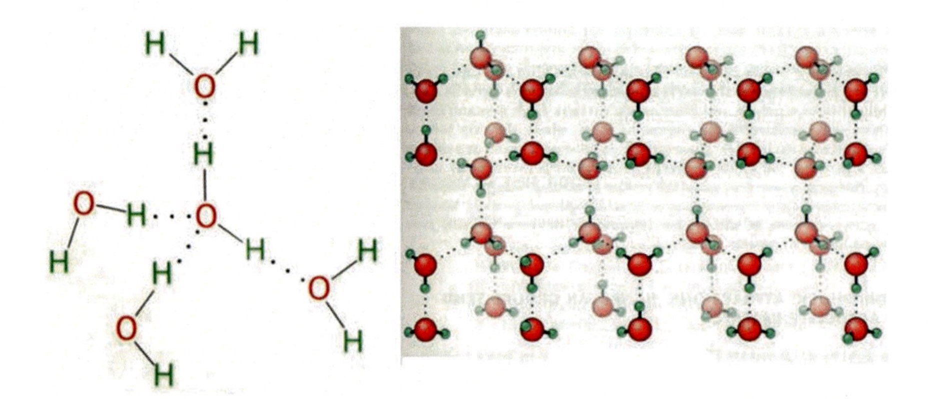

Fig. 2.6 Structure of a tetrahedral network consisting of hydrogen bonds in water or ice

Table 2.2 Some peculiar properties of water and explanation of hydrogen bonding and practical examples

Properties of water	Hydrogen bonding explained	Practical examples
High heat capacity	High energy required to break hydrogen bonds	Temperature stability of the earth and biological tissues
High evaporation energy	High energy required to break hydrogen bonds	Water evaporation reduces the temperature of body surface
Solids are less dense than liquids	Form hydrogen bond networks occupy space	Body surface ice protects rivers and lakes from freezing
Surface tension	Hydrogen bonds maintain the structure of adjacent molecules	Water is transported by capillary action in leaves, rhizomes

2.2.3 Understanding Water at the Molecular Scale

Water molecules plus the hydrogen bonds that connect them are like bricks plus cement in the material world, building the "edifice" of water structure. These two are the basis and guarantee of understanding water at the molecular scale. Several or even dozens of water molecules are connected by hydrogen bonds to form the cluster structure of water. In the last two decades, the structure of small clusters of water has been studied in detail. T. R. Dyke and R. J. Saykally and co-workers have determined the structure of dimer [9], trimer [10], tetramer [11], pentamer [12], and hexamer [13] of water by measuring the rotational energy levels of small clusters. From two to five, the water molecules are arranged in rings, but six water molecules form three structures with almost degenerate energy: rings, cages, and prisms [14, 15], showing the transition from a zero-dimensional cluster structure to a three-dimensional bulk structure.

Water molecules linked by hydrogen bonds to form ice in various bulk states are subject to the so-called BFP rules (Bernal-Fowler-Pauling Rules), the main contents of which are described as follows [16, 17]:

(1) Each oxygen atom is attached to two hydrogen atoms at a distance of about 0.96 Å, which are at an angle of about 105°.
(2) Each oxygen atom is attached to four surrounding oxygen atoms by hydrogen bonds, forming a tetrahedral structure.
(3) The interaction between two water molecules that are not neighbors is weak and insufficient to stabilize any of the bulk structures.
(4) There is one and only one hydrogen atom on each O–O linkage axis.

A look at the structure of bulk water (ice) on a molecular scale also reveals peculiar properties. Li and Ross [7] suggested that since the relative arrangement of the other three hydrogen atoms in the two water molecules forming hydrogen bonds in ice (connected by one hydrogen atom) can be divided into two classes according to symmetry, they form hydrogen bonds of two strengths with a 2:1 number ratio in ice, corresponding to the O–O vibrational peaks at 27 and 36 meV in the Inelastic Incoherent Neutron Scattering experiments. The linear temperature dependence of the average O–H spacing and bond angle of hydrogen bonds in water between 0 and 80 °C was obtained by K. Modig et al. from proton magnetic shielding tensor measurements [18]. The characteristics of hydrogen bond and water molecular dipole moments in water in the critical state (temperature 647 K, pressure 22.1 MPa, density 0.32 g/cm^3) [19], and the mechanism of partial decomposition of water into H_3O^+ and OH^- at high pressures (12–30 GPa) [20] were also understood in detail at the molecular level by molecular dynamics simulations.

Kropman and Bakker proposed a new method to be able to distinguish water molecules in several molecular layers around an ion in solution from other ordinary water molecules and thus directly observe the dynamics of water molecules closest to the ion [21]. They used femtosecond infrared region nonlinear spectroscopy to do the detection. The signal was first excited with a strong laser and then probed with another tunable laser beam delayed by a few picoseconds to measure the change in transmittance, and found that the signal decays through two channels, where the fast process corresponds to the bulk water in solution, and the slow process corresponds exactly to the water molecules around the ion. Using this method, they found that the layer of water molecules surrounding Cl^-, Br^-, and I^- is rigid. Molecular dynamics simulations have shown that the mechanism of transporting H_3O^+ and OH^- ions in water is also different, with three and four water molecules surrounding the two ions forming hydrogen bonds with them, respectively [22, 23].

The industry expects to use hydrogen as a clean energy source. However, to achieve this goal, it is necessary to use hydrogen separated from water. This is because today's industrial production of hydrogen from hydrides such as oil and natural gas already produces large amounts of greenhouse gases such as CO_2. The cost of using precious metals such as platinum as catalysts to produce hydrogen from water is too high, so people are turning to biomolecules. Hydrogenase, a catalyst, can easily break the OH bond of a water molecule and then combine the separated two protons and two electrons to form a hydrogen molecule (H_2) [24]. Successful exploitation of this reaction requires that we understand the details of the action of the water molecule and the enzyme molecule. Calculations have shown the presence of water molecule clusters on the inside or aside of biological macromolecules, such as DNA, with structures and dipole moment distributions that differ greatly from those of water in the bulk and gaseous states [25].

Understanding water at the molecular scale also includes understanding the electronic structure of water. First-principles calculations usually give the spatial distribution of electron clouds in water, orbital energy levels, etc., to help us make a microscopic description of water. Parrinello and co-workers have done extensive work in this area [22, 23, 25–28]. They drew the electronic density of states and the lowest unoccupied molecular orbit (LUMO) in liquid water [26]. They also gave the equivalence surface of the electron cloud distribution, which plays a key role in revealing the mechanism of H_3O^+ and OH^- ion transport in water [22, 23], and the interaction of water molecules with biomolecules [25]. The same information on the electron distribution appears in work studying the decomposition of water at high pressures [20].

Understanding water on a molecular scale often also takes into account the effects of quantum motion. This is because the water molecule has a very small mass of hydrogen, which has a significant quantum effect. The quantum effects in the $(H_3O_2)^-$ structure make the H double potential well structure between two oxygen atoms into a single potential well structure [27]. The quantum effect of hydrogen also plays a clear role in the phase transition process of high-pressure ice [28] and the transport of H_3O^+ and OH^- ions in water [22, 23].

2.3 Water on the Surface

Roughly speaking, water interacts with the outside world through "surface" contact, and the important role of water is mediated through the "surface". For example, the aforementioned applications of water in electrochemistry, catalysis, sensors, artificial rainfall, etc. all occur at the surface. To understand water at the molecular scale, we have to study the interaction between water and the surface in detail. At the molecular scale, this "surface" may be a single crystal platform, a step, a biological film, or a macromolecule. We view the interaction between water and the surface as confinement of water by the surface, both geometrically and physicochemically. In general, water in the confined system will exhibit very different properties from the bulk state, as can be seen in the following examples.

By type of surface, the study of water-surface interactions can be divided into research areas such as water and metal surfaces, semiconductor surfaces, oxide surfaces, other surfaces such as salts, organic surfaces, and biological surface interactions. The issues of concern and research interests are mainly focused on: (1) the adsorption structure of water and its electronic, vibrational, and reactive properties; (2) the co-adsorption of water with alkali metals, oxygen, CO, etc.; (3) the effect of temperature, pressure, and other conditions on the structure of water, the ordered-disordered transition of water at the surface; (4) the decomposition of water molecules at the surface; (5) the phenomenon of water immersion at the surface; (6) the actual behavior of water under non-high vacuum conditions and adsorption behavior on high tech materials; (7) behaviors of confined water in microporous, thin layer, biological systems, etc.

The adsorption of water on metal surfaces has been studied most [29–59], both because it was the starting point of this field of research (Döbereiner studied the "catalytic" effect of platinum in 1823 [1]) and because metal surfaces are relatively simple surfaces. Over the past hundred years, various experimental and theoretical studies have gradually led to the conclusion of basic rules for the adsorption of water on metal surfaces in general [60].

(1) At all coverages, water is adsorbed to the metal surface in an undecomposed molecular state.

(2) Even at very low coverages, water forms small clustered structures linked by hydrogen bonds.
(3) Dependent on the air pressure and growth temperature of the water, the water forms an ordered two-dimensional bilayer structure.
(4) The water forms a multilayer structure similar to that of bulk ice, which is independent of the constraints of the substrate.

However, this is only a rough generalization of the general law of water adsorption on metal surfaces, and the details have remained hotly debated recently. Feibelman has recently suggested that half of the water molecules are in a decomposed state on the surface of Ru(0001) [61]. Mitsui et al. observed that the diffusion of small clusters of water on the surface of Pd(111) is several orders of magnitude faster than that of individual water molecules [56]. Compared to a large number of focused experimental studies in this field, systematic theoretical work is still very lacking and increasingly urgent.

The study of water adsorption on semiconductors plays an important role in semiconductor industrial technology. Typically, water adsorption on Si surfaces causes oxidation of the Si surface to a stable SiO_2 surface, the so-called "wet oxidation" [62]. The water molecules will also decompose accordingly. However, whether the water adsorbed on Si(100)-(2 $\times$ 1) and Si(111)-(7 $\times$ 7) surfaces at room temperature decomposes or not is still debated in the literature [63–65]; structural models at the molecular level are still lacking. On Ge(100)-(2 $\times$ 1) and Ge(111) surfaces, the uniform conclusion is that water adsorbs in a decomposed state [66, 67]; water adsorption on the surfaces of other semiconductors such as GaAs, GaN, AlAs, InP, etc., follows a similar scenario. The adsorption of water usually does not change the reconstructed structure of the semiconductor surface.

The adsorption of water on oxide surfaces has also been carefully studied [68–77]. Water molecules are readily decomposed on oxide surfaces due to the competing effects of surface oxygen atoms. Calculations show that 1/3 of the water molecules in the adsorption structure of $p(3 \times 2)$ on the perfect surface of MgO(100) partially decompose due to collective effects (i.e., $H_2O \rightarrow H^+ + OH^-$) [69]. Even at low coverage, the molecular adsorption state of water on the α-Al_2O_3(0001) surface is sub-stable, and Hass et al. simulated the dynamics of water molecule decomposition by first-principles molecular dynamics [71]. The adsorption structure of water on the RuO_2(110) surface changes from a low-density water layer to an Ice X-like structure as the positive and negative voltages applied to the surface change [75]. On mica surfaces, water at room temperature still forms a two-dimensional ice structure that can be applied to artificial rainfall [76].

The interactions of water and salts [78–84] are generally weak and lattice matching is difficult, but they play an important role in natural phenomena, such as dissolution, etching, weathering, and meteorology. Low-energy electron diffraction (LEED) experiments have revealed an ordered phase of water on NaCl(100) surface $c(4 \times 2)$ [79]. X-ray diffraction was used to study the interface between liquid water and KDP crystals, and the two layers closest to the interface were found to be ordered and behave similarly to ice, the two more distant layers were less ordered, and the more distant ones behaved like bulk water [84].

The interaction between water and graphite has also been studied more. Graphite is often used as a representative of hydrophobic surfaces or as a simplified model of biological systems. Chakarov et al. observed the crystallization of an otherwise amorphous ice layer on a graphite surface under UV light irradiation [85]. Model simulations of water on graphite show that the contact angle of water droplets (> 0) is proportional to the adsorption energy of individual water molecules on graphite (< 0) [86], which builds a bridge to understanding macroscopic phenomena from the molecular-scale behavior of water.

IR absorption spectroscopy studies of water at hydrophobic organic hydrocarbon liquids as well as at CCl_4 surfaces (Fig. 2.7) show that water becomes weakly hydrogen-bonded and strongly oriented at the hydrophobic interface, contrary to the commonly held impression that water is strongly hydrogen-bonded at the hydrophobic interface [87]. X-ray reflectance measurements of paraffin wax floating on water show that the density of water at the interface decreases to about one less water molecule per 25–30 Å^2 of surface than in the bulk, with a minimum density of 0.9 times that of the bulk and a profile half-width of 8–10 Å for density defects. This is known as the "dewetting" effect [88].

Biological processes all occur in water, and water is present on the surface of both biological membranes and biomolecules. Recent X-ray diffraction studies have shown that a thin layer of water is sandwiched between two phospholipid bilayers (the main components of cell membranes) in the transition state structure of cell membrane fusion and division, and that this water layer plays an important role in this process [89]. K^+ and Na^+ form different hydrated structures at the opening of the K^+ ion channel protein KscA, causing selectivity of the channel for ion transport.

In general, Na^+ has to bind more water molecules, hindering its transport [90]. Water molecules are free to pass through the cell membrane while protons are not, also because hydrated protons cannot pass through the fine pores of water transport proteins [91]. The clusters of water bound to the small grooves of DNA are different from the bulk and affect the electronic structure of DNA [25].

Progress on the experimental aspects of water–solid surface interactions can be found in several review articles such as 1987, 2002, and 2009 [92–95].

Water is surface-constrained and usually exhibits properties different from those of the bulk state [96, 97]. Due to surface effects, water molecules are usually specifically oriented and ordered, and this order can be altered by changing the surface potential [48, 49]. Restricted water often behaves like supercooled water at a temperature 30 K lower [98]. Water behaves more uniquely under extended surface confinement, including two-surface intercalation and pore-like confinement. The water in the 1–2 layers between the two clay layers can remain liquid (supercooled water) at 228 K, which is not possible in the bulk state; relaxation dynamics studies show that water exists in a "strong" glassy state under these conditions and has a transition from a "fragile" state to a "strong" state [99]. Inside the carbon nanotubes, water molecules are connected to form peculiar tetragonal, pentagonal, hexagonal, and heptagonal tubes of ice with different diameters; more surprisingly, there is a critical point of

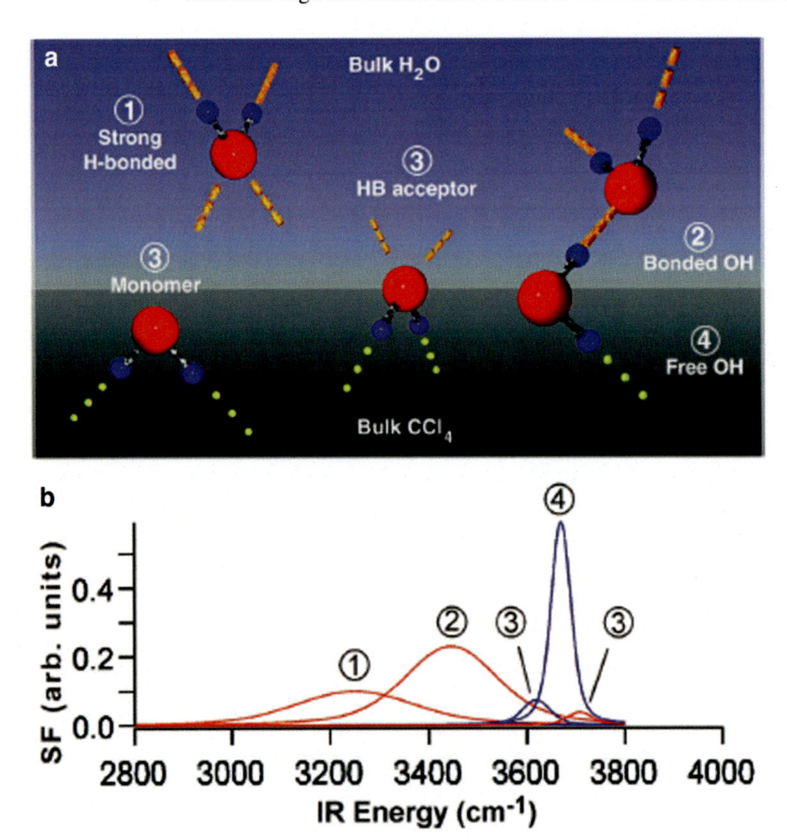

Fig. 2.7 Water on a hydrophobic CCl_4 surface. Different types of water molecules at the interface correspond to different OH vibrational IR absorption peaks, thus revealing that water exhibits weak hydrogen bonds and strong orientation effects at hydrophobic surfaces [46]

phase transition inside the finer tubes, above which there is no difference between ice and liquid water, and the phase transition between them is continuous. This is a completely different property from that of bulk water [100]. The transport of water within fine carbon nanotubes [101, 102] and protons along water chains within nanotubes [103, 104] have also been studied in careful simulations. In addition, the phenomena of water infiltration (i.e., hydrophobicity) and decomposition at the surface are typical examples of water-surface interactions that have received much attention from researchers recently.

The scientific approach to the study of the interaction between water and surfaces at the molecular level is briefly described here. The organic combination of experimental observations and theoretical simulations is an indispensable asset for understanding this interaction at the molecular scale. Almost all standard experimental analytical tools for surface science can be used to study water-surface interactions, such as X-ray photoelectron spectroscopy (XPS), ultraviolet photoelectron spectroscopy (UPS), X-ray diffraction (XRD), low-energy electron diffraction (LEED), helium atomic scattering (HAS), scanning tunneling microscopy (STM), high-resolution electron energy loss spectroscopy (HREELS), infrared absorption spectroscopy (IRAS), and temperature-controlled desorption spectroscopy (TPD), among others. Photoelectron spectroscopy can be used to observe the energy level shifts of the core and valence electrons of water molecules and the changes in the shape of the spectral peaks, and then infer information such as the molecular orientation of water molecules and whether they are decomposed. Based on the patterns of photon and electron diffraction and (helium) atom scattering, doing the Fourier transform one can easily obtain periodic information about the adsorption structure of water molecules on the surface, and can even fit the position of each atom. Scanning tunneling microscopy is becoming more and more widely used in the study of water adsorption on surfaces. It can not only directly observe various adsorption structures such as water clusters, chains, and bilayers in real space, but also observe the diffusion of water molecules and small clusters on the surface and the dynamics of water layer formation. Electron loss spectroscopy and infrared absorption spectroscopy can provide vibration information (energy and spectral shape) of the adsorbed water structure, including the translational, rotational, and oscillating motions of water molecules and the intramolecular OH stretching and HOH shearing motions. Temperature-controlled desorption measurements can tell the adsorption strength and desorption dynamics of water on the surface. Other experimental methods like atomic force microscopy (AFM) and neutron scattering are gradually applied to the study of the interaction between water and surfaces, such as studying the actual behavior of water under non-high vacuum conditions, the adsorption behavior on high-tech materials, and the confined behavior in microporous, thin-layer, and biological systems. The theoretical approach, except for the phenomenological simulation accompanied by experimental techniques (XRD, LEED), but more importantly is the first-principles calculations to study the structure, energy, and electronic states of water on surfaces; and first-principles-based molecular dynamics simulations, model-based classical molecular dynamics, and Monte Carlo simulations, the former being more accurate while the latter can simulate the structure and dynamics of water on larger scales and over longer periods.

As experimental and theoretical studies have advanced and their mutually promotion, there is increasing confidence in understanding, mastering, controlling, and applying water-surface interactions from the molecular scale. One example is that water molecules have been used as detectors of surface structure and properties. A large number of research papers are published each year in a variety of journals in physics, chemistry, biology, geography, meteorology, technology, and interdisciplinary disciplines. Almost every issue of Nature, Science, Phys. Rev. Lett. has 1–3 articles on the properties of water and its interaction with the outside.

The core concept of this book is to understand water at the molecular scale, including understanding its structure, vibrations, diffusion, phase transitions, relaxation, and various other properties. This is not only because the microscopic scale mechanism of water is fundamental to understanding its macroscopic peculiar properties, but also because water confined at this scale has many special properties that are different from its bulk state. Both of these points have very significant implications for the further use and manipulation of water resources that have almost unlimited potential. Here we restate the famous theoretical physicist R. P. Feynman's exciting atomic-scale view of the world [105], an idea that eventually led to the flourishing and far-reaching nanoscience and nanotechnology of today.

> What would the properties of materials be if we could really arrange the atoms the way we want them? I can't see exactly what would happen, but I can hardly doubt that when we have some control of the arrangement of things on a small scale we will get an enormously greater range of possible properties that substances can have. I can't see exactly what would happen, but I can hardly doubt that when we have some control of the arrangement of things on a small scale we will get an enormously greater range of possible properties that substances can have, and of different things that we can do.

References

1. J.W. Döbereiner, Schweigg. J. **39**, 1 (1823)
2. J.J. Berzelius, Jber. Chem. **15**, 237 (1837)
3. U. Raviv, P. Laurat, J. Klein, Fluidity of water confined to sub nanometer films. Nature **413**, 51 (2001)
4. D.D. Klug, Dense ice in detail. Nature **420**, 749 (2002)
5. O. Mishima, Y. Suzuki, Propagation of the polyamorphic transition of ice and the liquid-liquid critical point. Nature **419**, 599 (2002)
6. C.A. Tulk, C.J. Benmore, J. Urquidi, D.D. Klug, J. Neuefeind, B. Tomberli, P.A. Egelstaff, Structural studies of several distinct metastable forms of amorphous ice. Science **297**, 1320 (2002)
7. J. Li, D.K. Ross, Evidence for two kinds of hydrogen bond in ice. Nature **365**, 327 (1993)
8. W.L. Jorgensen, L. Salem, *The Organic Chemists' Book of Orbitals* (Academic Press, New York, 1973)
9. T.R. Dyke, K.M. Mack, J.S. Muenter, The structure of water dimer from molecular beam electric resonance spectroscopy. J. Chem. Phys. **66**, 498 (1977)
10. N. Pugliano, R.J. Saykally, Measurement of quantum tunneling between chiral isomers of the cyclic water trimer. Science **257**, 193 (1992)
11. J.D. Cruzan, L.B. Braly, K. Liu, M.G. Brown, J.G. Loeser, R.J. Saykally, Quantifying hydrogen bond cooperativity in water: VRT spectroscopy of the water tetramer. Science **271**, 59 (1996)
12. K. Liu, M.G. Brown, J.D. Cruzan, R.J. Saykally, Vibration-rotation tunneling spectra of the water pentamer: structure and dynamics. Science **271**, 62 (1996)
13. K. Liu, M.G. Brown, C. Carter, R.J. Saykally, J.K. Gregory, D.C. Clary, Characterization of a cage form of the water hexamer. Nature **381**, 501 (1996)
14. K. Liu, J.D. Cruzan, R.J. Saykally, Water clusters. Science **271**, 929 (1996)
15. J.K. Gregory, D.C. Clary, K. Liu, M.G. Brown, R.J. Saykally, The water dipole moment in water clusters. Science **275**, 814 (1997)
16. J.D. Bernal, R.H. Fowler, A theory of water and ionic solution, with particular reference to hydrogen and hydroxyl ions. J. Chem. Phys. **1**, 515 (1933)
17. L. Pauling, The structure and entropy of ice and of other crystals with some randomness of atomic arrangement. J. Am. Chem. Soc. **57**, 2680 (1935)
18. K. Modig, B.G. Pfrommer, B. Halle, Temperature-dependent hydrogen-bond geometry in liquid water. Phys. Rev. Lett. **90**, 075502 (2003)
19. M. Boero, K. Terakura, T. Ikeshoji, C.C. Liew, M. Parrinello, Hydrogen bonding and dipole moment of water at supercritical conditions: a first-principles molecular dynamics study. Phys. Rev. Lett. **85**, 3245 (2000)

20. E. Schwegler, G. Galli, F. Gygi, R.Q. Hood, Dissociation of water under pressure. Phys. Rev. Lett. **87**, 265501 (2001)
21. M.F. Kropman, H.J. Bakker, Dynamics of water molecules in aqueous solvation shells. Science **291**, 2118 (2001)
22. D. Marx, M.E. Tuckerman, J. Hutter, M. Parrinello, The nature of the hydrated excess proton in water. Nature **397**, 601 (1999)
23. M.E. Tuckerman, D. Marx, M. Parrinello, The nature and transport mechanism of hydrated hydroxide ions in aqueous solution. Nature **417**, 925 (2002)
24. J. Alper, Water splitting goes Au naturel. Science **299**, 1686 (2003)
25. F.L. Gervasio, P. Carloni, M. Parrinello, Electronic structure of wet DNA. Phys. Rev. Lett. **89**, 108102 (2002)
26. K. Laasonen, M. Sprik, M. Parrinello, "Ab initio" liquid water. J. Chem. Phys. **99**, 9080 (1993)
27. M.E. Tuckerman, D. Marx, M.L. Klein, M. Parrinello, On the quantum nature of the shared proton in hydrogen bonds. Science **275**, 817 (1997)
28. M. Benoit, D. Marx, M. Parrinello, Quantum effects on phase transitions in high-pressure ice. Comput. Mater. Sci. **10**, 88 (1998)
29. L.E. Firment, G.A. Somorjai, Low-energy electron diffraction studies of molecular crystals: the surface structures of vapor-grown ice and naphthalene. J. Chem. Phys. **63**, 1037 (1975)
30. L.E. Firment, G.A. Somorjai, Low energy electron diffraction studies of the surfaces of molecular crystals (ice, ammonia, naphthalene, benzene). Surf. Sci. **84**, 275 (1979)
31. B.A. Sexton, Vibrational spectra of water chemisorbed on Platinum (111). Surf. Sci. **94**, 435 (1980)
32. G.B. Fisher, J.L. Gland, The interaction of water with the Pt(111) surface. Surf. Sci. **94**, 446 (1980)
33. D. Doering, T.E. Madey, The adsorption of water on clean and oxygen-dosed Ru(001). Surf. Sci. **123**, 305 (1982)
34. P.A. Thiel, R.A. DePaola, F.M. Hoffmann, The vibrational spectra of chemisorbed molecular clusters: H_2O on Ru(001). J. Chem. Phys. **80**, 5326 (1984)
35. S. Andersson, C. Nyberg, C.G. Tengstål, Adsorption of water monomers on Cu(100) and Pd(100) at low temperatures. Chem. Phys. Lett. **104**, 305 (1984)
36. F.T. Wagner, T.E. Moylan, A comparison between water adsorbed on Rh(111) and Pt(111), with and without predosed oxygen. Surf. Sci. **191**, 121 (1987)
37. W. Ranke, Low temperature adsorption and condensation of O_2, H_2O, and NO on Pt(111), studied by core level and valence band photoemission. Surf. Sci. **209**, 57 (1989)
38. A.F. Carley, P.R. Davies, M.W. Roberts, K.K. Thomas, Hydroxylation of molecularly adsorbed water at Ag(111) and Cu(100) surfaces by dioxygen: photoelectron and vibrational spectroscopic studies. Surf. Sci. **238**, L467 (1990)
39. H. Ogasawara, J. Yoshinobu, M. Kawai, Water adsorption on Pt(111): from isolated molecule to three-dimensional cluster. Chem. Phys. Lett. **231**, 188 (1994)
40. G. Held, D. Menzel, The structure of the p($\sqrt{3}\times\sqrt{3}$)R30° bilayer of D_2O on Ru(001). Surf. Sci. **316**, 92 (1994)
41. M. Morgenstern, T. Michely, G. Comsa, Anisotropy in the adsorption of H_2O at low coordination sites on Pt(111). Phys. Rev. Lett. **77**, 703 (1996)
42. M. Morgenstern, J. Muller, T. Michely, G. Comsa, The ice bilayer on Pt(111): nucleation, structure and melting. Z. Phys. Chem. **198**, 43 (1997)
43. I. Villegas, M.J. Weaver, Infrared spectroscopy of model electrochemical interfaces in ultra-high vacuum: evidence for coupled cation-anion hydration in the Pt(111)/K^+ Cl^- system. J. Phys. Chem. **100**, 19502 (1996)
44. S. Smith, C. Huang, E.K.L. Wong, B.D. Kay, Desorption and crystallization kinetics in nanoscale thin films of amorphous water ice. Surf. Sci. **367**, L13 (1996)
45. P. Löfgren, P. Ahlström, D.V. Chakarov, J. Lausmaa, B. Kasemo, Substrate dependent sublimation kinetics of mesoscopic ice films. Surf. Sci. **367**, L19 (1996)
46. A. Glebov, A.P. Graham, A. Menzel, J.P. Toennies, Orientational ordering of two-dimensional ice on Pt(111). J. Chem. Phys. **106**, 9382 (1997)

47. G. Pirug, H.P. Bonzel, UHV simulation of the electrochemical double layer: adsorption of $HClO_4/H_2O$ on Au(111). Surf. Sci. **405**, 87 (1998)
48. X. Su, L. Lianos, Y.R. Shen, G.A. Somorjai, Surface-induced ferroelectric ice on Pt(111). Phys. Rev. Lett. **80**, 1533 (1998)
49. M.S. Yeganeh, S.M. Dougal, H.S. Pink, Vibrational spectroscopy of water at liquid/solid interfaces: crossing the isoelectric point of a solid surface. Phys. Rev. Lett. **83**, 1179 (1999)
50. H. Ogasawara, J. Yoshinobu, M. Kawai, Clustering behavior of water (D_2O) on Pt(111). J. Chem. Phys. **111**, 7003 (1999)
51. M. Nakamura, Y. Shingaya, M. Ito, The vibrational spectra of water cluster molecules on Pt(111) surface at 20 K. Chem. Phys. Lett. **309**, 123 (1999)
52. A.L. Glebov, A.P. Graham, A. Menzel, Vibrational spectroscopy of water molecules on Pt(111) at submonolayer coverages. Surf. Sci. **427**, 22 (1999)
53. M. Nakamura, M. Ito, Monomer and tetramer water clusters adsorbed on Ru(0001). Chem. Phys. Lett. **325**, 293 (2000)
54. K. Jacobi, K. Bedürftig, Y. Wang, G. Ertl, From monomers to ice – new vibrational characteristics of H_2O adsorbed on Pt(111). Surf. Sci. **472**, 9 (2001)
55. S. Haq, J. Harnett, A. Hodgson, Growth of thin crystalline ice films on Pt(111). Surf. Sci. **505**, 171 (2002)
56. T. Mitsui, M.K. Rose, E. Fomin, D.F. Ogletree, M. Salmeron, Water diffusion and clustering on Pd(111). Science **297**, 1850 (2002)
57. K. Morgenstern, J. Nieminen, Intermolecular bond length of ice on Ag(111). Phys. Rev. Lett. **88**, 066102 (2002)
58. H. Ogasawara, B. Brena, D. Nordlund, M. Nyberg, A. Pelmenschikov, L.G.M. Pettersson, A. Nilsson, Structure and bonding of water on Pt(111). Phys. Rev. Lett. **89**, 276102 (2002)
59. D.N. Denzler, C. Hess, R. Dudek, S. Wagner, C. Frischkorn, M. Wolf, G. Ertl, Interfacial structure of water on Ru(0001) investigated by vibrational spectroscopy. Chem. Phys. Lett. **376**, 618 (2003)
60. S.K. Jo, J. Kiss, J.A. Polanco, J.M. White, Identification of second layer adsorbates: water and chloroethane on Pt(111). Surf. Sci. **253**, 233 (1991)
61. P.J. Feibelman, Partial dissociation of water on Ru(0001). Science **295**, 99 (2002)
62. D.J. Elliot, *Integrated Circuit Fabrication Technology* (McGraw-Hill, New York, 1982)
63. W. Ranke, D. Schmeisser, Y.R. Xing, Orientation dependence of H_2O adsorption on a cylindrical Si single crystal. Surf. Sci. **152–153**, 1103 (1985)
64. Y.J. Chabal, S.B. Christman, Evidence of dissociation of water on the Si(100)2$\times$1 surface. Phys. Rev. B **29**, 6974 (1984)
65. R.A. Rosenberg, P.J. Love, V. Rehn, I. Owen, G. Thornton, The bonding of hydrogen on water-dosed Si(111). J. Vacuum Sci. Technol. A **4**, 1451 (1986)
66. F. Meyer, M.J. Sparnaay, The chemical and physical properties of clean Germanium and Silicon surfaces, in *Surface physics of phosphors and semiconductors*, ed. by C.G. Scott, C.E. Reed (Academic Press, London, 1975), chap. 6, pp. 321
67. R.H. Williams, I.T. McGovern, Adsorption on semiconductors, in the chemical physics of solid surfaces and heterogeneous catalysis, vol. 3, ed. by D.A. King, D.P. Woodruff (Elsevier, Amsterdam, 1984), chap. 6, pp. 267
68. W. Langel, M. Parrinello, Hydrolysis at stepped MgO surfaces. Phys. Rev. Lett. **73**, 504 (1994)
69. L. Giordano, J. Goniakowski, J. Suzanne, Partial dissociation of water molecules in the (3$\times$ 2) water monolayer deposited on the MgO(100) surface. Phys. Rev. Lett. **81**, 1271 (1998)
70. M. Odelius, Mixed molecular and dissociative water adsorption on MgO[100]. Phys. Rev. Lett. **82**, 3919 (1999)
71. K.C. Hass, W.F. Schneider, A. Curioni, W. Andreoni, The chemistry of water on alumina surfaces: reaction dynamics from first principles. Science **282**, 265 (1998)
72. P.J.D. Lindan, N.M. Harrison, Mixed dissociative and molecular adsorption of water on the rutile (110) surface. Phys. Rev. Lett. **80**, 762 (1998)
73. A. Vittadini, A. Selloni, F.P. Rotzinger, M. Gratzel, Structure and energetics of water adsorbed at TiO_2 anatase (101) and (001) surfaces. Phys. Rev. Lett. **81**, 2954 (1998)

74. R. Schaub et al., Oxygen vacancies as active sites for water dissociation on rutile $TiO_2(110)$. Phys. Rev. Lett. **87**, 266104 (2001)
75. Y.S. Chu, T.E. Lister, W.G. Cullen, H. You, Z. Nagy, Commensurate water monolayer at the $RuO_2(110)$ water interface. Phys. Rev. Lett. **86**, 3364 (2001)
76. M. Odelius, M. Bernasconi, M. Parrinello, Two dimensional ice adsorbed on mica surface. Phys. Rev. Lett. **78**, 2855 (1997)
77. L. Cheng, P. Fenter, K.L. Nagy, M.L. Schlegel, N.C. Sturchio, Molecular-scale density oscillations in water adjacent to a mica surface. Phys. Rev. Lett. **87**, 156103 (2001)
78. S. Fölsch, M. Henzler, Water adsorption on the NaCl surface. Surf. Sci. **247**, 269 (1991)
79. S. Fölsch, A. Stock, M. Henzler, Two-dimensional water condensation on the NaCl(100) surface. Surf. Sci. **264**, 65 (1992)
80. L.W. Bruch, A. Glebov, J.P. Toennies, H. Weiss, A helium atom scattering study of water adsorption on the NaCl(100) single crystal surface. J. Chem. Phys. **103**, 5109 (1995)
81. M. Foster, G.E. Ewing, An infrared spectroscopic study of water thin films on NaCl(100). Surf. Sci. **427**, 102 (1999)
82. A. Allouche, Water adsorption on NaCl(100): a quantum ab-initio cluster calculation. Surf. Sci. **406**, 279 (1998)
83. H. Meyer, P. Entel, J. Hafner, Physisorption of water on salt surfaces. Surf. Sci. **488**, 177 (2001)
84. M.F. Reedijk, J. Arsic, F.F.A. Hollander, S.A. de Vries, E. Vlieg, Liquid order at the interface of KDP crystals with water: evidence for icelike layers. Phys. Rev. Lett. **90**, 066103 (2003)
85. D. Chakarov, B. Kasemo, Photoinduced crystallization of amorphous ice films on graphite. Phys. Rev. Lett. **81**, 5181 (1998)
86. T. Werder, J.H. Walther, R.L. Jaffe, T. Halicioglu, P. Koumoutsakos, On the water–carbon interaction for use in molecular dynamics simulations of graphite. J. Phys. Chem. B **107**, 1345 (2003)
87. L.F. Scatena, M.G. Brown, G.L. Richmond, Water at hydrophobic surfaces: weak hydrogen bonding and strong orientation effects. Science **292**, 908 (2001)
88. T.R. Jensen, M.O. Jensen, N. Reitzel, K. Balashev, G.H. Peters, K. Kjaer, T. Bjornholm, Water in contact with extended hydrophobic surfaces: direct evidence of weak dewetting. Phys. Rev. Lett. **90**, 086101 (2003)
89. L. Yang, H.W. Huang, Observation of a memberane fusion intermediate structure. Science **297**, 1877 (2002)
90. L. Guidoni, V. Torre, P. Carloni, Potassium and sodium binding to the outer mouth of the K^+ channel. Biochemistry **38**, 8599 (1999)
91. K. Murata et al., Structural determinants of water permeation through aquaporin-1. Nature **407**, 599 (2000)
92. P.A. Thiel, T.E. Madey, The interation of water with solid surfaces: fundamental aspects. Surf. Sci. Rep. **7**, 211 (1987)
93. M.A. Henderson, The interation of water with solid surfaces: fundamental aspects revisited. Surf. Sci. Rep. **46**, 1 (2002)
94. A. Hodgson, S. Haq, Water adsorption and the wetting of metal surfaces. Surf. Sci. Rep. **64**, 381 (2009)
95. A. Verdaguer, G.M. Sacha, H. Bluhm, M. Salmeron, Molecular structure of water at interfaces: wetting at the nanometer scale. Chem. Rev. **106**, 1478 (2006)
96. N.E. Levinger, Water in confinement. Science **298**, 1722 (2002)
97. P. Ball, How to keep dry in water. Nature **423**, 25 (2003)
98. J. Teixeira, J.M. Zanotti, M.C. BellissentFunel, S.H. Chen, Water in confined geometries. Phys. B **234**, 370 (1997)
99. R. Bergman, J. Swenson, Dynamics of supercooled water in confined geometry. Nature **403**, 283 (2000)
100. K. Koga, G.T. Gao, H. Tanaka, X.C. Zeng, Formation of ordered ice nanotubes inside carbon nanotubes. Nature **412**, 802 (2001)

101. G. Hummer, J.C. Rasaiah, J.P. Noworyta, Water conduction through the hydrophobic channel of a carbon nanotube. Nature **414**, 188 (2001)
102. M.S.P. Sansom, P.C. Biggin, Water at the nanoscale. Nature **414**, 156 (2001)
103. C. Dellago, M.M. Naor, G. Hummer, Proton transport through water-filled carbon nanotubes. Phys. Rev. Lett. **90**, 105902 (2003)
104. D.J. Mann, M.D. Halls, Water alignment and proton conduction inside carbon nanotubes. Phys. Rev. Lett. **90**, 195503 (2003)
105. R.P. Feynman, There is plenty of room at the bottom, in *Annual Meeting of the American Physical Society*. 29th Dec, 1959. http://www.zyvex.com/nanotech/feynman.html

Chapter 3
Theoretical Approaches

3.1 A Brief History of Basic Water Research and Common Theoretical Methods

Da Fuenci was first fascinated by the behavior of water and engaged in thinking and exploring its modern scientific implications. Early studies of water were an organic combination of theoretical thinking and experimental discoveries. The catalytic effect was discovered through the study of water: Doebereiner found in 1823 that H_2 and O_2 on the surface of Pt produced water at room temperature [1], which also revealed the chemical composition of water. After 100 years of effort, starting with the experiments of Priestley and Cavendish in 1781 [2] to Morley's research in 1895 [3], that the ratio of hydrogen to oxygen in water was finally established at 2:1. In the 1930s, Mecker et al. determined that water molecules have a V-shaped structure through spectral analysis [4]. Summarizing various laws about the structure and binding of water molecules, Bernal, Fowler, and Pauling proposed a rule for the structure of water molecules in ice, called the BFP rule [5, 6]. This was the first systematic theoretical understanding of the laws of water. In 1957, Van Thiel et al. started to study the structure of small neutral clusters of water using molecular spectroscopy [7]. Since modern times, a large number of surface analysis techniques have been used for the experimental detection of water. The interactions of various metals (Pt, Pd, Au, Ag, Rh, Ru, Al), semiconductors (Si, Ge, GaAs, GaN, InP), oxides (MgO, TiO_2, Al_2O_3, mica), salts (NaCl), graphite, paraffin, and liquids (CCl_4), biological membranes, proteins, etc. with water have been studied in detail. Of these, the interaction between metal surfaces and water has been studied the most due to the relatively simple structure. Recently, synchrotron radiation techniques such as X-ray absorption, Raman scattering, terahertz spectroscopy, and nonlinear optical techniques, together with surface science techniques represented by scanning tunneling microscopy, have been used extensively to study bulk and interfacial water.

Theoretically, the first computer simulation of liquid water was performed by Barker and Watts in 1969 [8]. A large number of empirical potential models for water were proposed in the 1970s–1980s, including the commonly used ST2, SPC,

S. Meng and E. Wang, *Water*,
https://doi.org/10.1007/978-981-99-1541-5_3

TIP4P models, etc. [9]. In 1993, first-principles calculations based on density functional were used to simulate liquid water [10], and found that water molecules in liquid water have a very large dipole moment (~3.0D, compared to 1.6D for isolated water molecules). The quantum effects of hydrogen atoms in water were subsequently treated by path integration [11], and it was found that the quantum behavior leads to a reduction in the potential barrier for H transport. A large number of first-principles computational techniques have since been used to study bulk water, cluster water, and water adsorption on solid surfaces, and water surfaces. These include calculations by Peter J. Feibelman in 2002, which suggested that water has a semi-decomposed structure on the Ru(0001) surface [12]; in the same year, Meng Sheng et al. revealed the undecomposed structure of water on Pt(111) and discovered two types of hydrogen bonds in the surface water layer [13]; and in 2004, Ranea et al. found that quantum behavior leads to a water dimer molecule on the Pd(111) surface to diffuse at a very fast rate [14]; in 2010, theoretical calculations and scanning tunneling studies revealed the presence of a large number of five- and seven-membered rings in the molecular structure of the infiltrated water layer on the Pt(111) surface [15], etc.

The basic approach to theoretical simulation of water can be divided into three levels:

1. **Large-scale continuous medium simulations**

The earliest and simplest way to describe the structure and properties of water is to use the continuous medium model, i.e., to ignore the particulate microstructure of water and treat water as a medium distributed continuously in space, where the structure (e.g., density, etc.) and some properties (heat capacity, dielectric, etc.) of water can be expressed in the functions of Cartesian coordinates (x, y, z) of points distributed continuously in three-dimensional space. Using the continuous medium model, classical hydrodynamics was developed. It is very effective for understanding some of the macroscopic properties of water and is widely used when the behavior of water at the molecular scale is not emphasized.

Previously, a large number of simulations of fluids like water have been based on continuous medium models, such as those based on methods such as Finite Element and Lattice Boltzmann. It has been argued that continuous medium models can be used effectively at scales of more than 10+ water molecules. Their advantage is that they can simply deal with very large systems by ignoring details at the molecular level. They are mostly applied to simulations on large scales (microns and above) and are of great use in understanding the behavior of water in conventional fluids, microfluidic tubes, and bio-hygienic systems such as blood flow. However, due to the lack of molecular details, they cannot generally be used to explain water structure at the molecular level or interactions with other systems (biomolecules or surfaces). Due to the conventional nature of the methods, they will not be repeated here. A noteworthy trend is the organic combination of continuous medium simulations and molecular-level simulations based on the first principles or classical potential fields to model the behavior of water at multiple scales (multiscale modeling).

2. **Classical models**

Due to the limitations of early computer hardware, classical models are usually used in the literature to study the behavior of water. There are now about 100 various water models. Based on the type of charge, the two common types of models are the simple point charge model and the polarizable model, each of which has several common types. The former assigns a certain number of fractional charges to H and O atoms on a fixed water molecule structure to describe the polarity of the water molecule, such as ST2, SPC/E, TIP3P, TIP4P, TIP5P, and so on. Since water molecules are easily polarized and the polarity of water varies greatly in different environments, the treatment of fixed molecular structures (bond lengths and angles) and charges is often inadequate, so models with variable bond lengths and angles such as SPC/F, CF, NCF, etc., or allowing the value of the point charge to vary with the environment, such as ASP-W, PPC, POL5, etc., have been introduced. More complex models contain both structural and charge number variations such as DCF, CKL, SPC/FP, NCCvib, and POLARFLEX to name a few. The most complex water models contain more than 70 parameters, which is a huge difference from earlier models that simulated liquid water with a mixture of two bulk phases or replaced water molecules with point particles. These illustrate the complexity of water behavior and the limited and arbitrary nature of the model potential.

3. **First-principles simulations**

The first-principles approach describes the hydrogen–oxygen bonding in water, the hydrogen bonding between water molecules, the interaction between water and the outside, and the resulting structural properties and thermodynamic behavior of water at the atomic level by solving the electron Schrödinger equation. So first-principles calculations naturally give changes in bond lengths and bond angles of water molecules as the environment changes, and increases and decreases in the polarity of water, phenomena that are not easily described by simple models; and because it does not use any empirical parameters associated with water, it is highly accurate and overcomes the arbitrariness in classical models. These can help us to obtain data on the structure and stability of various kinds of water, especially for water on the nanoscale, which is almost the only valid approach. This is because the commonly used experimental methods at the nanoscale are extremely difficult to manipulate and the empirical parameters of water based on the bulk phase often fail. However, due to the huge amount of first-principles calculations and molecular dynamics calculations, the size of the water system simulated and the simulation time is one to three orders of magnitude smaller (less) than in classical simulations. Great success has been achieved in determining the structure of water on surfaces using first-principles calculations: by comparison with scanning tunneling microscopy images, the structure of water bilayers and water clusters, the kinetics of water decomposition, and the diffusion growth behavior of water on metal surfaces have been determined.

3.2 Classical Models of Water

Since 1933, when Bernal and Fowler published the first article on the empirical potential of water [5], many models of water have been developed. Most of these models rely on predetermined parameters and classical descriptions of water and hydrogen bonds, and thus they are empirical models. The water molecule is considered as a simple structureless point, or a multipoint particle with different charge pointings, to model the charge distribution of oxygen atoms, hydrogen atoms, or lone electron pairs. Directed electrostatic forces between point charges and non-directed dispersion forces between atoms [often in the form of the Lennard–Jones potential] represent the interactions between water molecules. These point charge models are on the basis that hydrogen bonds interactions contain mainly Coulomb interactions and a portion of the contribution of charge repulsion and dispersion forces [16]. Assuming a water molecule with m point charge sites and a Lennard–Jones site at the oxygen atom, the pairwise interaction potential can be written as

$$U = \sum_{i=1}^{m} \sum_{j=1}^{m} \frac{q_i q_j}{r_{i,j}} + \frac{A}{r_{OO}^{12}} - \frac{C}{r_{OO}^{6}},$$

where q_i and q_j are point charges at molecules i and j, and A and C are Lennard–Jones potential parameters. m, q, A, and C have model-dependent values. Table 3.1 lists the values of the relevant parameters in some typical models. The potential model incorporating electrostatic and charge repulsions can successfully explain the tetrahedral arrangement of water molecules in ice and liquid water. However, the Lennard–Jones interaction obeys a repulsion of $1/r^{12}$, which gives a large first peak of the oxygen–oxygen pair correlation function and an inadequate response to external pressure [17]. The weaker overlap integrals and Gaussian functions obtained based on quantum mechanical calculations can be used to describe these mutual repulsions well.

In general, the construction and development of each model should first be able to fit the experimentally determined structural and physical properties of water (e.g., pair correlation functions, diffusion coefficients, density anomalies or critical points, etc.) or be consistent with the results given by the exact ab initio potential energy (e.g., MCY, PPC, ASP-W models [18], etc.). Water is a deformable polar molecule. Models that do not reflect these characteristics are inaccurate and cannot be used to predict discoveries. In Table 3.1, the structures of water molecules described by different models, including bond lengths of hydrogen–oxygen bonds and hydrogen–oxygen-hydrogen bond angles, also vary, e.g., some water molecules are more rigid, some are more flexible, some have a fixed point charge, and some contain a polarized charge. In this way, water molecules with different properties are also described by applying the corresponding rigid, flexible, or charge-polarized water models. In general, rigid water models are used to describe water molecules qualitatively and are easier to implement, but charge-polarized water models have better transferability. We briefly discuss some typical models below. A review article on the development of water

Table. 3.1 Parameter settings for some typical models of water. The units are: 10^{-3} kcal·Å^{12}/mol (A), kcal·Å^6/mol (C), Å (length), degree (angle), and basic charge (e). M denotes the extra charge site

Model[a]	A	C	r_{OH}	HOH	q_O	q_H	r_{OM}	q_M
BF(1933,R)[18]	560.4	837.0	0.96	105.7	0.0	0.49	0.15	– 0.98
ST2(1973,R)[40]	238.7	268.9	1.0	109.47	0.0	0.2375	0.8	– 0.2375
SPC(1981,R)[41]	629.4	625.5	1.0	109.47	– 0.82	0.41	0.0	0.0
SPC/E(1987,R)[42]	629.4	625.5	1.0	109.47	– 0.8472	0.4238	0.0	0.0
SPC/HW(2001,R)[43]	629.4	625.5	1.0	109.47	– 0.8700	0.4350	0.0	0.0
PPC(1994,P)[44]	750.8	656.2	0.9430	106.00	0.0	0.5170	0.06	– 1.0340
TIP3P(1981,R)[45]	582.0	595.0	0.9572	104.52	– 0.834	0.417	0.0	0.0
TIP4P(1983,R)[45]	600.0	610.0	0.9572	104.52	0.0	0.52	0.15	– 1.04
TIP4P/FQ(1994,P)[46]	600.0	610.0	0.9572	104.52	0.0	0.63[b]	0.15	– 1.26[b]
TIP5P(2000,R)[47]	544.5	590.3	0.9572	104.52	0.0	0.2410	0.70	– 0.2410
SSD[c](2003)[48]	8296.1	11,022.7	–	109.47	–	–	–	–

[a] Abbreviations for water models. Year and type of presentation are in parentheses: R indicates rigid model; P indicates polarizable model
[b] Average
[c] Has only one action site distributed in the center of mass and a tetrahedral action potential

molecule models can be found in the literature [19]; some comments on the results obtained can be found in the literature [17].

(1) Two-state models

The two-state model considers water with two structural phases: ice-Ih and ice-II. In both phases, the water molecules exhibit similar tetrahedral coordination bonds, but the distances between the longer sub-nearest neighbors in the radial direction are 4.5 and 3.5 Å. In liquid water, the distribution and alternation of these two states determine the properties of water and its response to temperature and pressure. The two-state model has been used to explain water density anomalies and the effect of hydrogen bond networks on the thermodynamic behavior of water [20].

(2) Point charge models

The point charge model views water as a point charge or a point dipole. Hydrogen bonds are represented by a tetrahedral potential (e.g., the SSD model) or by simply counting the number of hydrogen bonds based on the orientation of the water molecules.

(3) Rigid models

The molecular model incorporates the intrinsic water structure and hydrogen bonding. The simplest type is the rigid model, which treats the water molecule as a rigid rotor with a fixed hydrogen–oxygen bond length and a hydrogen–oxygen-hydrogen bond angle. Point charges are distributed on these three atoms and the center of mass of the molecule or the central position of the lone pair of electrons. These positions correspond to the 3, 4, and 5-sites models, respectively. Usually, we choose the structural parameters and dipole distances of gaseous or liquid water. Most of the available empirical models are rigid models, such as the BF model, the popular ST2, SPC/E, TIP3P, TIP4P models, and the latest TIP5P and DEC models. Although rigid models are very simple and use very few parameters, these models reproduce well most of the properties of water molecules [17].

(4) Flexible models

Based on the rigid model, the flexible model relaxes the stiffness of the water molecule. Depending on the valence ionization potential of the water molecule, both the hydrogen–oxygen bond length and the hydrogen–oxygen-hydrogen bond angle inside it are allowed to be adjusted. This type includes CF, SPC/F, and NCF models, among others.

(5) Charge polarization models

Similarly, we can consider the polarization rates of water molecules as in the ASP-W, PPC, and POL5 models. These charges are not fixed and are determined by the water molecules in the immediate vicinity. Moreover, some advanced models that cover both elastic and polarization effects: such as DCF, CKL, SPC/FP, NCCvib, and POLARFLEX models [17], have emerged one after another.

These empirical models provide a good semi-quantitative description of the fundamental properties of water, such as liquid density, the heat of evaporation, diffusivity, structure, maximum density, critical parameters, dielectric constant, etc. These parameters in turn provide a great deal of information on the evolution of hydrogen bond networks in water. But analyzing the simulation results from about 50 different models, they all lack the consistency of the water simulations, exposing more or less the limitations of the different models. Even the ASP-W model, which contains as many as 70 fitted parameters, is no exception [18]. No model potential can reproduce all the properties of the water molecule simultaneously and completely. To give a specific example, the temperature corresponds to the maximum density of water. Figure 3.1 gives a plot of the temperature corresponding to the maximum density of water predicted by these classical models. Obviously, these results depend on the model itself. For the SPC/E model, the temperature corresponding to the maximum density is 235 K; for the TIP4P and TIP5P models, the temperatures corresponding to the maximum density are 248 K and 272 K, respectively; for the ST2 model, the temperature corresponding to the maximum density is at 304 K; while the SPC and TIP3P models have no maximum [17]. It follows that these values are all distinct from the temperature corresponding to the maximum density of real water at 4 °C (277 K). As we expected, the charge-polarized models (solid dots) are generally better than the non-polarized ones (circles), reflecting the importance of the polarization effect of water molecules.

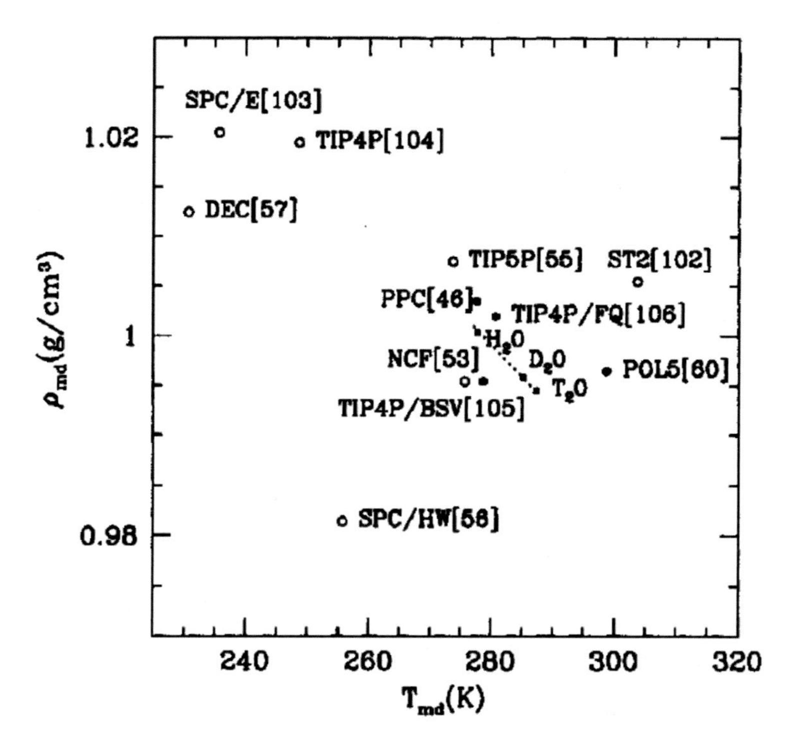

Fig. 3.1 The maximum density point of water (and its corresponding temperature) predicted by the classical model [17]. Circle: non-charge polarization model; Solid point: polarization model. The experimental values of H_2O, D_2O and T_2O are connected by dotted lines

Another example is that the popular TIP4P model underestimates the tetrahedral bonding environment of a water molecule and thus incorrectly predicts its dielectric constant. Also, models such as SPC/E, PPC, BSV, CC, DC, and TIP4P fail to correctly describe the oxygen–oxygen radial distribution function. For the hydrogen-bonded first coordination layer in liquid water, the results obtained from molecular dynamics simulations with these models are also quite different from those obtained from X-ray absorption spectroscopy. This result reveals the complexity of the behavior of water molecules and the need to develop more accurate classical models of water.

The limitations of these model potentials have several physical causes. First, the flexibility and polarization of water molecules are not well described in classical models, which makes it difficult to be accurately reproduced as well. Second, the electronic structure and dynamic charge transfer of water molecules are often inappropriately treated. The quantum nature of hydrogen is neglected in all models. Moreover, there are still many uncertainties in the numerical selection of the given physical quantities as fitting parameters. For example, the dipole distance of liquid water varies from 1.8 to 3.0D. However, the dipole moment is very sensitive in determining the structure, diffusivity, and dielectric constant of water. Ab initio calculations based on electronic structure provide a promising approach to overcome the shortcomings of these classical models and resolve the difficulties encountered. This approach can be used to describe the nature of the flexibility and charge polarization of water molecules, as well as many-body effects from the first principle of quantum mechanics. Moreover, it also succeeds in describing the interactions between different surfaces and between surfaces and molecules, often without relying on empirical parameters.

3.3 First-Principles Calculations and Density Functional Theory

First-principles calculation, also known as ab initio calculation, is a computational method for studying the structure and properties of matter from the perspective of electron motion since the emergence of quantum mechanics. This method is based on quantum mechanics to deal with the motion of electrons in a system, to obtain the wave function of electrons and the corresponding eigenenergy, so as to find the total energy of the system and the properties of bonding, elasticity, stability, etc. With the increasing computing power of contemporary computers, it has been increasingly applied to the study of solid, surface, macromolecular, and biological systems.

Since real systems usually have many electrons involved, strictly speaking, we need to deal precisely with the multi-electron Schrödinger equation.

$$H\psi(\vec{r},\vec{R}) = E\psi(\vec{r},\vec{R}) \tag{3.1}$$

where $\vec{r}$, $\vec{R}$ represents the coordinates of all electrons and nuclei, respectively. The Hamiltonian can be written in three parts, representing the kinetic energy of the electron, the energy of the nucleus, and the interaction of the electron with the nucleus, respectively.

$$\begin{aligned} H = H_e + H_N + H_{e-N} \\ &= -\sum_i \frac{\hbar^2}{2m_e}\nabla_i^2 \\ &\quad - \sum_n \frac{P^2}{2M_n} + \frac{1}{2}\sum_{m\neq n} V_N(\vec{R}_m - \vec{R}_n) \\ &\quad + \frac{1}{2}\sum_{i\neq j}\frac{e^2}{\left|\vec{r}_i - \vec{r}_j\right|} - \sum_{i,n}\frac{Z_n e^2}{\left|\vec{r}_i - \vec{R}_n\right|} \end{aligned} \tag{3.2}$$

Since the mass of the electron is much smaller than the nucleus, it moves much faster than the thermal motion of the nucleus, and the study of the electron motion can be done by considering the nucleus to be immobile, thus separating the motion of the nucleus from that of the equation, which is the Born-Oppenheimer approximation (adiabatic approximation) [21].

The crux of the problem is that the strict wave function of multiple electrons is difficult to express and calculate. Kohn even asserted that when the number of electrons is greater than 1000, the multi-electron wave function $\psi(\vec{r})$ is an unreasonable scientific concept that cannot be accurately calculated and recorded using today's computer and storage technology [22, 23]. Instead, a cubic meter of real solid has about 10^{27} electrons!

To solve this puzzle, various approximations have been made to the electron action. If the electron–electron interaction is completely neglected, the electron wave function can be written in the form of a product of single-electron wave functions: $\psi(\vec{r}) = \varphi_1(\vec{r}_1)\varphi_2(\vec{r}_2)\cdots\varphi_N(\vec{r}_N)$; then the many-body Schrödinger Eq. (3.1) for the electron reduces to the Hartree equation for the single electron [24].

$$\left[-\frac{\hbar^2}{2m_e}\nabla^2 + V(\vec{r}) + \sum_{j(\neq i)}\int d\vec{r}'\frac{|\phi_j(\vec{r}')|^2}{|\vec{r}-\vec{r}'|}\right]\phi_i(\vec{r}) = E_i\phi_i(\vec{r}) \tag{3.3}$$

where $V(\vec{r})$ is the potential field generated by the nucleus of the system. If the exchange antisymmetry of the electron wave function is taken into account, the Hartree–Fock equation is obtained by replacing the many-body wave function by the Slater determinant of the single-electron wave function [25].

$$\left[-\frac{\hbar^2}{2m_e}\nabla^2 + V(\vec{r}) + \sum_{j(\neq i)}\int d\vec{r}'\frac{|\phi_j(\vec{r}')|^2}{|\vec{r}-\vec{r}'|}\right]\phi_i(\vec{r}) - \sum_{j(\neq i)}\int d\vec{r}'\frac{\phi_j^*(\vec{r}')\phi_i(\vec{r}')}{|\vec{r}-\vec{r}'|}\phi_j(\vec{r}) = E_i\phi_i(\vec{r}) \tag{3.4}$$

Accurate quantum chemical calculations can be performed on individual water molecules or clusters of water using the Hartree–Fock equation, especially after considering perturbation corrections due to electron correlation effects. These methods include MP2, MP3, CCSD(T), CAS, CI, and quantum Monte Carlo methods. However, for larger or periodic aqueous systems, such calculations are too large or even impossible to perform. These methods are not described here; the interested reader is referred to quantum chemistry textbooks. To make some really meaningful studies for the water interaction system, other viable approaches must be found.

In 1927, Thomas and Fermi proposed the theory of using the electron density instead of the wave function to represent the energy [26, 27], which of course greatly reduced the variables in the calculation, but the results of this theory were crude and unreliable due to the lack of precise treatment of the kinetic energy term. In the 1960s, Kohn et al. successfully developed the idea of using the electron density as a system variable [28, 29], which is the modern Density Functional Theory (DFT).

3.3.1 Density Functional Theory

Density functional theory writes Hamiltonian as functional of the electron density. Hohenberg and W. Kohn formulated the famous Hohenberg–Kohn theorem in 1964 [28].

Theorem 1: The ground state energy and potential of the fully homogeneous fermionic systems without spin are uniquely determined by the particle number density $\rho(\vec{r})$.

Theorem 2: The energy functional $E[\rho]$ on particle density $\rho(\vec{r})$ takes the minimum value for the correct particle density $\rho(\vec{r})$ with constant particle number and is equal to the ground state energy.

The proofs of these two theorems are very simple and can be found in the literature [22, 28, 30]. According to the Hohenberg–Kohn theorem, the energy can be written as a functional of the electron density, and solving for the variance gives the ground state energy and the ground state wave function.

$$\begin{aligned} E[\rho] &= T[\rho] + U[\rho] + \int d\vec{r} v(\vec{r})\rho(\vec{r}) \\ &= T[\rho] + \frac{1}{2}\int\int d\vec{r}d\vec{r}'\frac{\rho(\vec{r})\rho(\vec{r}')}{|\vec{r}-\vec{r}'|} + E_{xc}[\rho] + \int d\vec{r} v(\vec{r})\rho(\vec{r}) \end{aligned} \tag{3.5}$$

But the form of the kinetic energy term $T[\rho(\vec{r})]$ is still unknown. And it is still a many-body equation. So Kohn and Sham (L. J. Sham) proposed [29]: to replace the actual kinetic energy $T[\rho(\vec{r})]$ by the kinetic energy functional of the multiparticle system without interaction $T[\rho(\vec{r})]$, attributing the difference to the unknown exchange–correlation term $E_{xc}[\rho(\vec{r})]$, thus transforming into a single-electron image. Namely:

$$\rho(r) = \sum_{i=1}^{N} |\phi_i(\vec{r})|^2 \tag{3.6}$$

$$T_0[\rho] = \sum_{i=1}^{N} \int d\vec{r}\phi_i * (\vec{r})(-\nabla^2)\phi_i(\vec{r}) \tag{3.7}$$

The variation on $\rho(\vec{r})$ can be reduced to the variation on $\phi_i(\vec{r})$

$$\delta\left\{E\left[\rho(\vec{r})\right] - \sum_{i=1}^{N} E_i\left[\int d\vec{r}\phi_i * (\vec{r})\phi_i(\vec{r}) - 1\right]\right\}/\delta\phi_i(\vec{r}) = 0 \tag{3.8}$$

Namely:

$$\left\{-\nabla^2 + v(\vec{r}) + \int dr\frac{\rho(\vec{r})}{|\vec{r}-\vec{r}'|} + \frac{\delta E_{xc}[\rho]}{\delta\rho}\right\}\phi_i(\vec{r}) = E_i\phi_i(\vec{r}) \tag{3.9}$$

This is the single-electron Kohn–Sham equation. Unlike the Hartree–Fock approximation, it is still strictly a single-electron equation. Unfortunately, the term $E_{xc}[\rho(\vec{r})]$ that generalizes all the complex interactions is unknown.

3.3.2 *Approximate Forms of Exchange–Correlation Functionals: Local Density Approximation and Generalized Gradient Approximation*

Since the exchange–correlation functional $E_{xc}[\rho(\vec{r})]$ is unknown, an appropriate approximation is required for the specific calculation. The Local Density Approximation (LDA) is the simplest and most effective approximation in practice [29]. It was first proposed and applied by Slater in 1951 [31, 32], even before density functional theory. This approximation assumes that the exchange–correlation energy at a location is only related to the density at that location and is equal to the exchange–correlation energy of a homogeneous electron gas of the same density.

$$E_{xc}^{LDA}[\rho] = \int d\vec{r}\rho(\vec{r})\varepsilon_{xc}^{unif}(\rho(r)) \tag{3.10}$$

The most commonly used local density approximation for exchange–correlation in specific calculations today is based on the Monte-Carlo method for homogeneous electron gases by Ceperley and Alder, [33, 34].

$$\left(r_s = \sqrt[3]{3/4\pi\rho}\right)$$

$$\varepsilon_x^{LDA}(r_s) = -0.9164/r_s \tag{3.11}$$

$$\varepsilon_c^{LDA}(r_s) = \begin{cases} -0.2846/\left(1 + 1.0529\sqrt{r_s} + 0.3334r_s\right) & (r_s \geq 1) \\ -0.0960 + 0.0622\ln r_s - 0.0232r_s + 0.0040r_s\ln r_s & (r_s \leq 1) \end{cases} \tag{3.12}$$

The LDA approximation has demonstrated great success in most materials calculations. Experience shows that LDA calculations of atomic free energy and molecular dissociation energy have errors in the range of 10–20%; for molecular bond lengths and crystal structures they can be accurate to about 1% [35]! However, LDA is not applicable for systems that are too far from a homogeneous electron gas or a slowly varying electron gas in space.

More precise considerations need to account for the effect of the charge density in the vicinity of a site on the exchange–correlation energy, for example by considering the contribution of a first-order gradient in density to the exchange–correlation energy,

$$E_{xc}^{GGA}[\rho] = \int d\vec{r} f_{xc}(\rho(r), |\nabla\rho(r)|) \tag{3.13}$$

the exchange energy can be taken as the modified Becke functional form [36, 37] ($x = |\nabla\rho|/\rho^{4/3}$, β is a constant).

$$E_x^{GGA} = E_x^{LDA} - \beta \int d\vec{r} \rho^{4/3} \frac{(1 - 0.55\exp[-1.65x^2])x^2 - 2.40 \times 10^{-4}x^4}{1 + 6\beta x \sinh^{-1} x + 1.08 \times 10^{-6}x^4} \tag{3.14}$$

This is called the Generalized Gradient Approximation (GGA) [38]. This approximation is semi-localized and, in general, gives more accurate energies and structures than LDA. It is more applicable for open electronic systems. Currently in common use are Becke [36], Perdew–Wang 91 [39, 40] and its more condensed form PBE [41], and BLYP [42].

Further, one can also take into account the approximation of higher-order gradients of density, which is called Meta-GGA or Post-GGA. This aspect has been studied but the accuracy of the calculations is not significantly improved compared to GGA, and there is still a need to find a sufficiently accurate and simple enough form. The exchange–correlation energy, as a function of the electron density, is in principle nonlocal. LDA and GGA make local and semi-local approximations, respectively, without considering nonlocal effects at all. Neither of them can correctly describe the non-local exchange–correlation action. For some weak effects, such as the Van der Waals force, non-local correlation effects need to be taken into account. Several works have attempted to describe the Van der Waals interactions within the framework of density functional theory, expecting more reliable binding energies, and some breakthroughs have been made [43, 44]. To improve the accuracy, in addition to the dependence of the total energy of the system on the electron density, one can also consider the dependence on the kinetic energy, i.e., the orbit-dependent exchange–correlation functional. In addition, the LDA or GGA approximation is actually equivalent to assuming weaker inter-electron correlations, so density functional theory is generally unable to handle strongly correlated systems, and this aspect requires combining LDA with dynamic mean-field theory (DMFT), and in some cases simplified methods such as LDA + U can be used.

Since the Hartree–Fock method gives a rigorous definition of the exchange energy, attempts have been made to introduce this rigorous exchange energy into the density functional method in place of local or semi-local approximations. But simply adding the Hartree–Fock exchange term to the density functional correlation term does not give satisfactory results. Essentially this is because our division between exchange and correlation energies is artificial, and it only makes physical sense to put correlation and exchange together (i.e., exchange–correlation terms). The so-called strict exchange energy in the Hartree–Fock method is actually only a strict form of mathematics, which cannot be matched directly with local correlation. In 1993, A. Becke proposed to replace the exchange term in the density functional by a mixture of the "strict" exchange energy and the local exchange energy in a certain ratio [45].

This is the hybrid functional approach. This method has been very successful, it is more accurate than the general GGA and LDA, especially B3LYP is generally better than other exchange–correlation forms in the calculation of chemical clusters. Much of the success of the hybrid density functional comes from the correction of self-interactions by "strict" exchange energy. Although this method is more accurate, the introduction of non-local "strict" exchange terms increases the computational effort, especially for periodic systems, and the energy converges slowly. For this reason, Jochen Heyd, Gustavo E. Scuseria, and Matthias Ernzerhof proposed the use of the screened Coulomb potential to calculate the exchange term, the HSE method [46]. This method speeds up the convergence of the system and still maintains accuracy. Even so, the hybrid density functional computation is still 10 to 100 times slower than the usual GGA computation.

3.3.3 Pseudopotential Methods

The actual solution of the Kohn–Sham equation is made difficult by the fact that the potential field terms $v(\vec{r})$ $1/r$ generated by the nucleus are divergent at the center of the atom, the wave function varies dramatically, and a large number of plane wave expansions are required. The usual practice is to divide the space region into Muffin-tin (pillbox) spheres, with the part of the wave function inside the sphere expanded by spherical waves (and their derivatives to energy), while setting the atomic potential field outside the sphere to be a constant zero. The energy band calculation methods developed and matured on this basis are the LAPW (Linearized Augmented Planewaves) and LMTO (Linear Combination of Muffin-tin Orbits) methods [47, 48]. It is also possible to use the pseudopotential method [30, 49–51], that is, by adding the coulomb attraction potential of the nucleus to a short-range repulsion potential, the sum of the two terms (pseudopotential) becomes relatively flat near the nucleus, while the resulting eigenenergy and valence electron wave function are the same as the real eigenvalue and the real wave function outside the nucleus (Fig. 3.2).

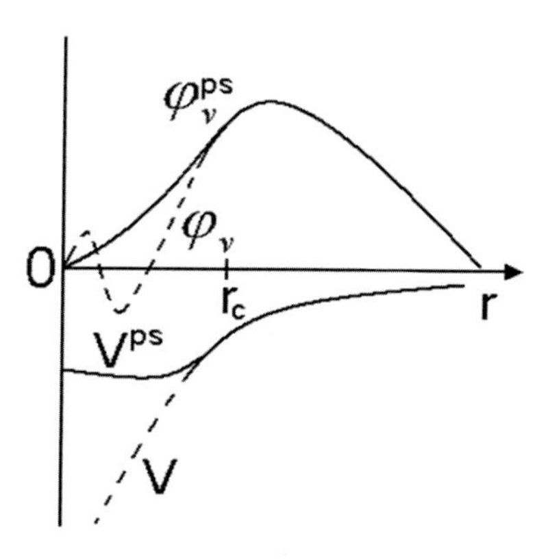

Fig. 3.2 Schematic diagram of the pseudopotential

The exact Hamiltonian H for the crystals is:

$$H|\phi_v\rangle = E_v|\phi_v\rangle, \quad H|\phi_c\rangle = E_c|\phi_c\rangle \tag{3.15}$$

where $|\phi_v\rangle$ and $|\phi_c\rangle$ are the strictly electronic wave functions of the valence states E_v and core states E_c, respectively. Considering $\langle\phi_c|\phi_v\rangle = 0$, constructing the pseudo-wave functions $|\phi_v^{Ps}\rangle$:

$$|\phi_v^{ps}\rangle = |\phi_v\rangle + \sum_c \langle\phi_c \mid \phi_v^{ps}\rangle|\phi_c\rangle \tag{3.16}$$

Then there are:

$$\begin{aligned}(H - E_v)|\phi_v^{ps}\rangle &= (H - E_v)\left(|\phi_v\rangle + \sum_c \langle\phi_c \mid \phi_v^{ps}\rangle|\phi_c\rangle\right) \\ &= (H - E_v)\sum_c |\phi_c\rangle\langle\phi_c \mid \phi_v^{ps}\rangle \\ &= \sum_c (E_c - E_v)|\phi_c\rangle\langle\phi_c \mid \phi_v^{ps}\rangle\end{aligned} \tag{3.17}$$

So there is:

$$\left(H + \sum_c (E_v - E_c)|\phi_c\rangle\langle\phi_c|\right)|\phi_v^{ps}\rangle = E_v|\phi_v^{ps}\rangle \tag{3.18}$$

Let: $H = T + V,\ V^{ps} = V + \sum_c (E_v - E_c)|\phi_c\rangle\langle\phi_c|$, then there is

$$(T + V^{ps})|\phi_v^{ps}\rangle = E_v|\phi_v^{ps}\rangle \tag{3.19}$$

This is the equation that the pseudo-wave function $|\phi_v^{Ps}\rangle$ satisfies, the V^{ps} is pseudo-potential.

The pseudopotential is non-local and energy level dependent. It is thus more difficult to construct. Empirical pseudopotentials, semi-empirical model pseudopotentials [52], and ab initio Norm-Conserving pseudopotentials [53–57] can be constructed by fitting atomic data. The latter does not have any additional empirical parameters and it satisfies that the corresponding wave function has the same shape amplitude outside the core region ($r > r_c$) with the real wave function, thus enabling the production of the correct charge density (Norm conservation). This is usually achieved by ensuring that the modulus-squared integration of the pseudo wave function in the core region ($r < r_c$) is consistent with the real wave function. The Norm-Conserving pseudopotentials have better transferability and can be used in different chemical environments, but often require a high plane-wave energy cutoff due to their strong locality. Ultrasoft pseudopotentials (USPP) proposed by

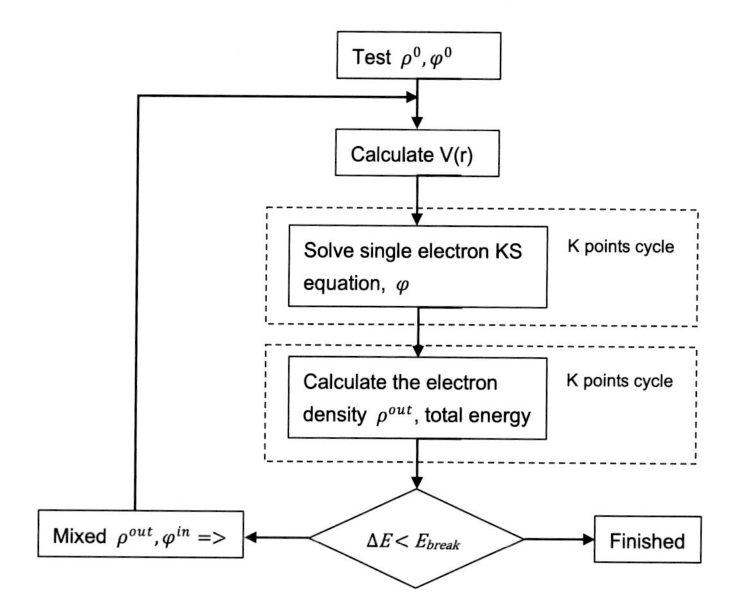

Fig. 3.3 Flow chart of the self-consistent solution of the Kohn–Sham equation

Vanderbilt [58] make the core region pseudo wave function also very smooth due to the abandonment of the Norm Conservation condition, thus solving this problem. The only caveat is that the wave function has to be projected back to obtain the correct charge density before it is calculated from the wave function. The Vanderbilt pseudopotentials greatly reduce the computational effort and enable the construction of pseudopotentials for first-row elements and transition metals with more localized electron distribution, which is widely used in practical calculations.

3.3.4 *Self-Consistent Calculations*

In practice, the Kohn–Sham equation is solved by a self-consistent calculation [59]. Its flow is shown in Fig. 3.3.

3.3.5 *Structural Optimization*

For a system with given atomic positions and elemental species, the total energy of the whole system in the multi-electron ground state is obtained by self-consistently solving the Kohn–Sham equation by density functional theory. The derivative of the total energy to the virtual microdisplacement of the system is the force on each atom (Hellmann–Feynman force). This provides a powerful way to predict the structure of matter in our theory. Since a stable structure in nature should have the lowest total energy, we simply vary the positions of the atoms according to their forces until the total energy of the whole system is at a minimum (all atoms have zero forces), i.e., a (global) minimum of the energy surface is found, then the corresponding structure of matter at that point is the most stable structure in nature. This process is known as structural optimization.

In order to ensure that the search for the minimum of the energy surface finds a global minimum rather than a local minimum, and to improve the efficiency of the whole search process, some powerful search algorithms are needed to move the atoms to the position of the most stable structure as fast as possible. The most commonly used methods are direct energy minimization, the deepest gradient (i.e., maximum force) method, the conjugate gradient method [60] (which takes into account whether the forces in the two steps before and after are in the same direction), quasi-Newton methods [61], damped dynamics methods, and so on.

First-principles calculations based on density founctional theory have achieved great success and remarkable developments within the last two decades, greatly contributing to the development of condensed matter physics, quantum chemistry, theoretical biology, and other disciplines. Kohn argued that DFT contributes to the study of multi-electron systems in two ways [22]: the first is the understanding of the basic physics, using only the charge density in three dimensions to get an accurate understanding of the connotation of multi-electron systems. The second is the practical aspect, where the conventional wave function approach can only deal with 10–20 electrons, while DFT can deal with systems of 10^2–10^3 atoms. For his outstanding contributions to DFT, Kohn shared the 1998 Nobel Prize in Chemistry. Looking ahead, Kohn believes that DFT and wave function methods should continue to develop in a complementary way in order to make our understanding of the electronic structure of the material world one step deeper [22].

3.3.6 Van Der Waals Interactions

None of the various localized exchange–correlation energy forms of the density functional theory can correctly describe van der Waals interactions. Post-Hartree–Fock methods such as quantum chemical MP2 are limited by finite basis vectors and finite cluster sizes, and are also often not suitable for describing periodic systems. Quantum Monte Carlo methods can in principle describe van der Waals interaction systems more accurately, but are too computationally intensive. Their statistical errors decay as a root number fraction of the computational time, and the statistical errors in quantum Monte Carlo for relatively large systems tend to be large. It is not yet possible to calculate atomic forces efficiently, and structural optimization is not possible. Also, the fixed-node approximation in the diffusion Monte Carlo method introduces a certain amount of error in the case of stronger correlations.

In order to balance computational accuracy with resource consumption, a common approach is to correct for errors in density functional methods in the van der Waals system, and one can then easily perform calculations and molecular dynamics simulations for large systems. Many kinds of methods have been proposed to modify the van der Waals interactions. For example, D. Langreth et al. proposed to calculate the van der Waals force by adding a non-local generalized form to the exchange–correlation energy; U. Rothlisberger et al. proposed to improve the density functional calculation by adding a correction term of the atomic center to the pseudopotential (DCACP); S. Grimme and many others used an empirical $1/R^6$ correction term to describe the van der Waals force (DFT-D); P. L. Silvestrelli proposed to calculate the van der Waals interaction between two molecules using the Wanier function obtained from density functional calculations; M. Scheffler et al. proposed an on-the-fly method to calculate the van der Waals interaction.

The correction term for the DFT-D method can be written as

$$E_{vdW} = \sum_{i>j} \frac{C_{6,ij}}{R_{ij}^6} f(R_{ij}),$$

where $C_{6,ij}$ is the strength of the van der Waals interaction between the i-th and j-th atoms, and R_{ij} is the distance between the i-th and j-th atoms. When we consider only the interaction between two molecules (or solids) and not the van der Waals interaction, then the i in this summation spreads over the atoms in the first molecule (or solid) while the j spreads over the atoms in the second molecule (or solid). Usually, the strength of the van der Waals interaction $C_{6,ii}$ between atoms of the same species can be calculated from the polarization rate of the atom.

$$C_{6,ii} = \frac{3}{4}\sqrt{N_i p_i^3}$$

where p_i is the polarizability of the i-th atom, as can be found in Miller's article, and N_i is the Slater-Kirkwood effective electron number. Halgren proposes an empirical algorithm to calculate this effective electron number. For hydrogen atoms, $N_i = 0.8$; and for other atoms

$$N_i = 1.17 + 0.33 n_i,$$

where n_i is the number of valence electrons of the i-th atom. Weitao Yang et al. proposed another method to obtain $C_{6,ii}$. They argued that the atomic polarizability p_i used in the above calculations were obtained by fitting them to experiments, and that different fits would give slightly different values of the polarizabilities. They used the strength of the van der Waals interactions between molecules directly to fit the atomic $C_{6,ii}$. In general, these methods give $C_{6,ii}$ values that do not differ too much. The van der Waals interaction strength $C_{6,ij}$ between two different atoms is obtained by a combination rule, which has been proposed by different authors. Grimme proposed

$$C_{6,ij} = \frac{2C_iC_j}{C_i + C_j},$$

where $C_{6,ii}$ is abbreviated as C_i. In another paper, he proposes another combinatorial approach

$$C_{6,ij} = \sqrt{C_iC_j}$$

Instead, M. Elstner et al. adopted the Slater-Kirkwood combination rule,

$$C_{6,ij} = \frac{2C_iC_jp_ip_j}{p_i^2C_j + p_j^2C_i}$$

With these combination rules and the atomic polarization rates, we can in principle obtain the parameters of the strength of the van der Waals interactions between all atoms.

The decay function f(R) has several different forms in the literature. The two most common forms are:

$$f(R) = \frac{1}{1 + \exp[-d(\frac{R}{R_0} - 1)]}$$

and

$$f(R) = (1 - \exp[-d(\frac{R}{R_0})^7])^4$$

where d is the parameter, generally taken it as 23 for the first form and 3 for the second, and R_0 can be found from the van der Waals radius of the atom. For the same kind of atom, we have

$$R_{0,ii} = 2R_{0,i}$$

where $R_{0,i}$ is the van der Waals radius of the i-th atom. In addition to this, Elstner et al. proposed a simpler approach: for all elements of the first period of the periodic table take $R_0 = 3.8$ Å, and for all elements of the second period take $R_0 = 4.8$ Å. For different kinds of atoms $R_{0,ij}$ can be obtained by a combination rule. Some have proposed to use a simple summation, i.e.

$$R_{0,ij} = R_{0,i} + R_{0,j}$$

It has also been proposed to use the cubic mean formula, i.e.

$$R_{0,ij} = \frac{R_{0,ii}^3 + R_{0,jj}^3}{R_{0,ii}^2 + R_{0,jj}^2}$$

Using these formulas above, we can very easily calculate the contribution of the van der Waals correction to the total energy as well as to the forces on the atoms.

In practical calculations, however, it has been found that if this energy correction term is added directly to the total density functional energy, the resulting energy profile tends to have a relatively large error. For this reason, Grimme proposed to introduce an overall scaling factor s_6, such that the van der Waals correction term is rewritten as

$$E_{vdW} = s_6 \sum_{i>j} \frac{C_{6,ij}}{R_{ij}^6} f(R_{ij})$$

This s_6 depends only on the different exchange–correlation forms in the density functional. Grimme fitted the value of the scaling factor under different exchange–correlation functionals by calculating the adsorption energies between different molecules. With the introduction of this scaling factor, this empirical correction method becomes more accurate. However, P. Jurečka et al. pointed out that this overall scaling factor, while improving the accuracy of the density functional calculation near the equilibrium position, also changes the decay coefficient of the van der Waals interaction at the long range. Since different exchange correlations have different scaling factors, different exchange correlations at the long range give different decay curves, which are clearly nonphysical. Therefore, they dropped this overall scaling factor s_6, and introduced another scaling factor s_R. This scaling factor changes the van der Waals radius of the atom (the R_0 term). The decay function f(R) is thus rewritten as.

$$f(R) = \frac{1}{1 + \exp[-d(\frac{R}{s_R R_0} - 1)]}$$

and

$$f(R) = (1 - \exp[-d(\frac{R}{s_R R_0})^7])^4$$

By the introduction of this scaling factor, the accuracy of the density functional can also be improved. Some later authors then introduced both scaling factors, making this empirical correction more reliable around the equilibrium position.

Recent breakthroughs have been made in the development of a general form of van der Waals functional, which are important for describing systems such as water, which are dominated by weaker hydrogen bond interactions. Langreth et al. invented van der Waals functional with general form [62]:

$$E_{xc}^{vdW} = E_x^{GGA} + E_c^{LDA} + E_c^{nl}$$

The van der Waals functional is composed of the exchange of the GGA plus the correlation term of the local density approximation (LDA) and the non-local correlation term; where the key non-local correlation term is written as

$$E_c^{nl} = \frac{1}{2}\iint drdr'n(r)\phi(r, r')n(r')$$

It is directly related to the density at r, r', and is linked by a mathematical relation ϕ of general form. Depending on the specific parameters in the GGA, ϕ can take slightly different forms.

3.4 Molecular Dynamics Simulations

Molecular dynamics simulations are one of the most effective methods for studying and understanding the material world, and especially its dynamical behavior, from a microscopic scale since the advent of computers [63–66]. This method calculates the forces on all the atoms in the system at a given moment from the interaction potential between atoms, and then uses the positions and velocities of the atoms at that moment to find the positions and velocities of all the atoms at the next moment (one time step apart) from the classical Newton equations; thus simulating the dynamical behavior of the whole system at a time interval from the molecular scale. It allows both to know the static properties of the whole system and to track the dynamical properties of the system. For a microcanonical esemble (where the system energy E, particle number N, and volume V are kept constant), the kinematic formulation is as follows.

$$\begin{cases} \vec{r}_{n,i+1} = \vec{r}_{n,i} + \vec{v}_{n,i}\Delta t \\ \vec{v}_{n,i+1} = \vec{v}_{n,i} + (\vec{F}_{n,i}/m_n)\Delta t \end{cases} \tag{3.20}$$

where $\vec{r}_{n,i}$, $\vec{v}_{n,i}$, $\vec{F}_{n,i}$ denote the position, velocity, and force of the nth particle (of mass m_n) at the moment $t_i = i\Delta t$, respectively. Δt is the time step.

The actual molecular dynamics simulation is divided into four stages: (1) input of initial coordinates, (2) structural optimization, (3) heating, convergence, and equilibrium, (4) commissioning, and data statistics and analysis. In general, the system is first structurally optimized by selecting the atomic equilibrium position as the initial position, and the initial velocity is selected according to the Boltzmann distribution. In order to bring the system to equilibrium, a convergence phase is crucial and essential, which often determines the success or failure of the whole simulation and the reliability of the results. The convergence can be done by rescaling the velocity to a specific requirement (equivalent to taking or making up energy from the system) at several steps. Once the system consistently gives determined values for the average kinetic energy and average potential energy, we consider that equilibrium has been reached. A formal simulation of the system can next be realized according to the kinetic equations, and the data on its system evolution is used for statistical analysis of the physics quantities. The general strategy for molecular dynamics simulations is shown in Fig. 3.4.

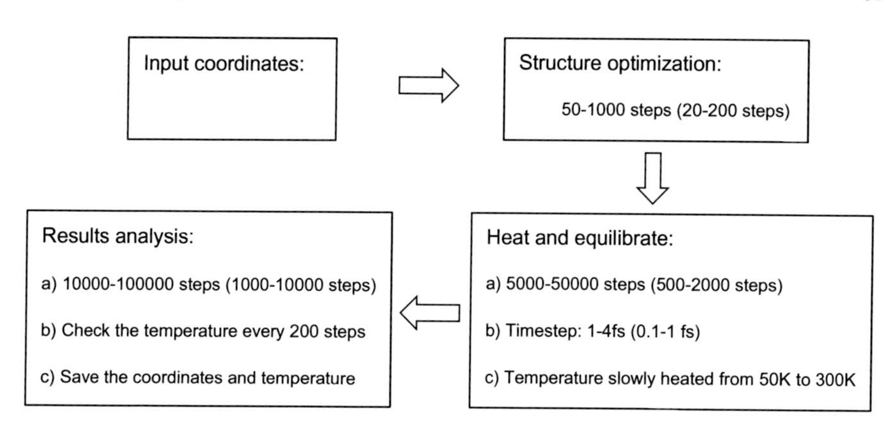

Fig. 3.4 A general strategy for molecular dynamics simulations. The numbers in parentheses correspond to first-principle molecular dynamics simulations

The use of molecular dynamics simulations has at least the following advantages: (1) it removes all the approximations in deriving analytical formulas; (2) it allows testing whether the proposed action model yields reasonable kinetic properties; (3) it provides systematic molecular-scale structural information and action rules that are most lacking in experiments, thus complementing experimental results and greatly facilitating the organic combination of theory and experiment. But the molecular dynamics approach also has its drawbacks. It is limited by the size of the computing power of the computer. First of all, the simulation time is limited, which may not be on the time scale of the actual physical process occurring, and the finite time-averaging also deviates from the ensemble averaging in phase space (ergodic hypothesis). Also, the finite system size may have an additional effect on the system. To minimize this effect, periodic boundary conditions are often taken in simulations. There are also errors introduced by finite time steps in the calculations and other cumulative errors that are worth noting.

3.4.1 Verlet Algorithm

The numerically more stable Verlet algorithm [67, 68] is often used in practice as a recursive description of the evolution of the system. It has the accuracy to fourth-order minima on the position $\vec{r}_{n,i}$. Assume that the position $\vec{r}(t+\Delta t)$ is expanded at t in time steps $\vec{r}_{n,i}$:

$$\begin{cases} \vec{r}(t+\Delta t) = \vec{r}(t) + \vec{v}(t)\Delta t + \frac{1}{2}\vec{a}(t)(\Delta t)^2 + \frac{1}{6}\vec{q}(t)(\Delta t)^3 + O((\Delta t)^4) \\ \vec{r}(t-\Delta t) = \vec{r}(t) - \vec{v}(t)\Delta t + \frac{1}{2}\vec{a}(t)(\Delta t)^2 - \frac{1}{6}\vec{q}(t)(\Delta t)^3 + O((\Delta t)^4) \end{cases} \tag{3.21}$$

The above two equations are added together to obtain:

$$\vec{r}(t+\Delta t) = 2\vec{r}(t) - \vec{r}(t-\Delta t) + \vec{a}(t)(\Delta t)^2 + O((\Delta t)^4) \quad (3.22)$$

We see that the position $\vec{r}$ is already exact to a fourth-order fraction of the time step Δt. But if we take the velocity $\vec{v}(t)$

$$\vec{v}(t) = [\vec{r}(t+\Delta t) - \vec{r}(t-\Delta t)]/2(\Delta t) \quad (3.23)$$

is only accurate to the third-order decimal of Δt.

Writing Eqs. (3.22) and (3.23) in a recursive form consistent with Eq. (3.20) gives:

$$\begin{cases} \vec{r}_{n,i+1} = 2\vec{r}_{n,i} - \vec{r}_{n,i-1} + \vec{F}_{n,i}(\Delta t)^2/m_n & (a) \\ \vec{v}_{n,i} = \left(\vec{r}_{n,i+1} - \vec{r}_{n,i-1}\right)/(2\Delta t) & (b) \end{cases} \quad (3.24)$$

This is commonly referred to as the Verlet algorithm. The steps for molecular dynamics simulations with this algorithm are:

(1) Pick the initial position $\vec{r}_{n,0}$ and initial velocity $\vec{v}_{n,0}$ of each particle.
(2) Calculate the force $\vec{F}_{n,0}$ on the particle at the initial moment. Let the position at the moment $t_1 = 1 \times \Delta t$

$$\vec{r}_{n,1} = \vec{r}_{n,0} + \vec{v}_{n,0}\Delta t + \vec{F}_{n,0}(\Delta t)^2/2m_n$$

(3) Calculate the force $\vec{F}_{n,i}$ on each particle at the moment $t_i = i\Delta t$.
(4) Calculate the position $\vec{r}_{n,i+1}$ at the moment $t_{i+1} = (i+1)\Delta t$ according to Eq. (3.24a).
(5) Calculate the velocity $\vec{v}_{n,i}$ at the moment t_i according to Eq. (3.24b).

The computation of position and velocity in Verlet's algorithm is not synchronized. The velocity Verlet algorithm has been proposed to solve this problem [69]:

$$\begin{cases} \vec{r}_{n,i+1} = \vec{r}_{n,i} + \vec{v}_{n,i}\Delta t + \vec{F}_{n,i}(\Delta t)^2/2m_n & (a) \\ \vec{v}_{n,i+1} = \vec{v}_{n,i}(t) + \left(\vec{F}_{n,i} + \vec{F}_{n,i+1}\right)\Delta t/2m_n & (b) \end{cases} \quad (3.25)$$

It is also accurate to third-order minima of Δt, and has better numerical stability.

3.4.2 Nose Thermostat

Usually, people are more interested in how to simulate a constant temperature system. In this system, the total energy is not conserved, and the temperature has a steady fluctuation around a central value. This can be achieved by introducing some constraints. The initial approach was to rescale the velocity at each step to a constant target value to achieve a constant temperature simulation. However, this approach does not imply that the system ensemble is a canonical ensemble at this time [70]. S. Nose proposed an Extended System (ES) approach that allows for a rigorous canonical ensemble simulation [71, 72]. This approach is achieved by introducing an additional degree of freedom s, as an exotic system coupled to a real system (N) containing N particles, or a "heat bath" (s), which controls the temperature regulation and fluctuations in the real system. The Hamiltonian quantities for the entire virtual system ($N + s$) are:

$$H = \sum_{n=1}^{N} \vec{p}_n^2/2m_n s^2 + \varphi(\vec{q}_n) + p_s^2/2Q + gkT \ln s \tag{3.26}$$

where the virtual parameters $(\vec{p}_n, \vec{q}_n, t)$ are transformed by

$$\vec{q}_n' = \vec{q}_n, \vec{p}_n' = \vec{p}_n/s, t' = \int^t \frac{dt}{s} \tag{3.27}$$

linked to the actual variable $\left(\vec{p}_n', \vec{q}_n', t'\right)$.

The total energy of the extended system ($N + s$) is conserved and its equilibrium distribution function is the distribution of the microcanonical ensemble. So the distribution function:

$$\begin{aligned} Z &= \int dp_s \int ds \int d\vec{p} \int d\vec{q}\delta\left[H_0(\vec{p}/s, \vec{q}) + p_s^2/2Q + gkT \ln s - E\right] \\ &= \int dp_s \int d\vec{p}' \int d\vec{q}' \int ds \cdot s^{3N}\delta\left[H_0(\vec{p}', \vec{q}') + p_s^2/2Q + gkT \ln s - E\right] \\ &= \frac{1}{gkT} \int dp_s \int d\vec{p}' \int d\vec{q}' \int ds \cdot s^{3N+1}\delta\left[s - \exp\left\{-\left[H_0(\vec{p}', \vec{q}')\right.\right.\right. \\ &\quad \left.\left.\left. +p_s^2/2Q - E\right]/ /gkT\right\}\right] \\ &= C \int d\vec{p}' \int d\vec{q}' \exp\left[-\left(\frac{3N+1}{g}\right) H_0(\vec{p}', \vec{q}')/kT\right] \end{aligned} \tag{3.28}$$

C is an irrelevant constant. From the above equation we see that if the appropriate potential function $gkT\ln s$ *is* chosen for s (here $g = 3N + 1$ can be taken, and in general g is an integer whose value depends on the specific simulation scheme), the equilibrium distribution function of the whole extended system ($N + s$) projected onto the physical system N is the distribution function of the strictly canonical ensemble, whether for the position space or the velocity space.

$$\rho(\vec{p}', \vec{q}') = \exp[-H_0(\vec{p}', \vec{q}')/kT] \tag{3.29}$$

We see that the Nose method actually achieves rescaling of the time variables by introducing a virtual degree of freedom s, thus bringing the physical system to the equilibrium distribution of the canonical ensemble. Where the virtual mass Q determines the strength of the coupling of the actual system to the virtual system. The value of Q *is* chosen carefully in the calculation, and the standard is the frequency corresponding to that value

$$\omega = \sqrt{2gkT/Q\langle s\rangle^2} \tag{3.30}$$

should be on the same order of magnitude as the phonon frequencies of the actual system; otherwise, the coupling of the two systems is disconnected and the fluctuation of the actual system deviates from the canonical system with finite time step during the simulation in finite time.

It can be shown that the constrained molecular dynamics approach proposed by Hoover et al. [73, 74] and Evans [75] is also a special case of this extended system approach, but it only maintains that the distribution of the system over the position space is canonical.

3.4.3 Vibrational Spectra Obtained from Molecular Dynamics Simulations

Defining the velocity autocorrelation function (VAF) in molecular dynamics simulations [76]:

$$C_{vv}(t) = \sum_{\alpha l\kappa} m_\kappa \langle v_{\alpha l\kappa}(t) v_{\alpha l\kappa}(0)\rangle \tag{3.31}$$

where $v_{\mathrm{alK}}(t)$ is the component of the velocity of the kth atom in the lth primitive cell at moment t on the axis α. m_K is the mass of the kth atom.

It can be shown that the Fourier transform of the velocity autocorrelation function gives the vibration spectrum of the whole system.

$$g(\varpi) = \sum_{j\vec{k}} \delta(\varpi - \varpi_j(\vec{k})) \tag{3.32}$$

This is because, if the velocity can be written in quadratic quantized operator form:

$$\begin{aligned} v_{\alpha l\kappa}(t) = {} & [\hbar/Nm_\kappa]^{1/2} \sum_{j\vec{k}} e^j_{\alpha\kappa}(\vec{k})\left[-i\varpi_j(\vec{k})\right] \\ & \cdot \exp\left[-i\varpi_j(\vec{k})t + i\vec{k}\cdot\vec{R}_l\right]\left[1/(2\varpi_j(\vec{k}))\right]^{1/2}\left[a_j(\vec{k}) + a_j^+(-\vec{k})\right] \end{aligned} \tag{3.33}$$

where $e^j_{\alpha\kappa}(\vec{k})$ is the polarization vector, N is the total number of atoms, and $a_j(\vec{k})$, $a_j^+(\vec{k})$ are the annihilation and production operators, respectively. Satisfying the correspondence.

$$\left[a_j(\vec{k}), a^+_{j'}(\vec{k}')\right] = \delta_{jj'}\delta(\vec{k}-\vec{k}'), \quad \left[a_j(\vec{k}), a_{j'}(\vec{k}')\right] = \left[a^+_j(\vec{k}), a^+_{j'}(\vec{k})\right] = 0 \tag{3.34}$$

Substituting into Eq. (3.31) we get:

$$\begin{aligned} C_{vv}(t) &= -\frac{(2\pi)^3}{V}\sum_{j\vec{k}} \exp\left[-i\varpi_j(\vec{k})t\right]\cdot\left\langle \hbar\varpi_j(\vec{k})(n_j(\vec{k})+1/2)\right\rangle \\ &= -\frac{(2\pi)^3}{V}k_B T\sum_{j\vec{k}}\exp\left[-i\varpi_j(\vec{k})t\right] \end{aligned} \tag{3.35}$$

Doing the Fourier transform, the right-hand side gives the vibrational spectral function: $-(2\pi)^3 k_B T/V \cdot g(\varpi)$.

In practical molecular dynamics simulations, the velocity correlation function is often defined as [65]:

$$C_{vv}(i\Delta t) = \frac{1}{NM}\sum_{j=1}^{M}\sum_{\kappa=1}^{N}\vec{v}_\kappa(j\Delta t)\cdot\vec{v}_\kappa(j\Delta t + i\Delta t) \tag{3.36}$$

If one wants to obtain a vibrational spectrum consistent with the intensity of the spectral absorption experimental peak, the Fourier transform of the dipole moment autocorrelation function of the entire primitive cell is done [77].

3.4.4 First-Principles Molecular Dynamics

Combining electronic ground state calculations of the system with traditional computer simulation methods gives rise to the first-principles molecular dynamics. Applying the Bonn-Oppenheimer approximation [21], it can be assumed that the corresponding electron distribution in each step of ion motion has reached equilibrium and the system is in the electronic ground state of that configuration. The energy of the system at this time can then be obtained by solving the Kohn–Sham equation using density functional theory [28, 29], and the derivative of the energy to the virtual microdisplacement of the system is the force on each ion. Thus the classical molecular dynamics approach can be transformed into a first-principles molecular dynamics approach by simply replacing the forces in the empirical model in the third step of Verlet's algorithm with the forces obtained from first-principles calculations. This is the Born–Oppenheimer first-principles molecular dynamics. Because of the inclusion of a precise treatment of electronic interactions, the simulation of the kinetic properties of the system by the first-principles molecular dynamics method appears to be more accurate. It should be noted that in what we usually call the first-principles molecular dynamics, the ion motion still follows the classical Newton's law, except that the interactions between the atoms are obtained by solving the quantum mechanical ground state of the multi-electron system. Molecular dynamics methods that take into account the quantum motion of ions (especially H-atoms) have emerged and it appears necessary to deal with some specific problems (e.g. the transport of H-ions [11, 78]).

The Car-Parrinello method was the first practical first-principles molecular dynamics method [79, 80] and is still widely used today in physical, chemical, and biological research and is still evolving. This method wonderfully imagines electrons as particles with a large mass (comparable to the mass of an ion), allowing the application of classical equations of motion to simulate both ion and electron motion in a system. The Lagrangian of the system is:

$$
\begin{aligned}
L &= T - V = T - E[\{\psi_i\}, \{\vec{R}_I\}, \{\alpha_v\}] \\
&\equiv \left(\sum_i \frac{1}{2}\mu_e \int d\vec{r} \left|\dot{\psi}_i\right|^2 + \sum_I \frac{1}{2} M_I \dot{\vec{R}}_I + \sum_v \frac{1}{2}\mu_v \dot{\alpha}_v^2 \right) \\
&\quad - \left(\sum_i \int d\vec{r} \psi_i^*(\vec{r}) \left[-(\hbar^2/2m_e)\nabla^2 \right] \psi_i(\vec{r}) + U[\rho(\vec{r}), \{\vec{R}_I\}, \{\alpha_v\}] \right)
\end{aligned}
\tag{3.37}
$$

where $\{\psi_i\}$, $\left\{\vec{R}_I\right\}$, $\{\alpha_v\}$ denote the electron wave function of the system, the nucleus coordinates, and the specific constraint coordinates (e.g., volume in a constant pressure simulation), respectively. U represents the total potential energy containing the electron Coulomb repulsion energy, the effective nuclear Coulomb repulsion energy, the energy of the electron in the nuclear potential field and the external field, the electron exchange–correlation energy, and the constraint term potential energy. So the equation of motion for the whole system (including ions and electrons) is (Λ is the operational matrix set up to satisfy the condition of orthogonal normalization of the wave function)

$$
\begin{cases}
\mu_e \ddot{\psi}_i(\vec{r}, t) = -\delta E / \delta \psi_i^*(\vec{r}, t) + \sum_k \Lambda_{ik} \psi_k(\vec{r}, t) \\
M_I \ddot{\vec{R}} = -\nabla_{R_I} E \\
\mu_v \ddot{\alpha}_v = -(\partial E / \partial \alpha_v)
\end{cases}
\tag{3.38}
$$

Of the above processes, only the dynamics of the ions have real physical meaning. The electron moves classically with the ion according to a virtual mass μ_e, and thus the simulated process is not a Bonn-Oppenheimer process, but Car-Parrinello shows that the simulated path deviates so little from the Bonn-Oppenheimer process that it can be ignored. Most importantly, although the wave function evolution is classical (spurious), the electronic ground state is indeed a rigorous solution to the Kohn–Sham equation [79].

Compared to classical molecular dynamics simulations that take the empirical atom–atom interaction potential, first-principles molecular dynamics can more accurately simulate the evolution of the system over time due to the use of electronic energy state calculations to rigorously represent the interaction forces between ions. Accordingly, it is also considerably more computationally intensive than classical simulations. Figure 3.4 also presents a comparison of the parameters of a first-principles molecular dynamics simulation with those of a classical molecular dynamics simulation, which can simulate 1–2 orders of magnitude less time than a classical simulation. However, with the rapid advances in computer technology, and especially the increasing computational power of massively parallel computing, first-principles molecular dynamics will be increasingly applied to research in various related fields such as condensed matter physics, chemical physics, molecular biology, and drug design. In particular, first-principles molecular dynamics will be an indispensable and irreplaceable theoretical tool to study the microscopic dynamics and interactions of water at the molecular scale, as described in this paper.

3.4.5 Path Integral Molecular Dynamics

The first-principles molecular dynamics approach takes into account the quantum mechanical effects of electrons, so that the interatomic forces are obtained from the electron interactions without the need for empirical parameters to describe the interatomic action potential. However, the motion of the nucleus is still described using classical Newtonian mechanics. This may not be accurate enough for lighter elements such as H. In particular, the quantum effects of H nuclei may be very significant at low temperatures and even at room temperature. A well-known example is the transport of H_3O^+ and OH^- in water where quantum effects greatly reduce the reaction barrier and accelerate the transport of H_3O^+ and OH^-. To describe the quantum effects of nuclei, path-integral molecular dynamics simulations can be performed using the Feynmann path-integral principle based on first-principles molecular dynamics.

According to the path integral principle, the quantum system partition function can be written as:

$$Z = \oint{}' \mathrm{D}\boldsymbol{R} \oint{}' \mathrm{D}\boldsymbol{r} \exp\left[-\frac{1}{\hbar}\int_0^{\hbar\beta} d\tau L_E\left(\left\{\dot{\boldsymbol{R}}_I(\tau)\right\}, \{\boldsymbol{R}_I(\tau)\}; \{\dot{\boldsymbol{r}}_i(\tau)\}, \{\boldsymbol{r}_i(\tau)\}\right)\right] \tag{3.39}$$

where

$$\begin{aligned} L_E &= T\left(\dot{\boldsymbol{R}}\right) + V(\boldsymbol{R}) + T(\dot{\boldsymbol{r}}) + V(\boldsymbol{r}) + V(\boldsymbol{R}, \boldsymbol{r}) \\ &= \sum_I \frac{1}{2} M_I \left(\frac{d\boldsymbol{R}}{d\tau}\right)^2 + \sum_{I<J} \frac{e^2 Z_I Z_J}{|\boldsymbol{R}_I - \boldsymbol{R}_J|} \\ &\quad + \sum_i \frac{1}{2} m_e \left(\frac{d\boldsymbol{r}_i}{d\tau}\right)^2 + \sum_{i<j} \frac{e^2}{|\boldsymbol{r}_i - \boldsymbol{r}_j|} \end{aligned}$$

$$-\sum_{I,i} \frac{e^2 Z_I}{|\boldsymbol{R}_I - \boldsymbol{r}_i|} \tag{3.40}$$

If we take the electron part of the partition function as

$$Z_e[\boldsymbol{R}] = \oint{}' \mathrm{D}\boldsymbol{r} \exp\left[-\int_0^\beta d\tau (T(\boldsymbol{r}) + V(\boldsymbol{r}) + V(\boldsymbol{R}, \boldsymbol{r}))\right] \tag{3.41}$$

then there is

$$Z = \oint{}' \mathrm{D}\boldsymbol{R} \exp\left[-\int_0^\beta d\tau (T(\boldsymbol{R}) + V(\boldsymbol{R}))\right] \cdot Z_e[\boldsymbol{R}] \tag{3.42}$$

Using the Born–Oppenheimer approximation, the total system partition function is

$$Z_{BO} = \oint \mathrm{D}\boldsymbol{R} \exp\left[-\int_0^\beta d\tau (T(\boldsymbol{R}) + V(\boldsymbol{R}) + E_0(\boldsymbol{R}))\right] \tag{3.43}$$

It is discretized using the Trotter decomposition method [81], i.e., the virtual time integral is transformed into the sum of P-segment paths as follows.

$$\begin{aligned} Z_{BO} = \lim_{P \to \infty} \prod_{S=1}^{P} \prod_{I=1}^{N} \left[\left(\frac{M_I P}{2\pi\beta} \right)^{\frac{3}{2}} \int d\boldsymbol{R}_I^{(S)} \right] \\ \times \exp\left[-\beta \sum_{S=1}^{P} \left\{ \sum_{I=1}^{N} \frac{1}{2} M_I \omega_P^2 \left(\boldsymbol{R}_I^{(S)} \right.\right.\right. \\ \left.\left.\left. - \boldsymbol{R}_I^{(S+1)} \right)^2 + \frac{1}{P} E_0\left(\left\{ \boldsymbol{R}_I^{(S)} \right\} \right) \right\} \right] \end{aligned} \tag{3.44}$$

where $\omega_P^2 = P/\beta^2$. Such a quantum system is equivalent to a classical system with the following effective potential:

$$\begin{aligned} V_{eff} = \sum_{S=1}^{P} \left\{ \sum_{I=1}^{N} \frac{1}{2} M_I \omega_P^2 \left(\boldsymbol{R}_I^{(S)} \right.\right. \\ \left.\left. - \boldsymbol{R}_I^{(S+1)} \right)^2 + \frac{1}{P} E_0(\{\boldsymbol{R}_I\}^{(S)}) \right\} \end{aligned} \tag{3.45}$$

Combined with the Car-Parrinello first-principles molecular dynamics approach, the Lagrangian of the system can be written as

$$
\begin{aligned}
L_{\text{AIPI}} = \frac{1}{P}\sum_{S=1}^{P}\Bigg\{&\sum_{i}\mu\left\langle\dot{\phi}_i^{(S)}|\dot{\phi}_i^{(S)}\right\rangle \\
&- E^{KS}\left[\{\phi_i\}^{(S)},\{\boldsymbol{R}_I\}^{(S)}\right] \\
&+ \sum_{ij}\Lambda_{ij}^{(S)}\left(\left\langle\phi_i^{(S)}|\phi_j^{(S)}\right\rangle - \delta_{ij}\right)\Bigg\} \\
&+ \sum_{S=1}^{P}\Bigg\{\sum_{I}\frac{1}{2}M_I^{'(S)}\left(\dot{\boldsymbol{R}}_I^{(S)}\right)^2 \\
&- \sum_{I=1}^{N}\frac{1}{2}M_I^{(S)}\omega_P^2\left(\boldsymbol{R}_I^{(S)} - \boldsymbol{R}_I^{(S+1)}\right)^2\Bigg\}
\end{aligned}
\tag{3.46}
$$

where $\left|\phi_j^{(S}\right\rangle$ is the electron wave function at the s-th imaginary time; $M_I{}'(S)$ is the virtual mass of the ion at the sth virtual time. The following equation of motion is obtained:

$$
\begin{aligned}
\frac{1}{P}\mu\ddot{\phi}_I^{(S)} &= -\frac{1}{P}\frac{\delta E\left[\{\phi_i\}^{(S)},\{\boldsymbol{R}_I\}^{(S)}\right]}{\delta\phi_i^{*(S)}} \\
&\quad + \frac{1}{P}\sum_{j}\Lambda_{ij}^{(S)}\phi_j^{(S)}
\end{aligned}
\tag{3.47}
$$

$$
\begin{aligned}
M_I'\ddot{\boldsymbol{R}}_I^S &= -\frac{1}{P}\frac{\delta E\left[\{\phi_i\}^{(S)},\{\boldsymbol{R}_I\}^{(S)}\right]}{\delta\boldsymbol{R}_I^{(S)}} \\
&\quad - M_I\omega_P^2\left(2\boldsymbol{R}_I^{(S)} - \boldsymbol{R}_I^{(S+1)} - \boldsymbol{R}_I^{(S-1)}\right)
\end{aligned}
\tag{3.48}
$$

Such path-integral molecular dynamics corresponds to motion in phase space and is essentially a sampling statistics method. In order to simulate quantum dynamical processes with real time significance, Voth et al. [82] developed a center-of-mass path-integral molecular dynamics method. The center-of-mass motion follows the quasi-classical Newtonian equations and its forces are derived from the mean potential generated by the non-center-of-mass modes, which evolve to represent the real dynamical evolution of the nucleus. A canonical transformation of the coordinates:

$$
\boldsymbol{u}_I^{(S)} = \frac{1}{\sqrt{P}}\sum_{S'=1}^{P}U_{SS'}\boldsymbol{R}_I^{S'}
\tag{3.49}
$$

where the matrix U is the unitary matrix after diagonalization of A($\boldsymbol{A}_{SS'} = 2\delta_{SS'} - \delta_{S,S'-1} - \delta_{S,S'+1}$). s = 1 for the center-of-mass mode and $\boldsymbol{u}_I^{(1)} = \frac{1}{P}\sum_{S=1}^{P}\boldsymbol{R}_I^S$. s > 1 for the non-center-of-mass mode. The Lagrangian of the system after the canonical transformation is:

$$
\begin{aligned}
L_{\text{AIPI}} = \frac{1}{P}\sum_{S=1}^{P}\Bigg\{ & \sum_{i}\mu\left\langle\dot{\phi}_i^{(S)}|\dot{\phi}_i^{(S)}\right\rangle \\
& - E\left[\{\phi_i\}^{(S)}, \left\{\boldsymbol{R}_I\left(\boldsymbol{u}_I^{(P)}\right)\right\}^{(S)}\right] \\
& + \sum_{ij}\Lambda_{ij}^{(S)}\left(\left\langle\phi_i^{(S)}|\phi_j^{(S)}\right\rangle - \delta_{ij}\right)\Bigg\} \\
& + \sum_{S=1}^{P}\Bigg\{\sum_{I}\frac{1}{2}M_I^{(S)}\left(\dot{\boldsymbol{u}}_I^{(S)}\right)^2 \\
& - \sum_{I=1}^{N}\frac{1}{2}M_I^{(S)}\omega_P^2\left(\boldsymbol{u}_I^{(S)}\right)^2\Bigg\}
\end{aligned}
\tag{3.50}
$$

Just take the virtual mass of each model

$$
\begin{aligned}
& M_I^{(1)} = M_I \\
& M_I{}'(s) = \gamma M_I^{(s)}, \quad (s = 2, \ldots, P)
\end{aligned}
\tag{3.51}
$$

In which $0 < \gamma \ll 1$, it is possible to make the non-center-of-mass modes move rapidly, achieving an adiabatic separation of the center-of-mass and non-center-of-mass modes with the center-of-mass taking quasi-classical motion. The center-of-mass path-integral molecular dynamics can simulate the process of nucleus motion in real time corrected by quantum mechanical effects and plays an important role in basic water research.

3.4.6 Combination of Molecular Dynamics and Electron Dynamics

When describing the kinetic processes in the electron excited states of bulk water or water clusters, it is necessary to consider the kinetic processes of the electrons in addition to the ion motion trajectories, since the electrons are not in the lowest energy ground state and it is possible for them to quantum transition between different energy levels. These kinetic processes can no longer be simulated using the Born–Oppenheimer approximation and can be simulated using the Ehrenfest theorem or the trajectory surface hopping approach [83]. For example, the Ehrenfest theorem can be used to calculate the forces on an electronic state at a giving moment (perhaps the average of several eigenelectron states) driving the ion motion:

$$M_J \frac{d^2 R_J^{cl}(t)}{dt^2} = -\nabla_{R_J^{cl}} \left[V_{ext}^J \left(R_J^{cl}, t\right) - \int \frac{Z_J \rho(r,t)}{\left|R_J^{cl} - r\right|} dr + \sum_{I \neq J} \frac{Z_J Z_I}{\left|R_J^{cl} - R_J^{cl}\right|} \right]$$

$$i\hbar \frac{\partial \phi_j(r,t)}{\partial t} = \left[-\frac{\hbar}{2m} \nabla_r^2 + \int \frac{\rho\left(r',t\right)}{|r - r'|} dr' + v_{xc}[\rho](r,t) \right.$$

$$\left. + v_{ext}(r,t) - \sum_I \frac{Z_I}{\left|r - R_I^{cl}\right|} \right] \phi_j(r,t) \tag{3.52}$$

Here the electron is described by the time-dependent density functional theory and is subject to an external field determined by the instantaneous nucleus coordinates; the nucleus partly follows the classical Newtonian mechanical path, but the forces include the effect of the time-dependent evolving electron density. The electronic state at this point is no longer the eigenstate of the system, and the effect of the excited state electrons on the rest of the electronic system is included. In recent years Meng Sheng et al. implemented and developed this first-principles-based hybrid ion–electron dynamics approach and used it to study surface ultrafast electron dynamics processes [84].

3.4.7 Current Problems

The scope and accuracy of current theoretical studies of water are still relatively limited. Theoretical and simulation studies of water still present at least the following problems, which deserve further discussion:

(1) Lack of simple and effective water models

This is especially true at the nanoscale and surface interfaces. Either the water models are too complicated and slow down the simulations, or the models are too coarse and make the accuracy of the results doubtful and cannot express the basic properties of water at the nanoscale. Developing models of water on the nanoscale is imperative.

(2) Lack of large-scale accurate first-principles approaches

Standard first-principles computational methods can handle up to hundreds of water molecules, and the molecular dynamics simulations time is about 1 ns, which is far from sufficient for studying water in nanoclusters or on the surface of biomolecules. It is a challenge to develop density functional methods to enable them to handle large systems. A promising direction is the development of the first-principles calculation of O(N) (N is the total number of electrons in the system).

(3) The quantum behaviors of H

The role of the H element in the water molecule is mysterious, the hydrogen bonds formed by it are the source of various specific properties of water and the source that guarantees the emergence of life activities on Earth. It is also the key to understanding the mechanism of photosynthesis and the operation of artificial photosynthetic systems. But is the commonly used approach of treating H as a point particle sufficient to explain the peculiar phenomena of water? This is a basic scientific question. Some recent studies imply that the quantum effects of the H element play a key role. This requires further development of simple and effective full-quantum theoretical methods, as well as further improvements in experimental techniques.

(4) Excited-state properties of water

The electronically excited-state properties of water have rarely been discussed, and corresponding theoretical and computational simulation work is even less available. This has had an extreme impact on understanding the atomic-scale processes of photolytic water and the physicochemical processes in the atmosphere and oceans. Indeed, experimental and theoretical understanding of these issues is extremely limited and very coarse.

(5) Consistency with experimental data

Although the theoretical calculations and the extensive experimental data match well in general, and the combination of the two reveals the molecular behaviors of water (a successful example is the theoretical and experimental determination of partially decomposed water layers on Ru(0001) subject to electron excitation), there are still data that are in dire need of theoretical explanation and understanding. For example, the XAS and XPS data of liquid water seem to show that the water has a chain structure with an average of two hydrogen bonds per water molecule. But this notion is greatly challenged by the fact that many studies suggest that liquid water adopts a traditional ice-like structure (about 3–4 hydrogen bonds per water molecule). These problems call for new developments and improvements in theory and experimental techniques.

References

1. J.W. Doebereiner, Schweigg. J. **39**, 1 (1823)
2. H. Cavendish, Experiments on air. Philos. Trans. R. Soc. **74**, 119–53 (1784)
3. E.W. Morley, Smithson. Contrib. Knowl. **980**, 111 (1895)
4. R. Mecke et al., Z. Physik **81**, 313, 445, 465 (1933)
5. J.D. Bernal, R.H. Fowler, A theory of water and ionic solution, with particular reference to hydrogen and hydroxyl ions. J. Chem. Phys. **1**, 515 (1933)
6. L. Pauling, The structure and entropy of ice and of other crystals with some randomness of atomic arrangement. J. Am. Chem. Soc. **57**, 2680 (1935)
7. M. Van Thiel, E.D. Becker, G.C. Pimentel, Infrared studies of hydrogen bonding of water by the matrix isolation technique. J. Chem. Phys. **27**, 486 (1957)
8. J.A. Barker, R.O. Watts, Structure of water; a Monte Carlo calculation. Chem. Phys. Lett. **3**, 144 (1969)

9. A. Rahman, F.H. Stillinger, Molecular dynamics study of liquid water. J. Chem. Phys. **55**, 3336 (1971)
10. K. Laasonen, M. Sprik, M. Parrinello, R. Car, Ab-initio liquid water. J. Chem. Phys. **99**, 9080 (1993)
11. D. Marx, M.E. Tuckerman, J. Hutter, M. Parrinello, The nature of the hydrated excess proton in water. Nature **397**, 601 (1999)
12. P.J. Feibelman, Partial dissociation of water on Ru(0001). Science **295**, 99 (2002)
13. S. Meng, L.F. Xu, E.G. Wang, S.W. Gao, Vibrational recognition of hydrogen-bonded water networks on a metal surface. Phys. Rev. Lett. **89**, 176104 (2002)
14. V.A. Ranea, A. Michaelides, R. Ramirez, P.L. de Andres, J.A. Verges, D.A. King, Water dimer diffusion on Pd{111} assisted by an H-bond donor-acceptor tunneling exchange. Phys. Rev. Lett. **92**, 136104 (2004)
15. S. Nie, P.J. Feibelman, N.C. Bartelt, K. Thuermer, Pentagons and heptagons in the first water layer on Pt(111). Phys. Rev. Lett. **105**, 026102 (2010)
16. W.L. Jorgensen, J. Chandrasekhar, J.D. Madura, R.W. Impey, M.L. Klein, Comparison of simple potential functions for simulating liquid water. J. Chem. Phys. **79**, 926 (1983)
17. B. Guillot, A reappraisal of what we have learnt during decades of computer simulations on water. J. Molecular Liquids **101**, 219 (2002)
18. C. Millot, J.C. Soetens, M.T.C. Martins Costa, M.P. Hodges, A.J. Stone, Revised anisotropic site potentials for the water dimer and calculated properties. J. Phys. Chem. A **102**, 754 (1998)
19. A. Wallqvist, R.D. Mountain, Molecular models of water: derivation and description. Rev. Comput. Chem. **13**, 183 (1999)
20. G.W. Robinson, C.H. Cho, J. Urquidi, Isosbestic points in liquid water: further strong evidence for the two-state mixture model. J. Chem. Phys. **111**, 698 (1999)
21. M. Born, K. Huang, *Dynamical Theory of Crystal Lattices* (Oxford University Press, Oxford, 1954)
22. W. Kohn, Nobel lecture: electronic structure of matter—wave functions and density functionals. Rev. Mod. Phys. **71**, 1253 (1998)
23. J.F. Jiang (江进福), Wavefunctions and density functionals. Physics **23**, 549 (2001). In Chinese
24. D.R. Hartree, The wave mechanics of an atom with a non-Coulomb central field part I theory and methods. Proc. Camb. Phil. Soc. **24**, 89 (1928)
25. V. Fock, Approximation method for the solution of the quantum mechanical multibody problems. Zeitschrift fur Phys. **61**, 126 (1930)
26. H. Thomas, Proc. Camb. Phil. Soc. **23**, 542 (1927)
27. E. Fermi, Accad. Naz. Lincei **6**, 602 (1927)
28. P. Hohenberg, W. Kohn, Inhomogeneous electron gas. Phys. Rev. **136**, B864 (1964)
29. W. Kohn, L.J. Sham, Self-consistent equations including exchange and correlation effects. Phys. Rev. **140**, A1133 (1965)
30. X.D. Xie (谢希德), D. Lu (陆栋), *Band Structure Theory of Solids* (Fudan University Press, Shanghai, 1998). In Chinese
31. J.C. Slater, A simplification of the Hartree-Fock method. Phys. Rev. **81**, 385 (1951)
32. J.C. Slater, *The Self-Consistent Field for Molecules and Solids* (McGraw-Hill, New York, 1974)
33. D.M. Ceperley, B.L. Alder, Ground state of the electron gas by a stochastic method. Phys. Rev. Lett. **45**, 566 (1980)
34. T.P. Perdew, A. Zunger, Self-interaction correction to density-functional approximations for many-electron systems. Phys. Rev. B **23**, 5048 (1981)
35. R.O. Jones, O. Gunnarsson, The density functional formalism, its applications and prospects. Rev. Mod. Phys. **61**, 689 (1989)
36. A.D. Becke, Density-functional exchange-energy approximation with correct asymptotic behavior. Phys. Rev. A **38**, 3098 (1988)
37. J.P. Perdew, in *Electronic Structure of Solids*, ed. by P. Ziesche, H. Eschrig (Akademic Verlag, Berlin, 1991), pp. 11
38. D.C. Langreth, J.P. Perdew, Theory of nonuniform electronic systems. I. Analysis of the gradient approximation and a generalization that works. Phys. Rev. B **21**, 5469 (1980)

39. J.P. Perdew, Y. Wang, Accurate and simple analytic representation of the electron-gas correlation energy. Phys. Rev. B **45**, 13244 (1992)
40. J.P. Perdew, J.A. Chevary, S.H. Vosko, K.A. Jackson, M.R. Pederson, D.J. Singh, C. Foilhais, Atoms, molecules, solids, and surfaces: applications of the generalized gradient approximation for exchange and correlation. Phys. Rev. B **46**, 6671 (1992)
41. J.P. Perdew, K. Burke, M. Ernzerhof, Generalized gradient approximation made simple. Phys. Rev. Lett. **77**, 3865 (1996)
42. C. Lee, W. Yang, R.C. Parr, Development of the Colle-Salvetti correlation-energy formula into a functional of the electron density. Phys. Rev. B **37**, 785 (1988)
43. Y. Andersson, D.C. Langreth, B.I. Lundqvist, Van der Waals interactions in density-functional theory. Phys. Rev. Lett. **76**, 102 (1996)
44. W. Kohn, Y. Meir, D.E. Makarov, Van der Waals energies in density functional theory. Phys. Rev. Lett. **80**, 4153 (1998)
45. A. Becke, Density-functional thermochemistry. 3. the role of exact exchange. J. Chem. Phys. **98**, 5648 (1993)
46. J. Heyd, G.E. Scuseria, M. Ernzerhof, Hybrid functionals based on a screened Coulomb potential. J. Chem. Phys. **118**, 8207 (2003)
47. D.J. Singh, *Planewaves, Pseudopotentials and the LAPW Method* (Kluwer Academic Publishers, Boston, 1993)
48. O.K. Andersen, Linear methods in band theory. Phys. Rev. B **12**, 3060 (1975)
49. W.E. Pickett, *Pseudopotential Methods in Condensed Matter Applications* (North-Holland, Amsterdam, 1989)
50. Z.Z. Li (李正中), *Theory of Solid State* (Higher Education Press, Beijing, 1985). In Chinese
51. S.J. Jie (解士杰), S.H. Han (韩圣浩), *Condensed Matter Physics* (Shandong Education Press, Jinan, 2001). In Chinese
52. S.G. Louie, J.R. Chelikowsky, M.L. Cohen, Ionicity and the theory of Schottky barriers. Phys. Rev. B **15**, 2154 (1977)
53. D.R. Hamann, M. Schlüter, C. Chiang, Norm-conserving pseudopotentials. Phys. Rev. Lett. **43**, 1494 (1979)
54. G.B. Bachelet, D.R. Hamann, M. Schlüter, Pseudopotentials that work: from H to Pu. Phys. Rev. B **26**, 4199 (1982)
55. D.R. Hamann, Generalized norm-conserving pseudopotentials. Phys. Rev. B **40**, 2980 (1989)
56. L. Kleinman, D.M. Bylander, Efficacious form for model pseudopotentials. Phys. Rev. Lett. **48**, 1425 (1982)
57. N. Troullier, J. Martins, Efficient pseudopotentials for plane-wave calculations. Phys. Rev. B **43**, 1993 (1991)
58. D. Vanderbilt, Soft self-consistent pseudopotentials in a generalized eigenvalue formalism. Phys. Rev. B **41**, 7892 (1990)
59. M.C. Payne, M.P. Teter, D.C. Allan, T.A. Arias, J.D. Joannopoulos, Iterative minimization techniques for ab initio total-energy calculations: molecular dynamics and conjugate gradients. Rev. Mod. Phys. **64**, 1045 (1992)
60. W.H. Press, B.P. Flannery, S.A. Teukolsky, W.T. Vetterting, *Em Numerical Recipes* (Cambridge University Press, New York, 1986)
61. P. Pulay, Convergence acceleration of iterative sequences. the case of scf iteration. Chem. Phys. Lett. **73**, 393 (1980)
62. M. Dion, H. Rydberg, E. Schroder, D.C. Langreth, B.I. Lundqvist, Van der Waals density functional for general geometries. Phys. Rev. Lett. **92**, 246401 (2004)
63. D.W. Heermann, *Computer Simulation Methods in Theoretical Physics* (Springer-Verlag, Berlin, 1989)
64. D.W. Heermann, Translated by K.C. Qin (秦克诚), *Computer Simulations in Theoretical Physics* (Beijing University Press, Beijing, 1996)
65. A.R. Leach, *Molecular modeling: principle and applications* (Addison Wesley Longman Limited, London, 1996)

66. C.L. Brooks, in *Computer Modeling of Fluids Polymers and Solids*, ed. by C.R.A. Catlow et al. (1990), pp. 289
67. A. Rahman, Correlations in the motion of atoms in liquid Argon. Phys. Rev. **136**, A405 (1964)
68. L. Verlet, Computer "experiments" on classical fluids. I. thermodynamical properties of Lennard-Jones molecules. Phys. Rev. **159**, 98 (1967)
69. W.C. Swope, H.C. Andersen, P.H. Berens, K.R. Wilson, A computer simulation method for the calculation of equilibrium constants for the formation of physical clusters of molecules: application to small water clusters. J. Chem. Phys. **76**, 637 (1982)
70. J.M. Haile, S. Gupta, Extensions of the molecular dynamics simulation method. II. Isothermal systems. J. Chem. Phys. **79**, 3067 (1983)
71. S. Nose, A unified formulation of the constant temperature molecular dynamics methods. J. Chem. Phys. **81**, 511 (1984)
72. S. Nose, A molecular dynamics method for simulations in the canonical ensemble. Mol. Phys. **52**, 255 (1984)
73. W.G. Hoover, A.J.C. Ladd, B. Moran, High-strain-rate plastic flow studied via nonequilibrium molecular dynamics. Phys. Rev. Lett. **48**, 1818 (1982)
74. A.C.J. Ladd, W.G. Hoover, Plastic flow in close-packed crystals via nonequilibrium molecular dynamics. Phys. Rev. B **28**, 1756 (1983)
75. D.J. Evans, Computer "experiment" for nonlinear thermodynamics of Couette flow. J. Chem. Phys. **78**, 3297 (1983)
76. C. Lee, D. Vanderbilt, K. Laasonen, R. Car, M. Parrinello, Ab initio studies on the structural and dynamical properties of ice. Phys. Rev. B **47**, 4863 (1993)
77. B. Guillot, A molecular dynamics study of the infrared spectrum of water. J. Chem. Phys. **95**, 1543 (1991)
78. M.E. Tuckerman, D. Marx, M. Parrinello, The nature and transport mechanism of hydrated hydroxide ions in aqueous solution. Nature **417**, 925 (2002)
79. R. Car, M. Parrinello, Unified approach for molecular dynamics and density-functional theory. Phys. Rev. Lett. **55**, 2471 (1985)
80. K. Laasonen, R. Car, C. Lee, D. Vanderbilt, Implementation of ultrasoft pseudopotentials in ab initio molecular dynamics. Phys. Rev. B **43**, 6796 (1991)
81. H. Kleinert, *Path integrals in quantum mechanics, statistics and polymer physics* (Word Scientific Publishing Company, Singapore, 1990)
82. S. Jang, G.A. Voth, Path integral centroid variables and the formulation of their exact real time dynamics. J. Chem. Phys. **111**, 2357 (1999)
83. J.C. Tully, Molecular dynamics with electronic transitions. J. Chem. Phys. **93**, 1061 (1990)
84. S. Meng, E. Kaxiras, Real-time, local basis-set implementation of time-dependent density functional theory for excited state dynamics simulations. J. Chem. Phys. **129**, 054110 (2008)

Chapter 4
Experimental Methods

4.1 Introduction to Experimental Methods of Basic Water Research

One of the most important aspects of basic research in water science is the study of the interaction between water molecules and other substances, and the interaction between water and the outside world is realized through the interface, which requires to study some properties of interfacial water and the interfacial properties of water itself (collectively referred to as "interfacial water" properties). And a deeper understanding of these interactions requires a microscopic, especially at the atomic-molecular level, to explore the microstructure and charge distribution, transfer of interfacial water. This is a frontier problem with complex content involving multidisciplinary fields. In recent years, due to the success and wide application of density functional theory methods, some progress has been made in theoretical treatment of the microstructures and interactions of interfacial water relatively; however, more accurate treatment of weak interactions such as hydrogen bonds and van der Waals forces, the establishment of models describing the structure of real water layers (liquid, several to tens of nanometers thick) at (solid, liquid) interfaces, and full quantization are still to be solved. Experimentally, the study of interfacial water is difficult, and simple methods for probing the structure and properties of interfacial water at the atomic-molecular level are lacking very much, and because the interactions of water-surfaces and between water molecules are weaker than normal chemical bonds (a fraction of the strength of normal chemical bonds), the interfacial structure can easily be artificially destroyed during experimental probing. There is a need to vigorously develop interface-sensitive, non-destructive experimental methods and tools, such as nonlinear optical methods such as sum-frequency vibrational spectroscopy, synchrotron radiation and terahertz studies, and novel scanning probe techniques.

Over the last three decades there has been a growing understanding of water—especially at the interfacial water molecular scale in detail—the most common and complex fundamental material system and its physicochemical properties through

S. Meng and E. Wang, *Water*,
https://doi.org/10.1007/978-981-99-1541-5_4

experimental and theoretical studies. However, due to the special characteristics of water, such as transparency, softness and weakness, existing experimental tools often have some inherent defects: experimental detection techniques (ion beams, electron beams, currents, etc.) are destructive to the detected water structure; there is a lack of interface-only sensitive detection techniques; experimental tools have limited spatial and temporal resolution, etc. In particular, due to the special "fragility" of the water molecule system (the strength of the hydrogen bonds connecting water is 1/10 to 1/5 of the normal chemical bonds) and complexity (water has more than 15 bulk phases and an uncountable number of "nano-phases"), it remains challenging to explore the fundamental issues of water and surface interaction using existing experimental techniques. This reality has led to, as the well-known German scientist D. Menzel put it: any discovery or insight into water is highly controversial and ambiguous. There is an urgent need to study the fundamental issues of water science and to vigorously develop core and new technologies with important challenges. Existing techniques for interfacial water such as sum-frequency vibrational spectroscopy and high-resolution surface energy spectroscopy should be further developed, and emphasis should be placed on developing a new generation of scanning probe techniques and ultrafast photoexcitation detection technology in complex environments (room temperature, non-vacuum, etc.) to study the physicochemical details of interfacial water structure and decomposition dynamics.

Scanning probe microscopy (SPM) techniques have irreplaceable applications in basic research in water sciences. Scanning probe microscopy is a general term for a range of microscopes, including scanning tunneling microscopy (STM), atomic force microscopy (AFM), scanning near-field optical microscopy (SNOM), and others. It is a class of microscopes for probing and characterizing surfaces, whose basic feature is to scan the surface of a local area by a very small probe tip to obtain various information within this local area, such as high-resolution topology, as well as information on electricity, magnetism, light, vibration, etc. In addition, SPM can also modify the local area range of a sample with high precision by applying forces and voltages to the surface by the tip. Therefore, SPM technology is the "eyes" and "hands" of scientific researchers to observe and transform the microscopic world.

Conventional atomic force microscopy in the field of interfacial water studies often fails to achieve reliable resolution at the nanometer/sub-nanometer level, and its resolving power and application range are greatly limited. The reason for this problem is that the small Young's modulus at the nano-aqueous interface deforms the sample, and the large viscosity coefficient in the aqueous solution environment reduces the quality factor and detection sensitivity of the probe. The development of high-resolution non-contact FM AFM systems and cryogenic scanning tunneling microscopes with high stability, high resolution, and inelastic tunneling spectroscopy is needed to address the critical scientific needs in water interface research. Once these new instruments are available, it will be possible to achieve nanometer/sub-nanometer spatial resolution for important systems such as molecules and ions adsorbed at the water interface, the generation and evolution of nano-bubbles at the water–solid interface, the nano-water layer at the gas–solid interface, and the microscopic mechanisms of water photolysis and electrolysis, and establish a method for imaging the water interface at the highest resolution level. The establishment and application of this technical approach could provide a powerful research tool for the important basic science field of water behaviour on solid surfaces and interfaces and advance a deeper understanding of this disciplinary problem.

Femtosecond laser spectroscopy, with its high temporal resolution, is a powerful tool for studying light-matter interactions and the induced various physicochemical processes; while STM, with its high spatial resolution, allows the study of the surface electronic states of samples at the single-atom scale. The combination of these two advanced techniques allows an in-depth study of the adsorption configurations of single water molecule, $H^+/OH^-/H$ and O atoms, H_2 and O_2 molecules, etc., the mechanisms of light absorption, photocatalytic chemical reactions, and processes such as light-induced surface desorption and electronic state jumping.

Sum-frequency vibrational spectroscopy (SFVS) has been widely used in recent years to study the interfacial structure of various kinds of water, and is a nonlinear laser spectroscopy technique with high surface resolution developed by Y. R. Shen's group at the University of California, Berkeley, two decades ago. So far, it is the only experimental technique that can provide a realistic surface (interface) vibrational spectra of water. SFVS vibrational spectroscopy allows one to obtain structural information of the surface (interface) at the molecular level. In particular, it can be applied to complex in situ environments, including various gases, liquids, solids and water interfaces, without demanding requirements on temperature, pressure, etc. Combined with other complementary techniques and theoretical calculations, SFVS can be used to investigate many important fundamental questions, including (i) the surface (interface) structure of water at the molecular level and how it changes under the influence of external factors; (ii) whether various molecules and ions adsorb at the surface (interface), what kind of adsorption kinetics and the relationship with the surface (interface) structure of water; and (iii) microscopic processes of water surface (interfacial) chemical reactions (e.g., photocatalytic reactions); (4) the kinetic processes of water surface (interfacial) reactions, etc. All these complex issues are directly related to the innovation of science and technology involving water surfaces (interfaces), such as water pollution treatment, green and clean energy, etc. SFVS technology also has its shortcomings, for example, it can only detect surfaces (interfaces) accessible by light, the spatial resolution is limited by the diffraction limit of light waves, and the experimentally measured spectra are difficult to solve without theoretical guidance.

Synchrotron radiation is an excellent performance light source and advanced synchrotron radiation techniques (e.g., angle-resolved photoelectron spectroscopy ARPES) are the prefered experimental tools for studying the electronic energy band structure of various materials and novel quantum phenomena. X-ray absorption spectroscopy, X-ray emission spectroscopy, and X-ray imaging are the main techniques for studying the microstructure of water (including surface, interface, and bulk phases) at the atomic and electronic levels, and are powerful tools for studying the hydrogen bond structure of water and its dynamics under normal and extreme conditions, and are beginning to play an irreplaceable and important role in the field of micro foundational research in water science such as confined systems, interfacial imaging, nanomaterials, and nano bio-composite systems. With the rapid advances in synchrotron radiation-based X-ray imaging technology, imaging methods with high density resolution, high spatial resolution, large field of view, and large depth of field are the coveted goals. The combination of waveband sheet imaging and grating phase scale imaging method is the most effective method to achieve this goal, the waveband sheet can provide high resolution, grating imaging can provide a large field of view, high scale, and therefore the development of waveband sheet-based grating phase scale imaging is the current development trend. At the same time, the development of soft X-ray nanoprobe CT methods, water window technology, and other related technology platforms will provide the most innovative approaches to water science research.

Each of these aspects is described in more detail in the following sections.

4.2 High Vacuum Surface Energy Spectrum Analysis

The experimental study of water on solid surfaces requires a clean and contamination-free environment. High vacuum environments can reduce the contamination by foreign impurities and signal interference as much as possible, and can make experimental conditions as simple and ideal as possible, and were therefore the first to be adopted to study water-surface interactions. Since the late 1960s, many traditional surface energy spectroscopy techniques have been developed in high vacuum environment, including photoelectron spectroscopy [mainly X-ray photoelectron spectroscopy (XPS) and ultraviolet photoelectron spectroscopy (UPS)], high-resolution electron energy loss spectroscopy, low-energy electron diffraction, temperature desorption spectroscopy, etc. Together with the recently developed scanning tunneling microscopy, these tools allow single-molecule-scale observations and studies of the coverage of water on surfaces, ordered structures, adsorption states, bonding properties of molecules and surfaces, and kinetic processes of water molecule decomposition on surfaces. The application of these methods is of great value to basic water science research.

The adsorption structure of water on the surface can first be observed with a low-energy electron diffractometer (LEED). Water is adsorbed on the surface of a single crystal sample in a vacuum environment, a beam of electrons is incident on the surface, the electrons dry diffract with the ordered structure on the surface and the outgoing beam shows diffraction spots. Since the incident low-energy electrons cannot penetrate the surface and only interact with the atomic structure of the two-dimensional surface layer, the diffraction spots are distributed in two-dimensional cycles, and the distance between the spots is determined by the lattice of the surface water adsorption structure, and the diffraction spots are also the inverted lattice of the two-dimensional surface structure. Using low-energy electron diffractometry (LEED), it may be possible to determine whether the structure of the water adsorbed layer is ordered and what the symmetry and period are. If the intensity of the diffraction spot is related to the incident electron kinetic energy, while the relationship is calculated using electron scattering theory for a surface with a known atomic structure model, the position of each atom within a single primitive cell on the surface can be approximately determined by fitting the two.

Temperature programmed desorption spectroscopy is a simple and effective means of measuring the adsorption energy and state of surface molecules. It is detected by gradually heating the surface sample with adsorbed molecules by a programmed method, so that adsorbates in different adsorption states are desorbed at different temperatures. In this way, the initial surface adsorbate can be inferred by studying the type and amount of desorption products as a function of temperature. For example, molecules that are weakly adsorbed on the surface through hydrogen bonds or van der Waals forces or that have multiple layers of adsorption will first desorb at a lower temperature (~ 100 K); molecular layers that have a strong direct interaction with the surface will desorb at another characteristic temperature; and molecules that decompose or undergo chemical reactions on the surface will desorb as reaction products from the surface at a higher characteristic temperature (~ 400–600 K). The desorption spectrum gives an idea of whether molecules adsorb as multilayers, whether they decompose, what the adsorption energy is approximately, and what their products are.

Photoelectron spectroscopy works as follows: when an X-ray incident source or an ultraviolet light source interacts with a sample, an electron at a certain energy level is excited, and the kinetic energy of this electron is measured to obtain information about the electronic structure in the sample. In practical analysis, the Fermi energy level is used as the zero point, and the binding energy of the electrons in the sample is measured, and the element under test can be determined. Since the electron binding energy of the element under test changes in relation to its surrounding chemical environment, the chemical binding state and valence of the element can be inferred from the measured energy position. The same approach can be used to analyze the Auger electron energy spectrum generated by the Auger process, as well as to determine the initial state of the surface element. Early laboratory work on photoelectron spectroscopy dates back to the 1940s. By the late 1960s, commercial X-ray photoelectron spectroscopy (XPS) instruments were available. With the development of vacuum technology, an ultra-high vacuum XPS instrument became available in 1972. During the development of XPS, Professor K. M. Siegbahn of the University of Uppsala, Sweden, made a special contribution, for which he was awarded the Nobel Prize in Physics in 1981.

H_2O and OH can be distinguished by photoelectron spectroscopy, and the OH groups produced when water is dissociated and adsorbed on a solid surface are measured, mainly based on the difference in the O 1s electron binding energy. Aziz gives the results of a study of the O 1s core electron energy levels and valence band spectra in water and OH groups [1], see Fig. 4.1.

High-resolution electron energy loss spectroscopy uses low-energy monochromatized electrons as the incident source, and when this beam of electrons interacts with the surface of the solid sample or with atoms or molecules adsorbed on the surface, the kinetic energy of the outgoing electrons is analyzed, and the energy difference with the incident electrons (i.e., energy loss) is the energy transferred to the surface

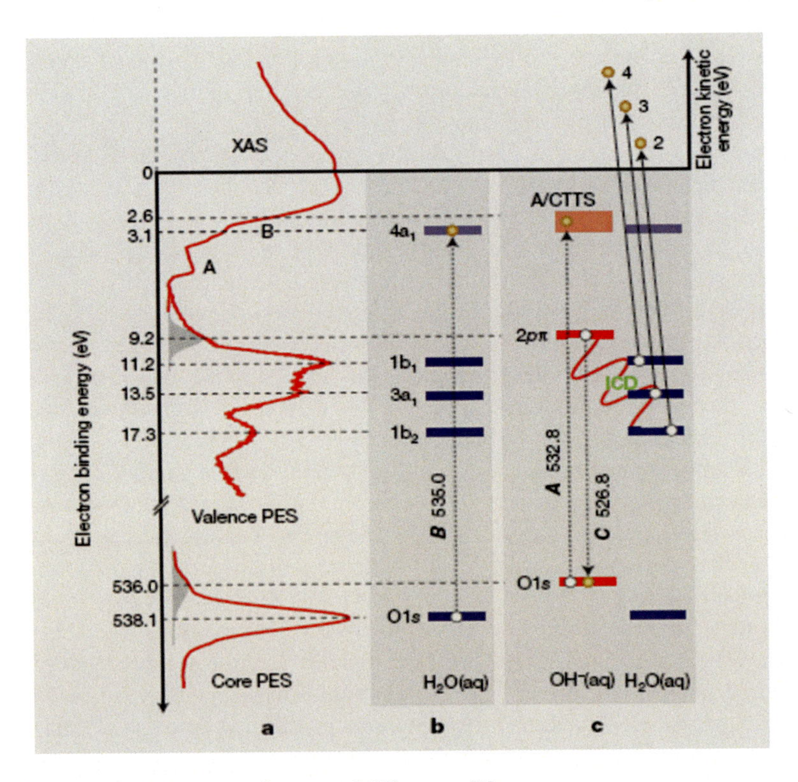

Fig. 4.1 Photoelectron spectra of water and OH groups [1]

vibrational freedom degrees, the intensity of which is related to the density of states of the vibrational model and the strength of the electron–phonon coupling. This allows the measurement of the vibrational spectrum of the surface and the surface adsorbate. The sample used in the study must be a single crystal-ordered surface. The advantage of this method is that the configuration orientation of the OH group on the surface can be detected, and the mechanism of H_2O dissociation can be investigated in more depth, contributing to the understanding of the H_2O adsorption and dissociation processes.

4.3 Scanning Probe Techniques for Microscopic Characterization

Scanning probe microscope (SPM) is a general term for a range of microscopes, including scanning tunneling microscope (STM), atomic force microscope (AFM), scanning near-field optical microscope (SNOM), and so on. SPM is a class of microscopes that probe and characterize surfaces, and its basic feature is to scan a local surface with a very small tip to obtain various information, such as high-resolution topology, electrical, magnetic, optical, vibration, and other information. In addition, by applying forces, voltages, etc., to the surface through the tip, SPM can also modify the sample with high precision in the local domain. SPM technology is thus the "eyes" and "hands" of scientific researchers to observe and transform the microscopic world. It not only brings the atomic and molecular world into view, but also provides the means to transform the microscopic world.

In 1982, G. Binnig and H. Rohrer, scientists at IBM's Zurich laboratory, invented STM using the electron tunneling effect in quantum mechanics [2]. Due to its extremely high spatial resolution, a wide range of applicability, and diversity of information obtained, STM became a landmark invention in the history of nanotechnology development. The basic function of STM is to obtain atomically resolved structural images of surfaces. However, for more than 20 years, STM technology has continued to evolve and the range of applications has expanded, and many advanced STM-based techniques have been implemented, making STM far more functional than simple atomic-resolution functions. For example, the development of scanning tunneling spectroscopy (STS) capabilities of STM has enabled access to information on surface local electronic structures and energy levels; inelastic tunneling spectroscopy (IETS) allows access to information on surface molecular vibrations, spin flips, and other excited states; spin-polarized STM allows access to information on surface magnetic structures and even single spin; STM tunnel combined with luminescence spectroscopy allows access to information on surface plasmons, single-molecule fluorescence, etc.; combined with STM manipulation of single atoms and molecules, various atomic structures can be constructed on the surface and their physicochemical properties can be studied, and even single-molecule chemical reactions on the surface can be manipulated artificially. Meanwhile, the application of STM is no longer limited to the surfaces of metals and semiconductors. For insulator surfaces, the surface can also be studied directly with STM by growing ultrathin insulator films on conducting substrates and using the tunneling effect. The development of these techniques provides strong conditions for basic research in water science.

The basic technology of STM is well established and the instruments are widely used commercially internationally. For example, Omicron in Germany, JEOL and Unisoku in Japan, RHK in the US, Createc in Germany, etc., have commercial room temperature, variable temperature, and low temperature strong magnetic field STM products. The limitations of these commercial products are: firstly, they are very expensive (a typical set of cryogenic strong-field STMs costs about 5 million RMB); secondly, the commercial instruments do not guarantee the implementation of some advanced technical features, such as inelastic tunneling spectral functions. More importantly, the latest features, such as spin-polarized STM, spin-resonance STM, STM-induced spectroscopy, etc., are still in the laboratory development stage and are not commercially available internationally.

In the past decade, STM has been widely used in high-resolution imaging of interfacial water, which can directly obtain the adsorption conformation of water molecules on the single-molecule layer surface, and combined with theoretical calculations can give the microstructure of the interfacial water hydrogen bond network. However, STM is difficult to image the internal structure of water molecules. The difficulty stems from the fact that water molecules are small molecules with closed-shell layers, and the frontier orbitals of water molecules are generally more than a few electron volts away from the Fermi energy level, making it impossible for STM to obtain the orbital information of molecules by resonant tunnelling, which is because small molecules become very unstable or even decomposed due to the excitation of tunnelling electrons at a bias of one electron volt. How to make it possible to image small molecules with stable orbitals within a small bias window is an important development direction for STM in interfacial water applications.

For the study of physicochemical properties of water on solid surfaces, in addition to the observation of atomic-scale structure of the adsorption configuration, dynamic processes, and decomposition and synthesis states of water using STM, advanced STM-based techniques can be used to obtain more in-depth information and greatly deepen one's understanding of the relevant physical problems. Inelastic tunneling spectroscopy (IETS) is a powerful tool for studying single-molecule vibrations with atomic-level resolution. It is also particularly suitable for the study of adsorption states of light atoms such as H and O. If STM is used in combination with IETS, the composition of water and its decompositions can be distinguished to avoid errors. Water on a solid surface often decomposes into hydrogen atoms and hydroxyl groups, which are not discernible from the atomic morphology alone. However, if inelastic tunneling spectroscopy is used, it is possible to target an atom or group and use tunneling currents to excite molecular vibrations to obtain its intrinsic vibrational frequency. The vibrations of light elemental groups such as hydrogen atoms and hydroxyl groups are the most characteristic and easily identifiable. Thus, their composition can be easily identified using inelastic tunneling spectroscopy, and this, together with the atomic resolution of the STM itself, can provide insight into the interpretation of specific adsorption sites of water on the surface, or decomposition and recombination processes generated by external excitation, providing insight into the catalytic decomposition reactions of water. Again, for example, if using fiber-optic introduction in STM, the change in the adsorption state of water is observed by irradiating the surface of adsorbed water with light pulses of specific wavelengths and polarizations, in combination with STM and IETS, a deeper understanding of the photocatalytic process can be obtained. However, reliable STM-IETS measurements of interfacial water have not been achieved experimentally so far, which stems from three main challenges: energy resolution, water molecular stability, and inelastic scattering cross section.

In the real environment, water often interacts with the solid surface of oxides and is even decomposed. To study the interaction of water with oxide surfaces, it is first necessary to clarify the atomic and electronic structure of the oxide surface. Oxide surfaces are a more difficult problem for traditional surface science, since most oxide samples are electrical insulators, and the introduction of electrons causes a charging effect on the sample surface, and the charged sample surface causes deflection of successive incident electrons, resulting in the inability to properly measure the sample's electron signal, which in turn limits the practical application of electron detection in basic research on oxide materials. Scanning tunneling microscopy requires conductive samples, while electron energy loss spectroscopy and photoelectron energy spectroscopy, for example, also require conductive samples to avoid the charging effect. Until recent years, it has been taken to (1) conductive oxides, such as oxygen vacancy-rich TiO_2; and (2) epitaxial growth of ultrathin oxide films on a conducting substrate such as a metal substrate to conduct using the tunneling effect of electrons. The ordered oxide film is prepared on the surface of a single-crystal metal substrate, and when the film thickness is 3–10 nm, it not only has the properties of the oxide bulk phase, but also electrons can penetrate the film to the metal substrate, thus eliminating the surface charge. Using these two methods, the atomic and electronic structures of the surfaces of some typical oxide substrates have been studied in detail, such as TiO_2, Fe_2O_3, MgO, $SrTiO_3$, etc. The study of water adsorption and decomposition on oxide surfaces from the atomic scale has not gained significant progress until recent years. On the other hand, when the thickness of oxide films is less than 2–3 nm in the low-dimensional state, they tend to exhibit some new physical and chemical behaviors, which differ from the bulk phase structure in surface chemistry reactions and physical properties and have many special properties.

Atomic force microscopy (AFM) was invented by Binnig et al. in 1986 [3]. With atomic-level resolution, AFM can be performed under vacuum, in air, or in liquids, and can scan directly on the surface of insulators, which is a unique advantage in water interface characterization and applications. AFM studies the interaction forces between the probe and the sample, and the accuracy of force detection can exceed nN level, and it can be transformed into the geometric morphology of sample surface with atomic resolution. The outstanding feature of AFM is that the sample does not need to be electrically conductive, and the experiment does not require special environments such as vacuum and low temperature, so the structure and properties of the material surface can be studied in situ and in real time at room temperature, in air and in water. Due to the excellent performance of AFM technology, it has become an extremely powerful tool for experimental research in condensed matter physics, chemistry, biology, materials, etc., creating a new situation.

A nanoscale thick water film covering a solid surface in atmospheric environment possesses unique physical/chemical properties due to the special interaction between the solid surface and water molecules and between water molecules. Applied research on AFMs working in an aqueous environment involves fundamental questions. For example, interactions between biomolecules, interactions between polymer molecules and substrates, dynamic processes of biological and chemical reactions, molecular changes induced by environmental factors, etc., providing very important tools for high-resolution imaging of macromolecules and their complexes in the near-physiological state, dynamic studies of biological processes, precise measurements of intermolecular interaction forces, single-molecule manipulation, protein non-adsorbed materials, etc.

In air, when an AFM probe comes into contact with a solid surface, water will condense between the two, forming a nano "water bridge". This nano "water bridge" can be used to transport chemical and biological molecules adhering to the probe to the surface and, due to the precisely controlled position of the AFM probe on the surface, can be used to form nanopatterns with specific chemical and biological functions. Such nano "water bridges" can also be used for nanomodification of the surface by electrochemical reactions. In addition, AFM studies of nanobubbles and gas layers verify the presence of nanoscale bubbles at the solid/water interface, and enable practically study the nanoscale properties of water at the solid/gas interface.

The characterization of interfacial water present on a solid surface will provide basic information about the nano-water film or water droplet, which is a very important method for our understanding of the fundamental properties of water, but a nondestructive, non-interfering, high-resolution method for interfacial water is lacking. For example, in the aforementioned AFM probe with bias (polarization force microscopy) method and tap-mode AFM imaging method, it inevitably applies electrostatic or mechanical forces to the interfacial water. Frequency Modulated Mode AFM (FM-AFM) can provide a nondestructive, non-interfering, high-resolution method for interfacial water. FM-AFM operates in a non-contact mode where the cantilever vibrates at a constant amplitude at a resonant frequency, and the force between the tip-sample is detected as a change in resonant frequency with a cantilever vibration amplitude of less than 1 nm (compared to 100 nm for tap AFM), thus improves the short-range force sensitivity and enables true atomic resolution. In addition, the small force between the tip and the sample in FM-AFM is reduced by three orders of magnitude compared to tap AFM, and basically enables nondestructive and non-interference imaging, which is of great significance for studying the nano-properties of water.

SPM technology has an irreplaceable application in the basic science of water, providing new information to our understanding of the important substance "water". The development of a high-resolution non-contact FM-AFM system and a cryogenic scanning tunneling microscope with high stability and high resolution, capable of stable inelastic tunneling spectroscopy (IETS), and use them to achieve nano/sub-nanoscale spatial resolution of important microscopic systems such as the water interface, can obtain high-resolution horizontal water interface imaging, which provides a powerful research tool for the behavior of water on solid surfaces and interfaces, an important area of basic science research.

4.4 Femtosecond Laser Detection

In reactions where solid materials are used as catalysts, the most important physical and chemical processes occur at the molecule-solid interface, including adsorption, diffusion of molecules on the surface, charge transfer, photon excitation, chemical bond breaking and recombination, diffusion and detachment of new product molecules, etc. To build a detailed picture of this process, we are challenged to simultaneously achieve (1) atomic-level spatial resolution, (2) energy resolution, and (3) time resolution in a catalytic reaction process. The catalytic decomposition of water on a surface is a model system in such reactions and is of great importance in clean energy research. We need to achieve atomic-level spatial resolution of water molecules with the catalyst surface during the reaction and accurately monitor the movement of water molecules on the catalyst surface during the reaction. These atomic spatially resolved data will provide the basic spatial coordinates for theoretical simulations of the reaction process. We also need to achieve energy-resolved spectroscopy analysis of the electronic states of the water molecules, the catalyst, and other molecules during the reaction. The energy-resolved spectral analysis will allow us to obtain the energy distribution of the electrons in the corresponding molecules or catalysts, which will be extremely helpful in understanding the catalytic reaction process. In addition, we need to achieve time resolution of fast processes such as excited states of the corresponding molecules in the reaction. Such time-resolved measurements will help us to understand the dynamics of catalytic reactions. It is important to emphasize that only if we achieve spatially resolved, energy resolved, and time-resolved measurements of the same water molecule in the reaction process, we can obtain a truly perfect image.

To achieve this goal, several schemes have been proposed. One option is based on scanning near-field optical microscopy techniques. The spatial confinement and enhancement effect of the tip on the light field is used to break through the laser diffraction limit and thus obtain the spatial resolution at the nanometer scale with a view to achieving both spatial and temporal resolution. Another option is based on optically coupled scanning tunneling microscopy (optically coupled STM). The discovery and application of femtosecond laser technology has made it possible to probe ultrafast physical and chemical processes, especially the intermediate transition states of chemical reaction processes. And the invention of STM has made it possible to resolve and image electron density and atomic structure at the atomic and submolecular scales. Combining the time-resolution of ultrafast lasers and the atomic-level spatial resolution of STM to achieve extreme detection and control in time and space is a current cutting-edge frontier topic in condensed matter physics and chemistry.

Since the mid-1990s, optically coupled STMs have been the goal of many international research groups [4]. Although there are various technical difficulties in the development of optically coupled STM, some research groups have made initial progress in recent years. The combined femtosecond laser + STM system was first realized by Tony Heinz's research group in the Department of Physics at Columbia University in 2004, where the diffusion of CO molecules adsorbed on the surface of Cu(110) was induced by laser through the combined utilization of STM function and femtosecond laser [5]. This instrument enables for the first time the optical coupling STM function, i.e., the combined utilization of femtosecond laser and STM. However, the STM shifts away when illuminated and resets after illumination, and only the initial state is detected without direct detection of the dynamic processes of electron and molecular excitation and diffusion, and in this sense no temporal resolution is achieved. In 2006, the Wilson Ho research group at the University of California, Irvine, for the first time achieved direct coupling between visible and near-infrared wavelength CW laser and tunneling electrons in the STM after years of efforts [6]. The principle of operation is to excite the electrons at the tip to the higher energy excited state (empty state), where the electrons tunnel to the empty state of the sample molecule, and the symmetry and spatial distribution of the electron wave function of the excited state of the molecule is determined by the measurement of the spatial distribution of the excited state tunneling current. This optically coupled STM system successfully achieves the coupling of continuous light photons and tunneling electrons. And the spatial distribution of the excited states of molecules was detected for the first time. In 2010, Hidemi Shigekawa's group at the University of Tsukuba, Japan, used two laser pulses to successively excite electron–hole pairs in a semiconductor material by the "photon pumping + electron detection" route, and monitored the magnitude of the tunneling current (average current) with the STM tip while the laser was irradiated. The variation of the tunneling current with the time delay between the two laser pulses was probed, and information such as the lifetime of the electron–hole pair inside the semiconductor material on the picosecond scale was obtained at the nanoscale for the first time [7]. Applying optically coupled STM to the fundamental study of water science problems, one will certainly be able to achieve spatially and temporally localized timing induction, detection, and quantum modulation of electronically excited states and catalytic reaction dynamic processes of surface-adsorbed water molecules.

4.5 Non-Linear Optical Techniques

Optical Sum Frequency Spectrum (SFS) is a technique developed in the last two decades to study surfaces and interfaces [8]. In particular, optical sum frequency spectrum has its own unique advantages for surface studies. It can be used to obtain surface spectra, especially surface and interfacial vibrational spectra of clean materials, as well as vibrational spectra of embedded interfaces. Moreover, optical sum frequency spectrum is particularly suitable for the study of water interfaces.

Figure 4.2 briefly depicts the process of sum frequency generation. Two incident laser pulses of frequencies ω_1 and ω_2, hitting the sample simultaneously, produce a sum-frequency reflected or projected output with a frequency $\omega = \omega_1 + \omega_2$. The sum-frequency signal can be expressed as:

$$S(\omega = \omega_1 + \omega_2) \propto \left| \widehat{e} \cdot \left(\overleftrightarrow{\chi}_S^{(2)} + \frac{\overleftrightarrow{\chi}_B^{(2)}}{i\Delta k} \right) : \widehat{e}_1 \widehat{e}_2 \right|^2 \quad (4.1)$$

where $X_S^{(2)}$ and $X_B^{(2)}$ represent the nonlinear polarization rates at the surface and in the body of the medium, respectively, the unit vector $\mathbf{e}_i$ describes the polarization of the *ith* beam, and Δk denotes the wave vector difference. As a second-order nonlinear optical process, the sum-frequency signal cannot be generated due to the central inversion symmetry of the electric dipole distribution in the bulk medium ($X_B^{(2)} \approx 0$), while at the surface and interface, the central inversion symmetry of

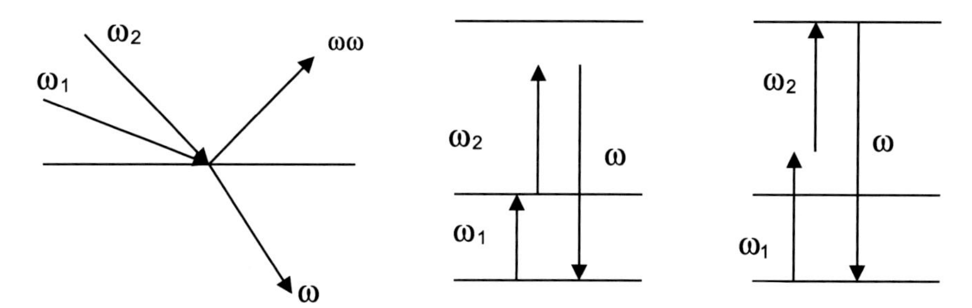

Fig. 4.2 Briefly depicts the process of sum-frequency generation. The corresponding energy diagram corresponds to the resonance of the vibrational and electron transitions

the electric dipole distribution is broken, allowing the sum-frequency signal to exist ($X_S{}^{(2)} \neq 0$). In simple terms, the surface and bulk structures of the medium differ, and the bulk material signal may be suppressed if a specific input/output polarization combination and sample orientation is chosen. Therefore, this means is particularly suitable for surface detection and research.

$X_S{}^{(2)}$ and $X_B{}^{(2)}$ of the medium contain both resonant and non-resonant contributions. For the ω_1 near-oscillatory transition (Fig. 4.2b), $X_S{}^{(2)}$ or $X_B{}^{(2)}$ can be expressed as

$$\overleftrightarrow{\chi}_{S,B}^{(2)} = \left[\overleftrightarrow{\chi}_{NR}^{(2)} + \sum_q \frac{\overleftrightarrow{A}_q}{(\omega_1 - \omega_q + i\Gamma_q)} \right]_{S,B} \tag{4.2}$$

where $\overleftrightarrow{\chi}_{NR}^{(2)}$ is the non-resonant contribution and $\overleftrightarrow{A}_q$, ω_q, and Γ_q represent the tensor magnitude, resonant frequency, and decay constant of the qth vibrational mode, respectively. From Eqs. (4.1) and (4.2), it follows that if all the oscillatory transitions are swept with ω_1, then the corresponding resonant enhancement of the SFG will produce a vibrational spectrum. Thus, the surface vibrational spectrum then corresponds directly to the surface structure of the material.

Similarly, by scanning the electron transitions with either ω or ω_2, we can obtain information on the electron spectrum. That is, if $\omega_1$and $\omega_2$are tunable, we are able to obtain resonances of both oscillatory and electronic transitions. This dual resonance SFS process is able to give a two-dimensional vibrational spectrum/electron spectrum information, detecting molecular species and structures more sensitively and selectively, and reducing the coupling between the chosen vibrational modes [9]. As an example, with different combinations of input/output polarizations it is possible to obtain surface vibrational spectra and provide information at the molecular level, including the molecule composition and the interface structure.

Besides, SFS is able to probe the semi-atomic layer. It has many advantages as a coherent laser spectroscopy technique, such as: high spectral resolution, high spatial resolution, high instantaneous resolution, etc. At the same time, it is capable of highly directional output, allowing remote in-situ detection of samples in the presence of hazardous environments. It is applicable to all interfaces through which light can be transmitted, and therefore has a wide and important range of applications, such as the ability to be applied to the study of liquid interfaces.

We briefly present the schematic diagram of a typical sum-frequency spectrum experiment [8], where a picosecond Nd:YAG laser is pumped onto the sample, producing two optical systems, each with a modulation interval from 420 nm to 8 μm, but with the ability to extend the modulation range from 210 nm to 16 μm. Modulating one or both incident beams, we can obtain single and double resonance SFG spectra. As shown in Fig. 4.3, the outgoing light passes through a filter and then passes through a gated light detection system where we can collect the sum-frequency signal. We can also apply a femtosecond laser with tunable optical parameters to obtain sum-frequency spectra with high temporal resolution.

Although the method of sum-frequency spectrum is widely used to study the interface between water and other substances, there are still some limitations to the method. For example: (1) it cannot be used to study places where light cannot reach; (2) if the material absorbs light, then a weaker light intensity must be used to avoid damage to the material by the laser, in which case the sum-frequency signal may be too weak to be measured. (3) The wavelength of the currently tunable infrared portion of the light is up to 16 μm, and it is very expensive to generate longer infrared light. (4) The current sensitivity of SFS detection is about a fraction of a monolayer, and smaller concentrations of ions or molecules at the interface can not be detected. (5) Spatial resolution is determined by the light wave and is relatively low. (6) Spectral understanding and theoretical models are cumbersome, especially for complex interfaces.

Some of these drawbacks can be addressed in conjunction with other detection techniques, such as X-ray absorption, STM or AFM, and there is scope for further development of SFS itself, such as the development of dual resonance techniques [i.e., using tunable infrared and visible light to excite vibrational and electronic energy levels separately], or the use of time-resolved femtosecond laser techniques.

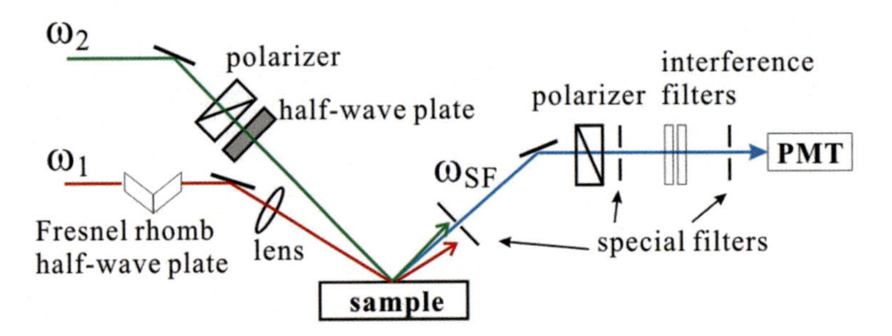

Fig. 4.3 Schematic diagram of the sum-frequency spectrum measurements

4.6 Synchrotron Radiation and Neutron Scattering

Synchrotron radiation is an excellent performance light source, and advanced synchrotron radiation techniques (e.g., angle-resolved photoelectron spectroscopy ARPES) are the preferred experimental tools for the study of various materials and systems as well as novel quantum phenomena. X-ray absorption spectroscopy (XAS), photoelectron spectroscopy (PES), ultraviolet photon emission spectroscopy (UPS), X-ray Raman scattering spectroscopy (XRS), X-ray emission spectrum (XES), X-ray imaging, neutron diffraction and scattering are the main techniques for studying the microstructure of water (including surface, interface and bulk phases) from the atomic and electronic levels, and are powerful tools for studying the hydrogen bond structure of water and its dynamics under normal and extreme conditions, and are beginning to play an irreplaceable and important role in the microfoundational research field of water science, such as confined systems, interfacial imaging, nanomaterials and nanobio-composite systems. Neutron sources are sensitive probes of atomic nuclei, especially for light elements, while synchrotron X-rays are sensitive to electronic structures, and the two are a well-coordinated complement to each other. The organic combination of neutron scattering and diffraction techniques with synchrotron X-ray techniques will be a tremendous contribution to the development of water science research.

Water has been considered to be a molecular form with random and disordered networks connected by hydrogen bonds, whose vibration fluctuates rapidly, and the hydrogen bonds breaking and reforming on roughly femtosecond to picosecond scales. Water and its hydrogen bond interactions with condensed matter result in their unique atomic and electronic structures under normal and extreme conditions, which exhibit different properties (e.g., rapid density reduction at temperatures below 277 K, aqueous solvent channels in biological cells, etc.). The core in water science is thus the microstructure of hydrogen bonds. Addressing this fundamental question helps advance the understanding of many physical, chemical, and biological processes.

Since 2004, the use of synchrotron spectroscopy techniques to study the hydrogen bond structure of water has received general attention and has also generated some controversy. For example, using XAS and XRS methods, Wernet et al. found that most water molecules in liquid water contain two hydrogen bonds and form a ring or chain network, whereas in ice there is a tetrahedral structured network of four hydrogen bonds [10]. Immediately afterwards, Smith et al. measured XAS spectra at variable temperature (−22 °C to 15 °C) and found a strong temperature dependence in the absorption spectrum characteristics of water, suggesting that water in the bulk phase exists as tetrahedra containing four hydrogen bonds [11]. The views of the two research groups were conflicting. Since then, several research groups have investigated the structure of water from a time-resolved perspective to further explore and validate the hydrogen bond model of water. However, basically the subsequent studies support the traditional tetrahedral theory, for example, Hermann et al. pointed out that the traditional tetrahedral structure is still considered reasonable [12]. Odelius et al. used XES to monitor the core-state decay process to obtain information such as the lifetime of the 1S electrons of O in water and heavy water in the femtosecond scale, and tried to estabilish a new hydrogen bond model from the perspective of dynamics [13]. Using high-resolution XAS and XES spectroscopy, Fuchs et al. observed that liquid water exhibits strong isotopic effects, and studied the hydrogen bond configuration and the ultrafast decay time scale of the 1S core state electrons of O [14]. In addition to the variable temperature experiments, Fukui et al. compared the variation of O–K edge XRS in water at different pressures, and found that the number of hydrogen bonds showed a trend of increasing and then decreasing with pressure, which helps to understand the changing properties of hydrogen bonds under extreme conditions [15].

Being able to image the water interface in three dimensions is very promising. Current research methods for water science are mainly based on physical and chemical methods, focusing on the study of its chemical components and functions. The shortcoming of these methods is that they cannot visually give the true three-dimensional structure image of water in various forms. The structure of a substance has an important influence on the properties of the substance, so the research of the structure of a substance is a hot topic that has emerged in recent years. The study of structure is important for the understanding of the properties and synthesis of substances. X-ray CT technology provides us with the possibility to obtain the three-dimensional structure of water in various forms non-destructively. Conventional X-ray imaging techniques are mainly based on absorption contrast and geometric-optical approximation. Absorption contrast arises from the difference in X-ray absorption coefficients of the sample structure and composition. However, for light elemental substances such as C, H, and O, their absorption of X-rays is relatively small, making it difficult to obtain clear images. With the application of synchrotron light sources, X-ray phase contrast imaging technology, water window technology, and neutron imaging technology have been developed rapidly, making it possible to obtain the three-dimensional structure of water in various forms without damage.

Since X-ray phase contrast imaging is mainly used to enhance boundary and interface information of materials with different refractive coefficients or large thickness gradients, this unique capability is particularly suitable for studying water–gas and water–solid interfaces. A large number of related studies have been conducted internationally, such as Scheel et al. used X-ray imaging to study the morphology of the water–sand interface and the surface tension of water [16], Wang et al. used ultrafast X-ray phase imaging to study the water fusion process and the dynamics of thick liquid jets [17], and Fezzaa et al. used ultrafast X-ray phase contrast imaging to study the interface fusion process of two drops of water [18], Weon et al. used X-ray imaging to study the tension at the water–air interface as a function of X-ray irradiation time [19], as well as the existing form of water in biological macromolecules and the distribution of trace elements in water pollution, etc., which have laid the foundation for the study and application of synchrotron X-ray imaging in the water life sciences. The resolution of X-ray nanomicroscopy is now better than 30 nm, allowing the observation of the three-dimensional morphological distribution and dynamic processes of nanobubbles in aqueous solutions, effectively observing their evolution. This is a completely new field, and many valuable explorations require the urgent development of new synchrotron radiation techniques and methods.

With the improvement of synchrotron radiation experimental conditions, time-resolved XAS has been able to be applied well, making it possible to study the dynamic changes throughout the catalytic process. For example, Kleifeld et al. used time-resolved XAS to observe changes in the distance between water molecules and Zn ions to reflect the conformational changes of the active center, while combining quantum chemical calculations to reveal the key role of water on bacterial alcohol dehydrogenation protease catalysis [20].

Neutron scattering is one of the effective methods to probe the atomic structure of aqueous solutions. The experimentally obtained scattering data can be correlated with the distribution function g(r) by Fourier transform, which gives structural information such as atomic spacing and coordination number. The advantage of the neutron technique is that it interacts with the nucleus and even isotopes can be distinguished from the experimental data. For water science research, a lot of research work has also been carried out on neutron source techniques [21, 22]. Synchrotron radiation and neutron source techniques are powerful tools for studying the hydrogen bond structure of water and its dynamics from the atomic and electronic levels under both normal and extreme conditions. The neutron imaging technique developed in recent years with the combination of neutron source and phase contrast imaging, the distribution of gas pores and liquid droplets inside high density materials is even more powerful for the study of material related properties.

The use of synchrotron X-ray spectroscopy and techniques such as neutron diffraction and scattering enables the study of the microstructure of water, especially the hydrogen bond network, and the fundamental properties of water at the nanoscale under conventional, confined and extreme (high temperature, high pressure, low temperature, electric and magnetic fields) conditions, at the atomic and electronic levels. The experimental technique of synchrotron radiation using a combination of high spatial resolution, high temporal resolution, and high energy spectral resolution, combined with neutron diffraction, is an important tool for studying the structure and dynamics of the electron excited states on photocatalytic surfaces and the mechanism of the surface photocatalytic hydrogen production process, the structure as well as the dynamics of the surface electron excited states, the effect of surface defects on surface photochemistry, the migration dynamics of surface atomic molecules, and the dynamis of photocatalytic hydrogen production.

References

1. E.F. Aziz, N. Ottosson, M. Faubel, I.V. Hertel, B. Winter, Interaction between liquid water and hydroxide revealed by core-hole de-excitation. Nature **455**, 89 (2008)
2. G. Binnig, H. Rohrer, C. Gerber, E. Weibel, Tunneling through a controllable vacuum gap. Appl. Phys. Lett. **40**, 178 (1982)
3. G. Binnig, C.F. Quate, C. Gerber, Atomic force microscopy. Phys. Rev. Lett. **56**, 930 (1986)
4. S. Grafstrom, Photoassisted scanning tunneling microscopy. J. Appl. Phys. **91**, 1717 (2002)

5. L. Bartels, F. Wang, D. Moller, E. Knoesel, T.F. Heinz, Real-space observation of molecular motion induced by femtosecond laser pulses. Science **305**, 648 (2004)
6. S.W. Wu, N. Ogawa, W. Ho, Atomic-scale coupling of photons to single-molecule junctions. Science **312**, 1362 (2006)
7. Y. Terada, S. Yoshida, O. Takeuchi, H. Shigekawa, Real-space imaging of transient carrier dynamics by nanoscale pump–probe microscopy. Nat. Photonics **4**, 869 (2010)
8. Y.R. Shen, in *Frontier in Laser Spectroscopy*, ed. by T.W. Hansch, M. Inguscio (North Holland, Amsterdam, 1994), pp. 139–165
9. M.A. Belkin, Y.R. Shen, Doubly resonant IR-UV sum-frequency vibrational spectroscopy on molecular chirality. Phys. Rev. Lett. **91**, 213907 (2003)
10. P. Wernet, D. Nordlund, U. Bergmann, M. Cavalleri, M. Odelius, H. Ogasawara, L.A. Nalund, T.K. Hirsch, L. Ojamae, P. Glatzel, L.G.M. Pettersson, A. Nilsson, The structure of the first coordination shell in liquid water. Science **304**, 995 (2004)
11. J.D. Smith, C.D. Cappa, K.R. Wilson, B.M. Messer, R.C. Cohen, R.J. Saykally, Energetics of hydrogen bond network rearrangements in liquid water. Science **306**, 851 (2004)
12. A. Hermann, W.G. Schmidt, P. Schwerdtfeger, Resolving the optical spectrum of water: coordination and electrostatic effects. Phys. Rev. Lett. **100**, 207403 (2008)
13. M. Odelius, H. Ogasawara, D. Nordlund, O. Fuchs, L. Weinhardt, F. Maier, E. Umbach, C. Heske, Y. Zubavichus, M. Grunze, J.D. Denlinger, L.G.M. Pettersson, A. Nilsson, Ultrafast core-hole-induced dynamics in water probed by X-Ray emission spectroscopy. Phys. Rev. Lett. **94**, 227401 (2005)
14. O. Fuchs, M. Zharnikov, L. Weinhardt, M. Blum, M. Weigand, Y. Zubavichus, M. Bär, F. Maier, J.D. Denlinger, C. Heske, M. Grunze, E. Umbach, Isotope and temperature effects in liquid water probed by X-Ray absorption and resonant X-Ray emission spectroscopy. Phys. Rev. Lett. **100**, 027801 (2008)
15. H. Fukui, S. Huotari, D. Andrault, T. Kawamoto, Oxygen K-edge fine structures of water by x-ray Raman scattering spectroscopy under pressure conditions. J. Chem. Phys. **127**, 134502 (2007)
16. M. Scheel, R. Seemann, M. Brinkmann, M. Di Michiel, A. Sheppard, B. Breidenbach, S. Herminghaus, Morphological clues to wet granular pile stability. Nat. Mater. **7**, 189 (2008)
17. Y.J. Wang, X. Liu, K.S. Im, W.K. Lee, J. Wang, K. Fezzaa, D.L.S. Hung, J.R. Winkelman, Ultrafast X-ray study of dense-liquid-jet flow dynamics using structure-tracking velocimetry. Nat. Phys. **4**, 305 (2008)
18. K. Fezzaa, Y.J. Wang, Ultrafast X-Ray phase-contrast imaging of the initial coalescence phase of two water droplets. Phys. Rev. Lett. **100**, 104501 (2008)
19. B.M. Weon, J.H. Je, Y. Hwu, G. Margaritondo, Decreased surface tension of water by hard-X-ray irradiation. Phys. Rev. Lett. **100**, 217403 (2008)
20. O. Kleifeld, A. Frenkel, J.M.L. Martin, I. Sagi, Active site electronic structure and dynamics during metalloenzyme catalysis. Nat. Struc. Bio. **10**, 98 (2003)
21. T. Strässle, A.M. Saitta, Y. Le Godec, G. Hamel, S. Klotz, J.S. Loveday, R.J. Nelmes, Structure of dense liquid water by neutron scattering to 6.5 GPa and 670 K. Phys. Rev. Lett. **96**, 067801 (2006)
22. A.K. Soper, Joint structure refinement of x-ray and neutron diffraction data on disordered materials: application to liquid water. J. Phys.: Condens. Matter **19**, 335206 (2007)

Chapter 5
Water Molecules, Small Clusters and Bulk Water

5.1 First-Principles Calculations of Free Water Molecule and Water Dimers

We first perform first-principles calculations for the free individual water molecule H_2O and the water dimer $(H_2O)_2$ and compare the results with experiments to test the reliability and accuracy of this approach when dealing with water-related problems.

5.1.1 The Magic of Density Functional Theory: Accurately Predicting the Structure of Water Molecules

It has long been known that water is made up of two elements, hydrogen (H) and oxygen (O). The chemical formula of water is H_2O. But what is the atomic structure of the water molecule, and what are the OH bond lengths and bond angles? In particular, does it have a linear structure like CO_2 or have a specific bond angle? This problem can now be easily solved by calculating the electronic orbitals of the three bonded atoms, i.e., by performing the first-principles calculation.

We use density functional theory for first-principles calculations, take six valence electrons ($2s^2 2p^4$) of O and one electron for each of the two H's. The action of the atomic nuclei on the electrons is replaced by the Vanderbilt ultra-soft pseudopotentials [1] (USPP). The Kohn–Sham equation for this triatomic system is solved using the exchange–correlation form of PW91 [2] and a plane wave expansion [3]. The energy cutoff of the plane wave is taken as 300 eV. The structural optimization is performed using the VASP (Vienna ab initio Simulation Package) software [4–6]. A 10 Å edge length periodic primitive cell was used in the calculations. A single Γ point in the Brillouin zone was taken instead.

As shown in Fig. 5.1, the initial configuration of the water molecule is taken to be nearly linear, with an HOH angle of 175° and an OH distance of 1.0 Å. A conjugate gradient (CG) search algorithm [7] was used to optimize the structure

S. Meng and E. Wang, *Water*,
https://doi.org/10.1007/978-981-99-1541-5_5

until the force on the three atoms was zero (< 0.01 eV/Å). We were amazed to see that after 38 ionic steps, a stable V-shaped water molecule structure was obtained: the OH bond length was 0.972 Å and the angle was 104.53°. This agrees well with the experimentally values (OH = 0.957 Å, HOH = 104.52°) [8], although the initial configuration deviates far from this value, fully demonstrating the magic-like power of electronic orbital calculations with density functional theory.

Figure 5.2 shows the calculated electronic density of states for the water molecule, with small differences (< 0.1 eV) between the energy gaps of each orbital and the experimental results of photoelectron spectroscopy [9]. Table 5.1 lists the results of the calculations for the free water molecule. There is little change in the structure of water molecules calculated with a higher energy cut-off (400 eV) and with LDA. The FLAPW (full-potential augmented plane waves) method [10] test gives bond lengths and bond angles of 0.972 Å and 104.40°, respectively, which are consistent with the pseudopotential results.

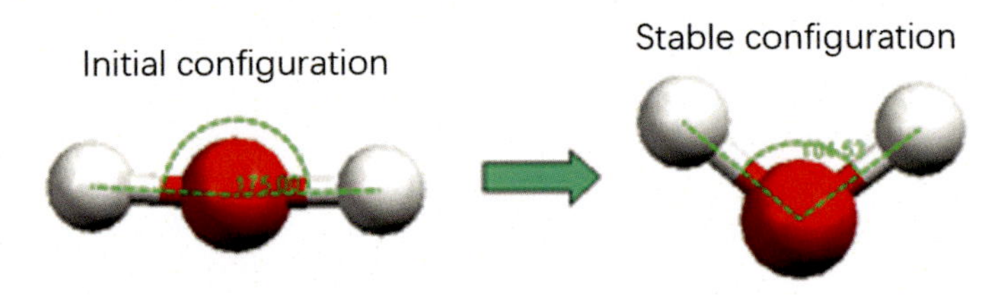

Fig. 5.1 Calculation of stable configurations of water molecule

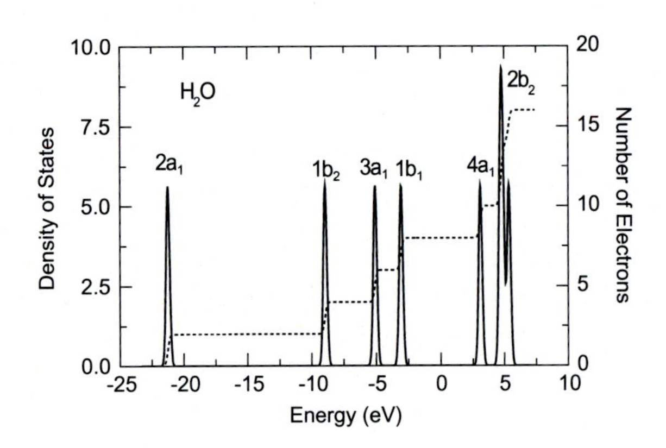

Fig. 5.2 Density of states of free water molecule

Table 5.1 Calculated results for the structure of the free water molecule

(OH, ∠HOH)	LDA	GGA
E_{cut} = 300 eV	(0.973 Å, 105.69°)	(0.972 Å, 104.53°)
E_{cut} = 400 eV	(0.975 Å, 105.66°)	(0.973 Å, 104.62°)

5.1.2 Molecular Structure of Water Dimers

Similarly, we perform a first-principles calculation for $(H_2O)_2$. The calculation reveals that the configuration in which two water molecules maintain the symmetry of C_s and are connected by a hydrogen bond in the middle is the most stable (as in Fig. 5.3).

First, we check the convergence of ab initio calculations on the energy cutoff and the primitive cell size. As shown in Fig. 5.4, the horizontal axes are the energy cutoff and the primitive cell side length, respectively, and the vertical axis is the binding energy of $(H_2O)_2$, i.e., the energy of a single hydrogen bond. The ultra-soft pseudopotential (USPP) and projector augmented plane wave (PAW) pseudopotential were used, respectively [11]. We see that for USPP, a cutoff of 300 eV already converges the binding energy; while for PAW it has to increase to 500 eV, in agreement with the usual trend. The binding energy converges for both at a primitive cell size of 12 Å.

The energy cut-off is chosen to be 400 eV, the primitive cell size is 12 Å, and the exchange–correlation with the ultra-soft pseudopotential and PW91 is used. Figure 5.5 shows the potential energy surface (PES) of $(H_2O)_2$ near the energy minimum for the structural parameters R_{OO}, α, β, which represent the OO distance

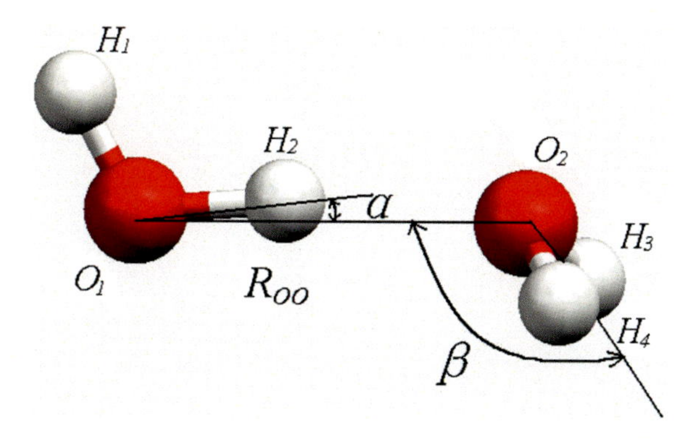

Fig. 5.3 The most stable structure of $(H_2O)_2$

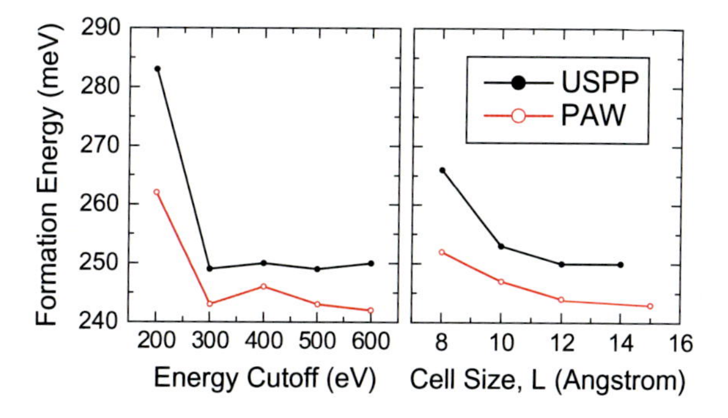

Fig. 5.4 Convergence of the binding energy of $(H_2O)_2$ for the energy cut-off and primitive cell size (L is the side length of the cubic primitive cell)

(R_{OO}), the hydrogen bond angle (α), and the angle between the proton acceptor plane and the OO linkage (β), respectively. As seen in Fig. 5.5, the relative energy takes the form of a Lennard–Jones potential for R_{OO} and a parabolic curve in a small range near the energy minimum for α, β, although with ups and downs.

The parameters we obtained for the most stable $(H_2O)_2$ structure are R_{OO} = 2.856 Å, $\alpha = 2.79°$, and $\beta = 126.35°$, which agree well with the experiment [12]. A comparison of these parameters with experimental data is presented in Table 5.2, along with results for PAW(+PW91), (USPP+) LDA, MP2 [13], and model

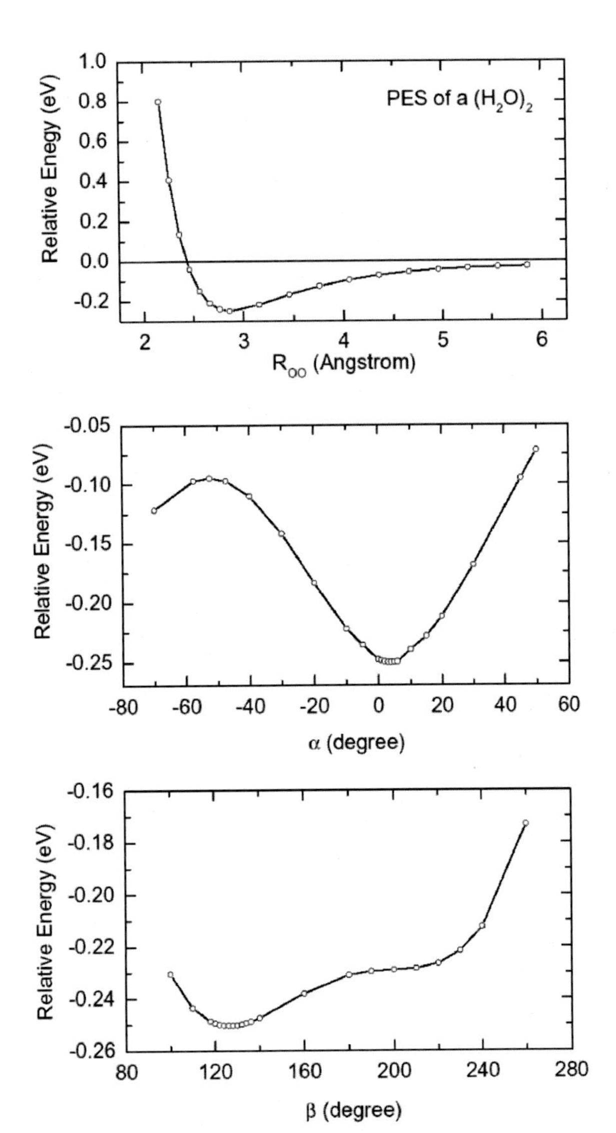

Fig. 5.5 Potential energy surface (PES) of $(H_2O)_2$

calculations (TIP3P, CT-PC) and averaged over 2 ps of molecular dynamics (MD) simulations at 15 K.

USPP calculations and PAW yield nearly identical steady-state structures with approximately the same hydrogen bond energy (USPP: 250 meV, PAW: 243 meV), both of which are consistent with the experimental value (236 meV). Although PAW is slightly more accurate than USPP, PAW requires a higher energy cut-off and a larger primitive cell, which is more computationally intensive. The R_{OO} in the LDA results is too small, while the binding energy is as high as 400 meV, and thus very unreliable. This suggests that GGA is necessary when dealing with hydrogen-bonded systems. Therefore, we conclude that using USPP + PW91 for water and hydrogen bonding is a very accurate and efficient method.

5.1.3 Vibrational Spectra of Water Molecule and Dimer

We also did molecular dynamics simulations for free H_2O and $(H_2O)_2$, respectively. The time step was taken as 0.5 fs and the energy cut-off was 300 eV, simulated in a 12 Å × 12 Å × 12 Å orthorhombic primitive cell, with the k-point taken as a single Γ-point. The exchange–correlation energy with the ultra-soft pseudopotential and PW91 is used. The equilibrium is typically 0.5 ps at 15 K, and then simulated for 2 ps using the microcanonical system synthesis. The average structural data for the 4000 $(H_2O)_2$ configurations in the molecular dynamics simulations are also listed in Table 5.1, which is closer to the experimental values than the statically calculated structures. This is because the data in the experiments is the results averaged over time.

The velocity autocorrelation function is calculated using the simulated data in the time channel, and the Fourier transform is done to obtain the vibrational spectrum of the molecule. For a single free water molecule, three frequencies are obtained as 1595, 3727, 3858 cm^{-1}, and they correspond to three modes of HOH shear motion, symmetric OH stretching vibration, and asymmetric OH stretching vibration, respectively. This is in agreement with the experimental data of 1595, 3657, 3756 cm^{-1}[14]. The test of force constants method in CPMD software [15] (USPP + PBE, energy cut-off 408 eV) resulted in 1588, 3739, 3841 cm^{-1}, which is essentially the same as the results given by VASP.

Figure 5.6 illustrates the vibrational spectrum of $(H_2O)_2$. Table 5.3 lists the detailed data. ν_1, ν_2, ν_3 are the three intramolecular vibrational modes of the proton acceptor (A) and proton donor (D), respectively. ν_7, ν_8, ν_9 represent the motion between the two molecules: the motion of the H forming hydrogen bonds perpendicular to the donor HOH plane (ν_7) and in the HOH plane (ν_8), and the OO stretching motion (ν_9). ν_{10}, ν_{11}, ν_{12} are the rotation and rolling modes of the two molecules. The OH stretching vibration frequency of the donor molecule is red-shifted, indicating the effect of hydrogen bonds. Compared to other theoretical approaches, the results

Table 5.2 Calculated results for the structure of the $(H_2O)_2$ molecule

	USPP	PAW	LDA	TIP3P	CT-PC	MD	MP2[a]	Expt.[b]
d_{O1H1} (Å)	0.972	0.970	0.974	0.9572*	0.9572*	0.971	–	–
d_{O1H2} (Å)	0.985	0.983	0.993	0.9572*	0.9572*	0.981	–	–
$\angle H_1O_1H_2$ (°)	104.98	104.80	106.10	104.52*	104.52*	105.03	–	–
$d_{O2H3,4}$ (Å)	0.974	0.972	0.976	0.9572*	0.9572*	0.972	–	–
$\angle H_3O_2H_4$ (°)	105.53	105.26	106.42	104.52*	104.52*	105.26	-	-
R_{OO} (Å)	2.856	2.869	2.708	2.745	2.976	2.935	2.949(6)	2.976(0–0.03)
α(°)	2.79	3.08	3.33	-3.93	2.02	4.80	5.3(2)	-1 ± 6
β(°)	126.53	124.30	125.67	158.47	130.01	127.0	124.8(2)	123 ± 6
E_{form} (meV)	250	243	400	282	222	242	207	236

Note *Model parameters, [a]Literature [13], [b]Literature [12]

we obtained are closer to the experimental values, especially for the low frequencies rotation part. This is a good indication of the reliability of our approach to the hydrogen-bonding system.

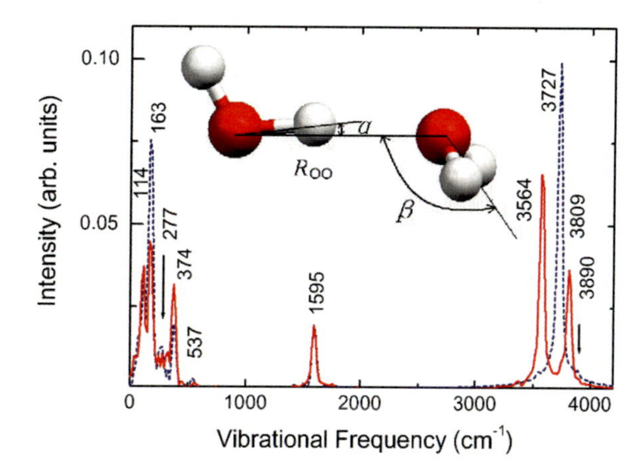

Fig. 5.6 Vibrational spectrum of $(H_2O)_2$

Table 5.3 Vibrational frequencies of $(H_2O)_2$ molecules (A: Proton acceptor, D: proton donor. Hb: H-bonded)

Frequency	BOMD[a]	VWN[b]	MP2[c]	VASP	Expt.[d]
ν_1 (A)	3750	3686	3745	3727	3626
ν_2 (A)	1565	1563	1693	1595	1600
ν_3 (A)	3860	3786	3882	3890	3714
ν_1 (D)	3580	3394	3358	3564	3548
ν_2 (D)	1590	1574	1720	1595	1618
ν_3 (D)	3825	3744	3860	3809	3698
ν_7 (Hb)	710	785	715	537	520[e]
ν_8 (Hb)	410	464	398	374	320[e]
ν_9 (O–O)	225	271	220	277	243[e]
ν_{10}	150	174	193	163	
ν_{11}	150	163	178	163	155[e]
ν_{12}	150	151	155	114	

Note [a]Literature [16], [b]Literature [17], [c]Literature [18], [d]Literature [19], [e]Literature [20]

5.2 From Water Dimer to Water-Water Interaction Model Potentials

First-principles calculations, while accurate, are quite computationally intensive. Due to the limitations of computer hardware, usually, first-principles simulations are one to two or three orders of magnitude smaller (less) than classical simulations in terms of system size and simulation time. In order to simulate larger systems and behavior over longer times, classical models are also often used in the literature to study water. The two main types of models commonly used are the simplest point-charge models and polarizable models, each of which has several common types. All of these models, however, are based first and foremost on the simplest case of water-water interactions: the dimer molecule of water.

The two-body interaction of water with water is through hydrogen bond between two molecules. Analysis shows that hydrogen bond mainly originates from electrostatic interactions [21]. A simple model of water-water interaction is the appropriate assignment of a point charge of a certain size to the rigid structure of the water molecule, describing hydrogen bond in terms of the Coulomb potential energy between them together with the inter-OO Lennard–Jones interaction.

$$\mathrm{E} = \sum_{i}^{m}\sum_{j}^{n}\frac{q_i q_j}{r_{i,j}} + \frac{A}{r_{oo}^{12}} - \frac{C}{r_{oo}^{6}} \tag{5.1}$$

The most commonly used point charge models [22] are SPC/E, TIP3P, TIP4P, etc. and their model parameters are shown in Table 5.4. Where the mass and charge distribution is shown as an example in Fig. 5.7.

Table 5.4 Parameters of the simple model of water

	SPC	SPC/E	TIP3P	BF	TIP4P	ST2	CT-PC
r_{OH} (Å)	1.0	1.0	0.9572	0.96	0.9572	1.0	0.9572
∠HOH (°)	109.47	109.47	104.52	105.7	104.52	109.47	104.52
$A \times 10^{-3}$ (kcal Å^{12}/mol)	629.4	629.4	582.0	560.4	600.0	238.7	2307.0
C (kcal Å^6/mol)	625.5	625.5	595.0	837.0	610.0	268.9	4760.0
q_O	– 0.82	– 0.8472	– 0.834	0.0	0.0	0.0	– 0.66
q_H	0.41	0.4238	0.417	0.49	0.52	0.2375	0.33
q_M	0.0	0.0	0.0	−0.98	−1.04	−0.2375	0.0
r_{OM} (Å)	0.0	0.0	0.0	0.15	0.15	0.8	0.0
δq	0.0	0.0	0.0	0.0	0.0	0.0	0.045

M is a virtual point on the angular bisector of the water molecule. For the ST2 model, M represents the center of the two lone pairs of electrons. See the literature [22]

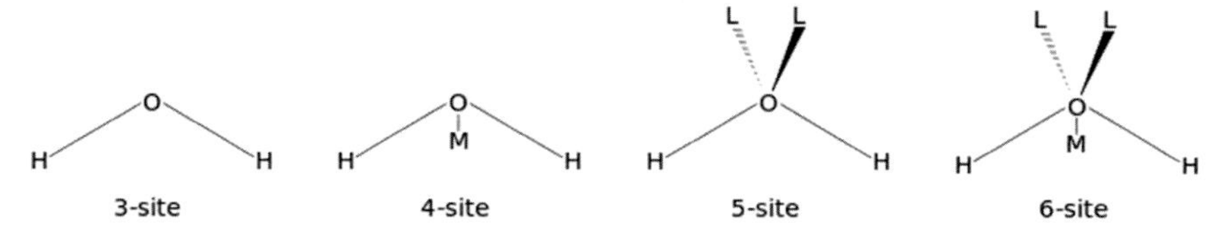

Fig. 5.7 Distribution of interaction sites in the water classical model

The simple model of water built on two-body interactions is designed to simulate the properties of liquid water. Their parameters should be chosen in such a way that reproduce experimental data on bulk water, such as pair correlation functions, diffusion coefficients, or reproduce the results of recent high-precision first-principles calculations. It has been found that these models, although simple, are essentially better at describing bulk water. Adopting the TIP3P model, we simulated ice using the AMBER software [23, 24], taking 384 water molecules in a 22.6 Å × 22.6 Å × 22.6 Å cubic primitive cell for 10 ps (time step 1 fs), and obtained a hexagonally arranged Ice Ih structure consistent with experiment.

Curiously, however, these simple models of the water-water interaction do not deal with the structure of small clusters of water molecules accurately enough. The structures and action energies of $(H_2O)_2$ that we obtained with the TIP3P model are also listed in Table 5.2, and they are far from the experimental values. The reason is that in order to represent the bulk nature of water, these models overemphasize the average nature of a water molecule in the bulk state, whereas in clustered and space-constrained water systems, the nature of the water molecule is inhomogeneous and differs from the bulk average result.

5.3 A New Water-Water Two-Body Interaction Model

We present a new simple model for water-water interactions which is suitable for dealing with systems of water molecules with unsaturated hydrogen bonds, such as small clusters, one-dimensional water chains, water at surface interfaces, etc. For small molecules of water dimers, structures that agree well with experiments and first-principles calculations are given, greatly improving the results of models such as TIP3P.

The basic consideration is the charge transfer effect caused by the addition of hydrogen bond. Analysis shows that hydrogen bond comes from charge transfer in a small fraction except mainly from electrostatic interactions [21]. Roughly 90% of the electron transfer is from proton acceptor to proton donor (as in Fig. 5.8). To include the effects of charge transfer, our approach is very simple: on the basis of the TIP3P model, the charge on the H providing the hydrogen bond and the O accepting the hydrogen bond is redistributed so that H gets a little more charge, i.e., the transferred electron δq. The dependence of the structural parameters R_{OO}, α, and β on δq is

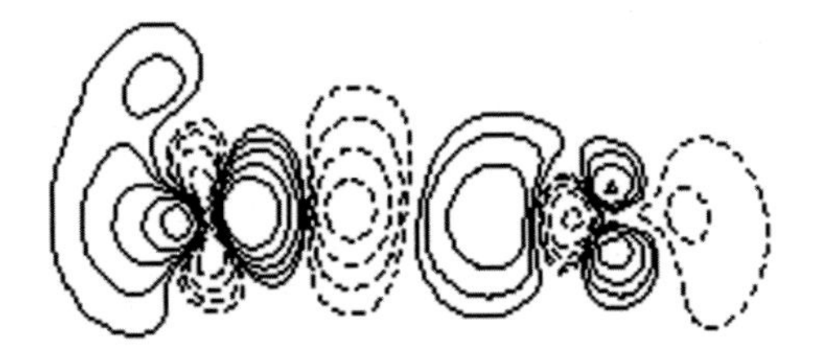

Fig. 5.8 Charge transfer for $(H_2O)_2$. The solid line corresponds to the electron-gaining region and the dashed line to the electron-losing region

shown in Fig. 5.9. Detailed experiments show that the parameter $\delta q = 0.047e$ is most reasonable. We call this model the charge transfer-point charge interaction model (CT-PC). Its parameters are listed in Table 5.4. Its predicted structure ($R_{OO} = 2.954$ Å, $\alpha = 4.4°$, $\beta = 116.5°$) and energy (228 meV) of $(H_2O)_2$ are listed in Table 5.2. Compared to the TIP3P model, the results of this model agree surprisingly well with experimental and first-principles calculations. Moreover, this model yields an $(H_2O)_2$ dipole moment of 2.72 D (1D = 1/4.80 e·Å), which is in better agreement with the experimental value of 2.60 D [25], unlike the TIP3P model (3.85 D), which differs considerably.

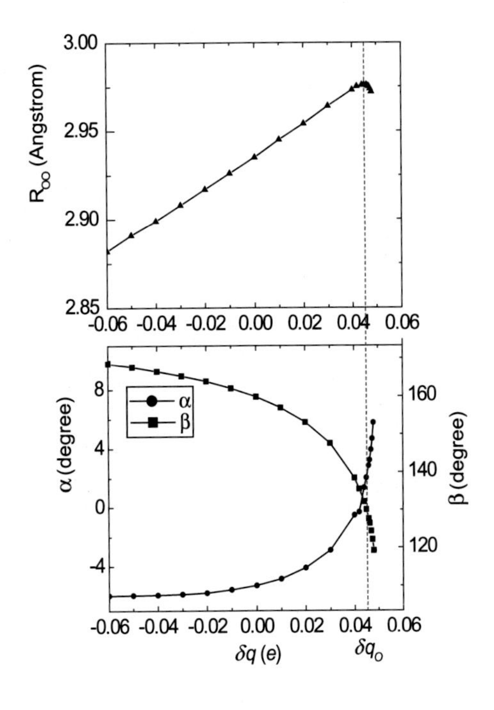

Fig. 5.9 Dependence of the structural parameters of $(H_2O)_2$ on charge transfer in the CT-PC model. $\delta q_O = 0.045e$ is the optimal parameter

5.4 Experimental Measurements of Water Cluster Structures

More water molecules are clustered together to form a cluster structure of water. The atomic structure of gaseous water clusters can be measured by vibration–rotation tunneling spectroscopy (VRT spectroscopy). The principle of the measurement is to use ultrasonic expansion to prepare various mixtures of pure water molecular clusters and measure their absorption by a far-infrared laser beam (50–100 cm^{-1}), where absorption lines of different intensities with equal spacing of positive and negative symmetry appear near a specific frequency, corresponding to the light absorption due to the quantum vibration and rotation of a specific structure of a particular water cluster. By comparing the theoretical fit with the structure of the water cluster predicted by the model, key information such as the structure of the water cluster, the electrode moment, the structural transformation, the strength of the interaction, and the potential energy surface of the interaction can be obtained. This has a key role in understanding the hydrogen bonding of water, developing water models, and studying atmospheric physical phenomena.

Using this method, J. S. Muenter and R. J. Saykally et al. measured the molecular structure of water multimers. Based on the well-known structure of a single water molecule, J. S. Muenter et al. first determined the structure of a water dimer [12], as shown in Fig. 5.10. The measured structure of the water dimer is mirror-symmetric, with one water molecule lying entirely within that mirror, providing a proton attached to the oxygen of the other water molecule, forming a hydrogen bond, hence the two are called the hydrogen bond donor (H-bond donor) and acceptor (H-bond acceptor), respectively. The hydrogen bond acceptor has its central axis in the plane and its two protons are symmetrically distributed about the plane. The structural parameters of the water dimer were measured as follows: the OO bond length $R_{OO} = 2.976$, the angle θ_d between the centerline of the donor and the OO line is 51°; the angle θ_a between the centerline of the acceptor and the OO line is 57°. It is generally accepted that hydrogen bonding is linear, i.e., the angle X–H…Y between the hydrogen bond and the covalent bond is 180°. However, in water dimers this hydrogen bond angle is not strictly 180°, where the donor OH bond deviates from the OO linkage by an angle of $-1° \pm 6°$ (i.e., $\theta_d - 104.5/2°$).

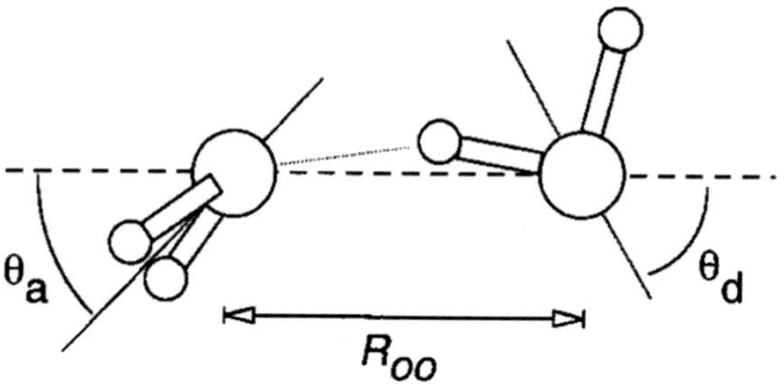

Fig. 5.10 Structure of the water dimer

The structure of the water trimer as determined by R. J. Saykally et al. using VRT spectroscopy [26] is shown in Fig. 5.11. The water trimer cluster has a cyclic structure with each water molecule providing and accepting a hydrogen bond. The average OO bond lengths are 2.97, 2.97, and 2.94 Å. The difference in OO bond lengths comes from the distribution of OH in the vertical direction: each of the three water molecules has ahanging OH bond, with two of them pointing upwards and one pointing downwards on the vertical paper surface. Due to the difference in symmetry, the hydrogen bonds between the water molecules that both have an OH pointing upward are shorter; while the hydrogen bonds between the two water molecules that each have an OH pointing in a different direction are longer, forming the structure in Fig. 5.11. In the experiment, two equivalent structures exist, namely the two OH pointing up or the two OH pointing down structures. They can be transformed between them at will, simply by rotating one of the water molecules OH that not hydrogen-bonded by 180°. This conversion can be easily observed in experiments [26].

The tetrameric and pentameric structures of gaseous water measured by VRT spectroscopy are also cyclic [27, 28]. Similar to the trimer structure, each water molecule provides one hydrogen bond and accepts one hydrogen bond, forming a hydrogen-bonded ring. The number of hydrogen bonds for the entire cluster is four and five, respectively. In the water tetramer, two free OHs facing upward and two facing downward, and the water molecules with OHs pointing upward and downward are arranged intermittently; this structure is the most stable, and the electric dipole moment of the entire cluster is zero. In pentamers, the number of free OHs pointing upward and downward is unequal, similar to trimers, who are in dynamic equilibrium with three free OHs pointing upward or downward, respectively. The average OO bond length in the pentamer is 2.862 Å. See Fig. 5.12.

The case of clusters composed of six water molecules varies. The gaseous water hexamer cluster has a variety of structures, and in addition to forming a ring-like structure similar to a tri-pentamer, it can also form a variety of three-dimensional steric

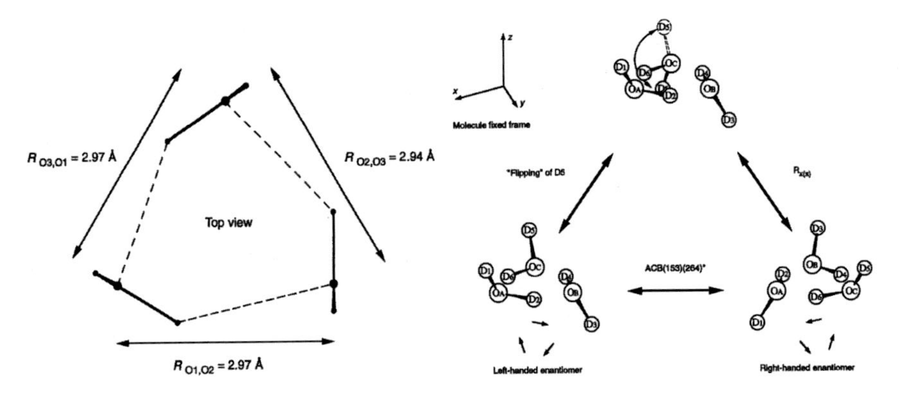

Fig. 5.11 (Left) Structure of a water trimer, (Right) Transformation between the two structures of the water trimer

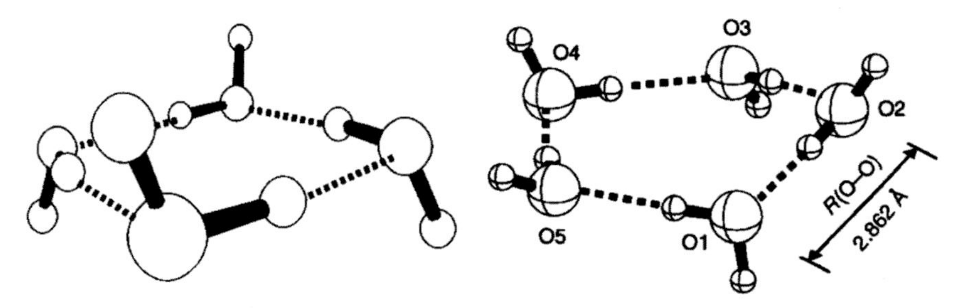

Fig. 5.12 The tetrameric and pentameric structure of water

structures [25]. The most stable water hexamer structures are shown in Fig. 5.13, where the cage-like three-dimensional structure exhibits the highest stability due to the formation of more hydrogen bonds (eight), and the trigonal hexamer cluster, although containing nine hydrogen bonds, is slightly less stable than the cage-like structure due to the large distortion of its structure relative to the ideal hydrogen bond configuration and the lower strength of these hydrogen bonds.

Comparing these small clusters of water, from dimer to hexamer, there is a structural transition from forming a chain or ring hydrogen-bonded structure to a three-dimensional structure. The larger clusters of the six extra water molecules all form a three-dimensional hydrogen-bonded network, indicating a slow transition of water

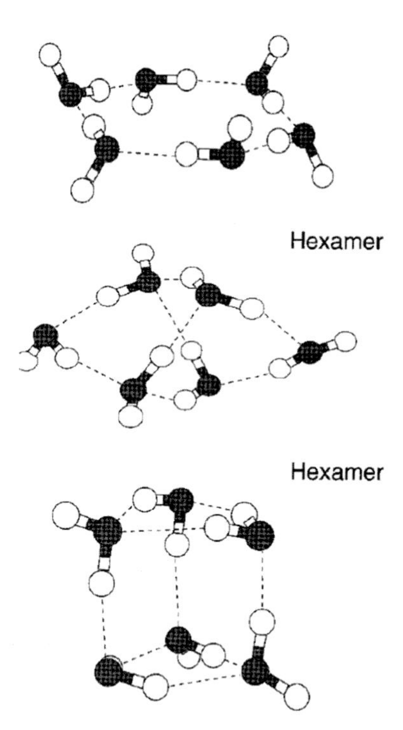

Fig. 5.13 Three different dimeric structures

from clusters to a bulk-phase structure. As the average number of hydrogen bonds per water molecule increases, the structural characteristics and strength of the hydrogen bonds change. For example, the R_{OO}-spacing becomes progressively smaller as the number of water molecules n in the (D2O)n cluster increases, from 2.98 to 2.75 Å (see Fig. 5.14), indicating a gradual increase in hydrogen bond strength. This is due to the fact that as the number of hydrogen bonds increases, the water is more strongly polarized, resulting in stronger hydrogen bonds. The calculated electric dipole moments of individual water molecules in different clusters as a function of cluster size are also shown in Fig. 5.14. It is evident from the figure that the dipole moment of water increases from 1.85 D in isolated molecules to 2.7 D in octamers.

The molecular structures measured by R. J. Saykally and co-workers from water dimers to several different small hexameric clusters are summarized in Fig. 5.15 [25]. From two to five, the water molecules are arranged in rings, but six water molecules form three structures with almost concise energies: rings, cages, and prisms, showing

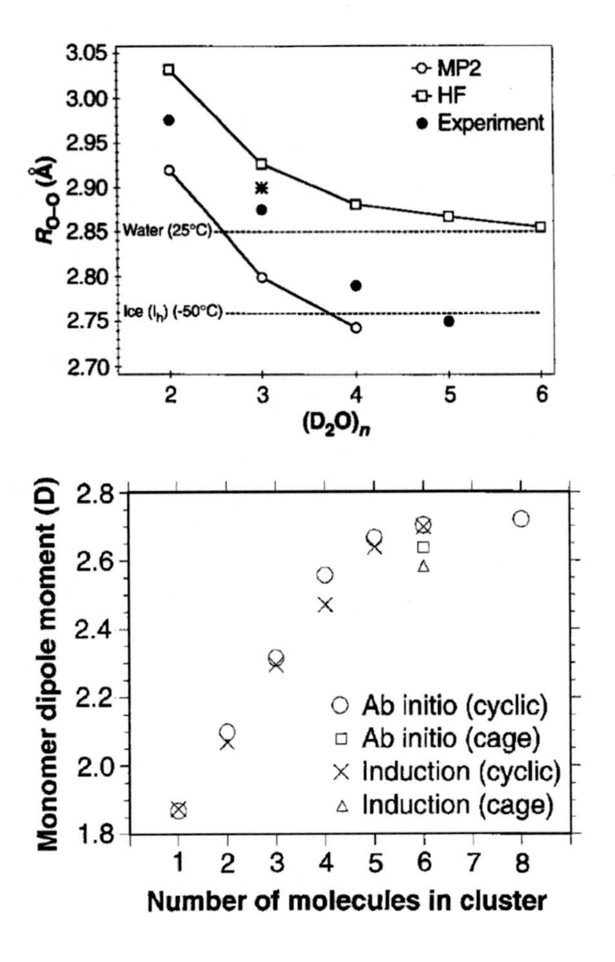

Fig. 5.14 (Top) Change in O–O spacing with the number of water molecules in a water multimer [27]. (Bottom) Variation of the electric dipole moment with the number of water molecules in a water multimer [25]

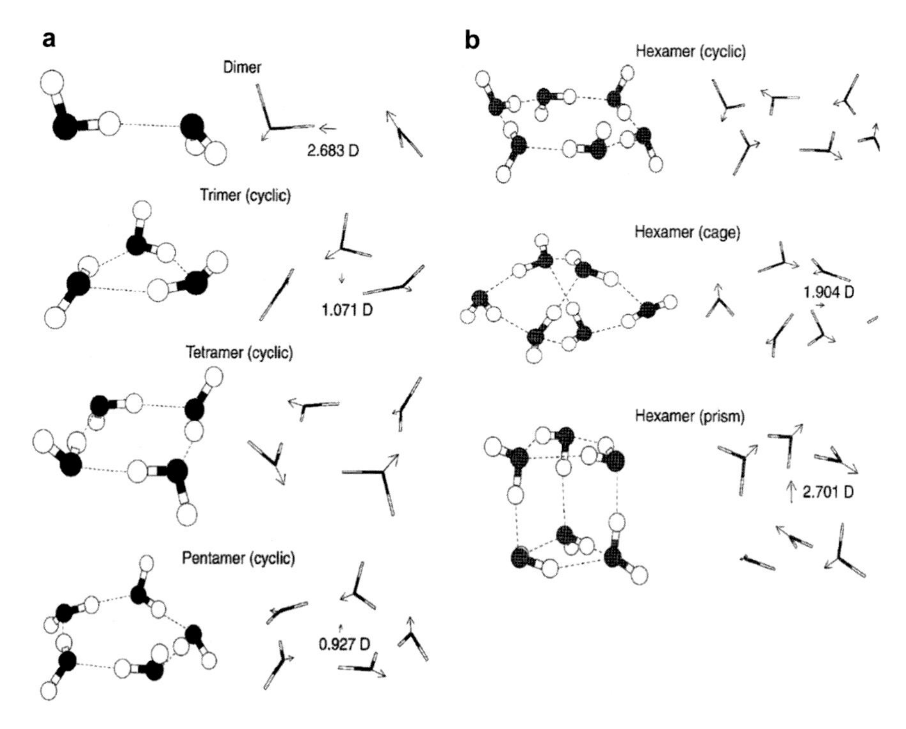

Fig. 5.15 Small clusters of di-, tri-, tetra-, pentameric and hexameric molecules of water [25]. Among these hexameric clusters are ring, cage, and prismatic structures. Their structures and dipole moments are shown in the figure

the transition from a zero-dimensional cluster structure to a three-dimensional bulk structure.

5.5 Structure of Bulk Water

At increasing humidity or decreasing temperature, these small clusters of water further coalesce to form small droplets on the nanoscale to the macroscopic scale. One can study the structure of liquid water using synchrotron radiation or neutron diffraction, usually by measuring the radial distribution function of O, H, and its variation with temperature, pressure, solute, and other factors. Recently, the structure and bonding around that liquid water molecule are also often inferred by measuring the change in the core electron energy level of O.

The precise structure of liquid water, especially the quantitative description of the hydrogen bond network and its variation, is highly controversial and still unclear. It is common to use the phenomenological two-state model, which treats liquid water as a mixture of two water phases with different structural property parameters, to describe the variation of thermodynamic properties such as volume and heat capacity of water with temperature and pressure. Based on the latent heat absorbed when the ice melts, it is commonly inferred that about 5–7% of the hydrogen bonds in liquid water under ordinary conditions are broken, that is, 20–30% of the water molecules have hydrogen bonds broken, leaving about 70–80% of the water molecules still maintaining a tetrahedral configuration similar to that in ice, with four hydrogen bonds formed around them.

In 2004, using X-ray absorption and emission spectroscopy measurements (Fig. 5.16) and fitting them to spectral lines of water clusters with known structures, Nilsson et al. inferred the surprising conclusion that 80% of the water molecules in liquid water may provide and accept a hydrogen bond, thus forming a chain structure [29].

This is currently inconsistent with the results of both classical and first-principles molecular dynamics simulations, as in Table 5.5. Some scientists believe that this conclusion may exaggerate the number of OH roots in liquid water that are not

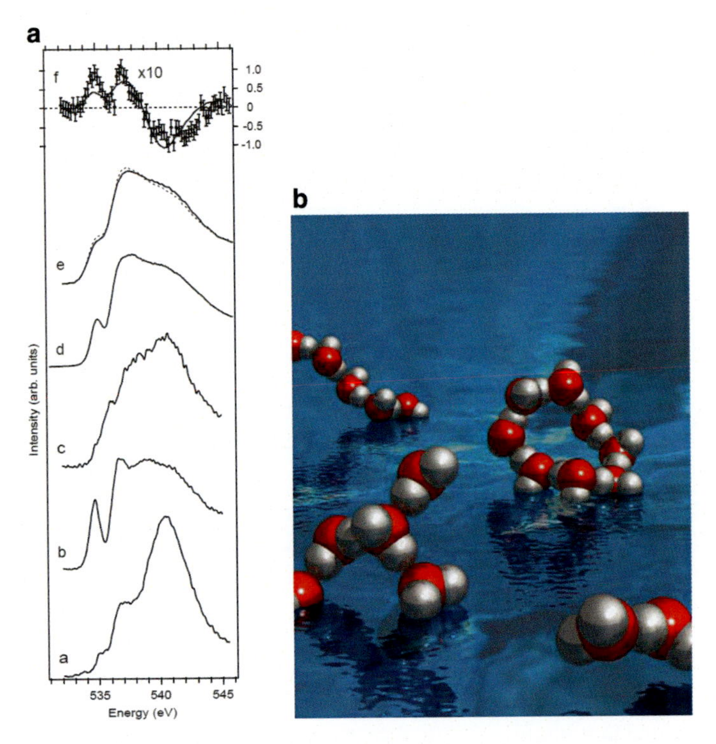

Fig. 5.16 (Left) X-ray absorption spectra (XAS) of various water structures [29]. **a** Ih bulk ice (XAS quadratic electron spectrum); **b** Ih ice surface (XAS auger electron spectrum); **c** ice Ih surface covered by NH_3; **d** room temperature water, taken from literature [30]; **e** water at 25 °C (solid line) and 90 °C (dashed line); **f** difference in spectral lines between water at 25 °C and bulk ice (solid line); spectral line difference (multiplied by a factor of 10) between water at 90 °C and 25 °C. (Right) A visualization of the chain structure in water [31]

Table 5.5 The percentages of various classes of water molecules in liquid water obtained by various methods

	Method			
Type	EXP + FIT	SPC	MCYL	CPMD
25°C				
DD	15^{+25}_{-15}	70	50	79
SD	80 ± 20	27	41	20
ND	5 ± 5	3	9	1
90 °C				
DD	10^{+25}_{-15}	56	39	63
SD	85^{+15}_{-20}	37	47	34
ND	5 ± 5	7	14	3

DD water molecules as two-proton donors; *SD* water molecules as single-proton donors; *ND* water molecules that do not become proton donors. *SPC* and *MCYL* are results of molecular dynamics simulations of the classical model of water. *CPMD* is a result using Car-Parrinello first-principles molecular dynamics simulation

hydrogen-bonded [32] and that a more reasonable number maybe around 5% by direct measurement of the IR absorption surface of free OH [33].

At low temperatures, water coalesces into beautiful ice crystals and snowflakes. Water presents more than a dozen different crystal phases depending on temperature and pressure, as shown in Fig. 2.2. The most common of these is the hexagonal ice Ih, whose simplified structural unit [34] is shown in Fig. 5.17. Snowflakes vary in shape, and it has been claimed that "no two snowflakes are exactly alike". Some photographs of snowflakes are collected in Fig. 5.18. At 50–130 K, water also forms a variety of amorphous forms, such as high-density amorphous (HDA) ice and low-density amorphous (LDA) ice; the phase transition between them is also an important area of study. The formation of liquid or amorphous ice between 136 and 232 K is experimentally extremely difficult and little is known about it, hence the name "no man's land".

It should be noted that under surface adsorption or confined conditions, water exhibits a dozen new ice phases, and the structure and stability of small clusters of water differs from the free cluster case above, and further descriptions are covered in later chapters.

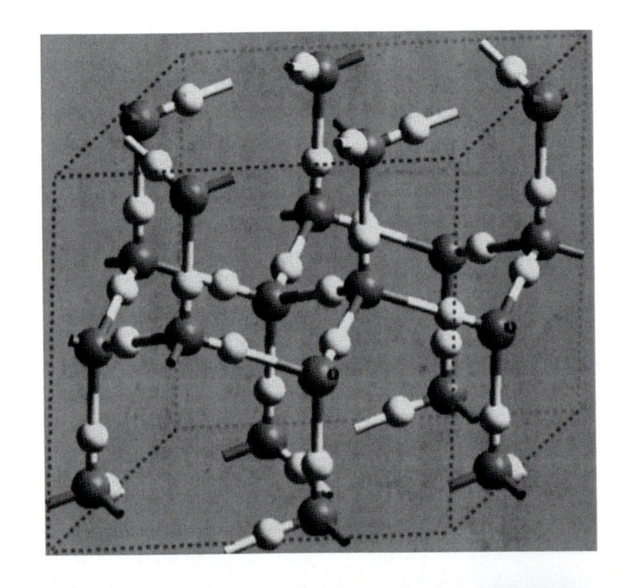

Fig. 5.17 Schematic diagram of the atomic structure of the most common ice phase Ih [34]. The black balls indicate oxygen atoms and the white balls indicate hydrogen atoms

Fig. 5.18 Different snowflake morphologies

References

1. D. Vanderbilt, Soft self-consistent pseudopotentials in a generalized eigenvalue formalism. Phys. Rev. B **41**, 7892 (1990)
2. J.P. Perdew, Y. Wang, Accurate and simple analytic representation of the electron-gas correlation energy. Phys. Rev. B **45**, 13244 (1992)
3. W. Kohn, L.J. Sham, Self-consistent equations including exchange and correlation effects. Phys. Rev. **140**, A1133 (1965)
4. G. Kresse, J. Hafner, Ab-initio molecular-dynamics simulation of the liquid-metal amorphous-semi -conductor transition in Germanium. Phys. Rev. B **49**, 14251 (1994)
5. G. Kresse, J. Furthmüller, Efficiency of ab-initio total energy calculations for metals and semiconductors using a plane-wave basis set. Comput. Mat. Sci. **6**, 15 (1996)
6. G. Kresse, J. Furthmüller, Efficient iterative schemes for ab initio total-energy calculations using a plane-wave basis set. Phys. Rev. B. **54**, 11169 (1996)
7. W.H. Press, B.P. Flannery, S.A. Teukolsky, W.T. Vetterting, *Em Numerical Recipes* (Cambridge University Press, New York, 1986)
8. W.S. Benedict, N. Gailer, E.K. Plyler, Rotation-vibration spectra of deuterated water vapor. J. Chem. Phys. **24**, 1139 (1956)
9. D.W. Turner, C. Baker, A.D. Baker, C.R. Brundle, *Molecular Photoelectron Spectroscopy* (Wiley-Interscience, London, 1970)
10. D.J. Singh, Planewaves, *Planewaves, Pseudopotentials and the LAPW Method* (Kluwer Academic Publishers, Boston, 1993)
11. P.E. Blöchl, Projector augmented-wave method. Phys. Rev. B **50**, 17953 (1994)
12. T.R. Dyke, K.M. Mack, J.S. Muenter, The structure of water dimer from molecular beam electric resonance spectroscopy. J. Chem. Phys. **66**, 498 (1977)
13. J.G.C.M. van Duijneveldt-van de Rijdt, F.B. van Duijneveldt, Convergence to the basis-set limit in ab initio calculations at the correlated level on the water dimer. J. Chem. Phys. **97**, 5019 (1992)
14. K. Kuchitsu, Y. Morino, Estimation of anharmonic potential constants .I. linear XY2 molecules. Bull. Chem. Soc. Jpn. **38**, 805 (1965)
15. J. Hutter et al. CPMD. Copyright IBM Zurich Research Laboratory and MPI für Festkörperforschung 1995–2001.
16. R.N. Barnett, U. Landma, Born-oppenheimer molecular-dynamics simulations of finite systems - structure and dynamics of $(H_2O)_2$. Phys. Rev. B **48**, 2081 (1993)
17. F. Sim, A. St-Amant, I. Papai, D.R. Salahub, Gaussian density functional calculations on hydrogen-bonded systems. J. Am. Chem. Soc. **114**, 4391 (1992)
18. M.J. Frisch, J.E.D. Bene, J.S. Binkley, H.F. Schaefer III., Extensive theoretical studies of the hydrogen-bonded complexes $(H_2O)_2$, $(H_2O)_2H$ +, $(HF)_2$, $(HF)_2H$+, F_2H–, and $(NH_3)_2$. J. Chem. Phys. **84**, 2279 (1986)
19. A.J. Tursi, E.R. Nixon, Matrix-isolation study of the water dimer in solid nitrogen. J. Chem. Phys. **52**, 1521 (1970)
20. R.M. Bentwood, A.J. Barnes, W.J. Orville-Thomas, Studies of intermolecular interactions by matrix isolation vibrational spectroscopy: Self-association of water. J. Mol. Spectrosc. **84**, 391 (1980)
21. H. Umeyama, K. Morokuma, The origin of hydrogen bonding: an energy decomposition study. J. Am. Chem. Soc. **99**, 1316 (1977)
22. W.L. Jorgensen, J. Chandrasekhar, J.D. Madura, R.W. Impey, M.L. Klein, Comparison of simple potential functions for simulating liquid water. J. Chem. Phys. **79**, 926 (1983)
23. D.A. Case et al., *AMBER 6* (University of California, San Francisco, 1999)
24. D.A. Pearlman, D.A. Case, J.W. Caldwell, W.S. Ross, T.E. Cheatham, S. Debolt, D. Ferguson, G. Seibel, P. Kollman, Amber, a package of computer programs for applying molecular mechanics, normal mode analysis, molecular dynamics and free energy calculations to simulate the structural and energetic properties of molecules. Comp. Phys. Commun. **91**, 1 (1995)

25. J.K. Gregory, D.C. Clary, K. Liu, M.G. Brown, R.J. Saykally, The water dipole moment in water clusters. Science **275**, 814 (1997)
26. N. Pugliano, R.J. Saykally, Measurement of quantum tunneling between chiral isomers of the cyclic water trimer. Science **257**, 1937 (1992)
27. J.D. Cruzan, L.B. Braly, K. Liu, M.G. Brown, J.G. Loeser, R.J. Saykally, Quantifying hydrogen bond cooperativity in water: VRT spectroscopy of the water tetramer. Science **271**, 59 (1996)
28. K. Liu, M.G. Brown, J.D. Cruzan, R.J. Saykally, Vibration-rotation tunneling spectra of the water pentamer: structure and dynamics. Science **271**, 62 (1996)
29. P.H. Wernet, D. Nordlund, U. Bergmann, M. Cavalleri, M. Odelius, H. Ogasawara, L.A. Nalund, T.K. Hirsch, L. Ojamae, P. Glatzel, L.G.M. Pettersson, A. Nilsson, The structure of the first coordination shell in liquid water. Science **304**, 995 (2004)
30. S. Myneni, Y. Luo, L.Å. Näslund, M. Cavalleri, L. Ojamäe, H. Ogasawara, A. Pelmenschikov, P.H. Wernet, P. Väterlein, C. Heske, Z. Hussain, L.G.M. Pettersson, A. Nilsson, Spectroscopic probing of local hydrogen-bonding structures in liquid water. J. Phys. Condens. Matter **14**, L213 (2002)
31. Y. Zubavicus, M. Grunze, New insights into the structure of water with ultrafast probes. Science **304**, 974 (2004)
32. J.D. Smith, C.D. Cappa, K.R. Wilson, B.M. Messer, R.C. Cohen, R.J. Saykally, Energetics of hydrogen bond network rearrangements in liquid water. Science **306**, 851 (2004)
33. K. Lin, X.G. Zhou, S.L. Liu, Y. Luo, Identification of free OH and its implication on structural changes of liquid water. Chin. J. Chem. Phys. **26**, 121 (2013)
34. D.R. Hamann, H_2O hydrogen bonding in density-functional theory. Phys. Rev. B **55**, R10157 (1997)

Chapter 6
Experimental Studies of Water-Surface Interactions

The experimental work on the interaction between water and various types of surfaces is very extensive, and only some fundamental and representative work is briefly described in this chapter, with a view to provide the readers with a general impression.

6.1 Individual Water Molecules and Small Clusters on the Surface

A single water molecule on the surface is very difficult to observe because it diffuses extremely easily due to its small diffusion potential, and usually there is only evidence of the presence of a single water molecule at sufficiently low temperatures (< 40 K).

On the simplest high-symmetry surface of transition metal, Salmeron et al. directly observed a single water molecule on the Pd(111) surface using scanning tunneling microscopy (STM) at 40 K [1]. In the STM image shown in Fig. 6.1, the single water molecule is a small bright spot and its internal structure cannot be distinguished. They also simultaneously observed the diffusion process of a single water molecule on the Pd surface. A single water molecule collide with other single water molecules in the diffusion pathway and gradually coalesce into water dimers, trimers, and other multimers. In particular, Salmeron et al. observed that water dimers form and then diffuse extremely rapidly, at a rate 10,000 times faster than that of single water molecules. This is explained by the fact that the quantum motion of the hydrogen nucleus of the water molecule in the dimer plays a major role in the rotation process [2].

Ertl et al. obtained the vibrational spectrum of a single water molecule on Pt by electron energy loss spectroscopy (EELS) measurements [3]. They found that the energy of a single water molecule vibrating perpendicular to the Pt surface is 15 meV and the rotation energy of water molecules confined by the surface is 28 and 36 meV.

Recently, on a thin layer of NaCl(001) surface, Ying Jiang's group [4] has not only observed individual water molecules, but also obtained information on the internal

S. Meng and E. Wang, *Water*,
https://doi.org/10.1007/978-981-99-1541-5_6

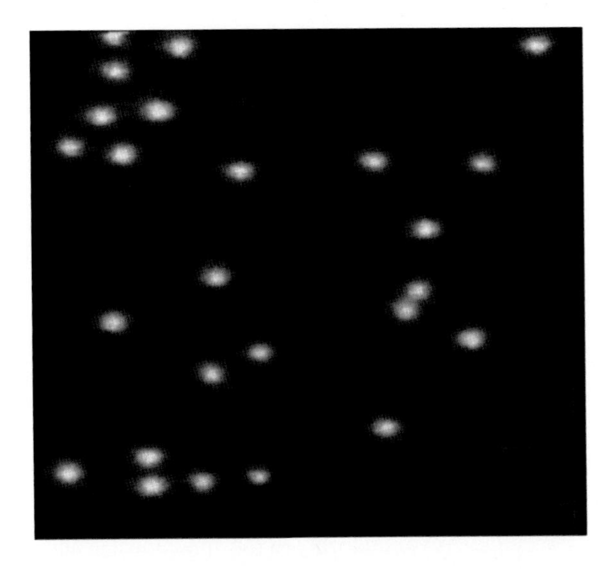

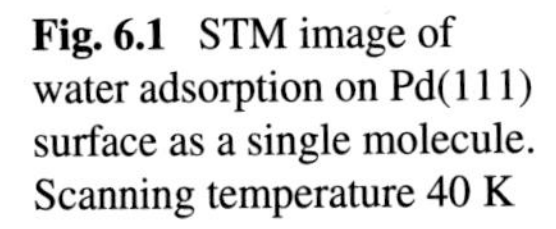

Fig. 6.1 STM image of water adsorption on Pd(111) surface as a single molecule. Scanning temperature 40 K

degrees of freedom of individual water molecules, i.e., information on the distribution of electron orbitals. These experimental observations are further supported by theoretical works based on the first-principle calculations. First, they deposited a thin layer of NaCl on an Au(111) substrate and obtained a perfect NaCl(001) bilayer structure. Further water molecules were deposited on it and found to be isolated single molecules adsorbed dispersedly on the surface at low temperature and low coverage, as in Fig. 6.2. The stripe-like structure in the figure reflects the reconstruction of the surface of Au(111) substrate under the thin NaCl layer.

A closer look at the bright spots in Fig. 6.2 shows that unlike the bright spots deposited directly onto the metal surface, the bright spots on the surface of the thin NaCl layer have an internal structure. The internal structure of these bright spots is also changing under different bias pressures. Detailed analysis shows that this internal structure corresponds to the electronic orbitals of individual water molecules, i.e., the highest occupied molecular orbital (HOMO) and the lowest unoccupied molecular orbital (LUMO), or a mixture of both. When the substrate is under positive bias, the water molecule presents its HOMO orbital, which is characterized by a distribution of electron clouds as two flaps of mirror image symmetry, with the dark line in the middle as the symmetry plane. This corresponds to the HOMO orbital of the water molecule ($1b_1$), which consists mainly of the p_z-orbitals of O. The water molecule plane is the region of wave function nodes. When the substrate is under negative bias, the orbitals are characterized by an "egg-shaped" ellipsoid when viewed in the direction normal to the surface. This corresponds to the LUMO orbital of the water molecule ($4a_1$), with the large end of the ellipsoid corresponding to the position of the O atom and the pointy end to the position of the H atom. Under certain bias pressures, the observed water molecule orbitals are a mixture of these two cases. In this way, Ying Jiang et al. not only observed the internal structure of the water molecule (i.e.,

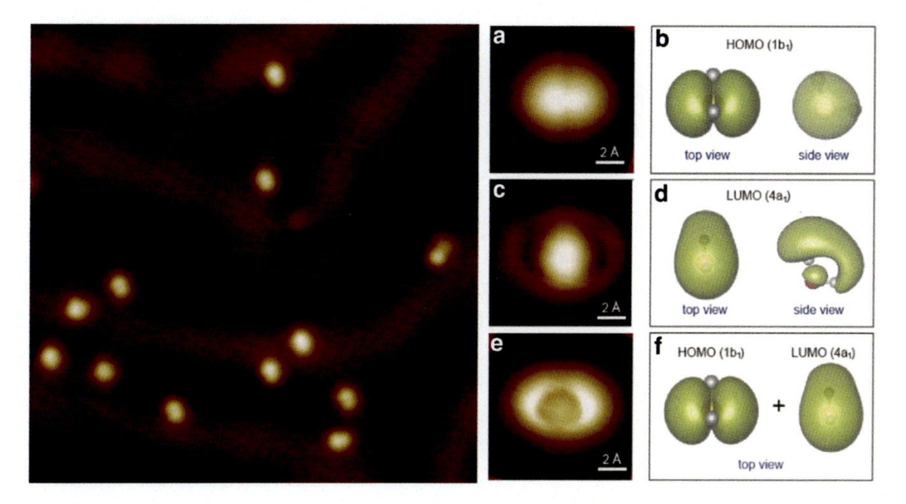

Fig. 6.2 (Left) STM images of single water molecules adsorbed on two thin layers of NaCl grown on the Au(111) surface. (Middle) Magnified images at different bias voltages. Substrate is positively biased at 100 mV (A); − 100 mV (C); − 50 mV (E). (Right) Calculated electronic orbital shapes of water molecules [4]

the orientation of the OH bond) directly from experiment for the first time, but also gave information on the electronic orbitals of the single water molecule!

In this work, the structure and orbital information of individual water molecules can be observed for two reasons: (1) the NaCl insulator film inserted between the water molecule and the metal substrate makes the water molecule decouple from the substrate, thus allowing the intrinsic orbital information of the water molecule to be retained intactly; (2) using the modified scanning tunneling microscope tip as the top gate pole, controlling distance and coupling strength between the tip and the water molecule with pico-metric precision, artificially spreading and moving the leading orbitals of water molecules near the Fermi surface, and modulating the distribution of the density of states for molecular orbitals near the Fermi energy level, thus greatly enhancing the signal-to-noise ratio of imaging. In fact, in this experiment, the water molecule is closer to the STM tip than to the metal substrate due to the isolation effect of the NaCl insulating layer, and thus the coupling strength between the water molecule and the tip is much stronger than that between the water molecule and the metal substrate, which explains that the HOMO orbital of the water molecule is observed instead when the substrate is positively biased! Using an insulating layer to attenuate the electronic coupling between the molecules on the surface and the substrate would be an effective and unique new way to observe the geometry and electronic structure of small molecules.

By increasing the temperature of the substrate above 40 K, individual water molecules will begin to diffuse and coalesce into small clusters. Experimentally, the water dimer on the Pt(111) surface was imaged directly with STM and the image in Fig. 6.3d was observed. The image presents a six-valve symmetric distribution, which is explained by the fact that the water dimer is constantly rotating around the water molecule with lower absorption position during the imaging process. The higher water molecule spends more time in the top position of the six surrounding Pt atoms during the rotation, resulting in the hexagonal symmetric distribution of the image.

Another cluster of water that forms a hexagonal structure is the hexamer of water. Among the small clusters, it is the most stable structure. Unlike the free hexamer cluster in the gaseous state, which has a three-dimensional structure, the cluster on the surface is a ring-like quasi-planar structure, with each water molecule providing and accepting a hydrogen bond from the preceding water molecule, forming a symmetric six-membered ring. Its image is also shown in Fig. 6.3g.

Based on the hexameric structure, water can further form clusters of larger size. The heptamers, octamers, and ninamers formed by water molecules are shown in

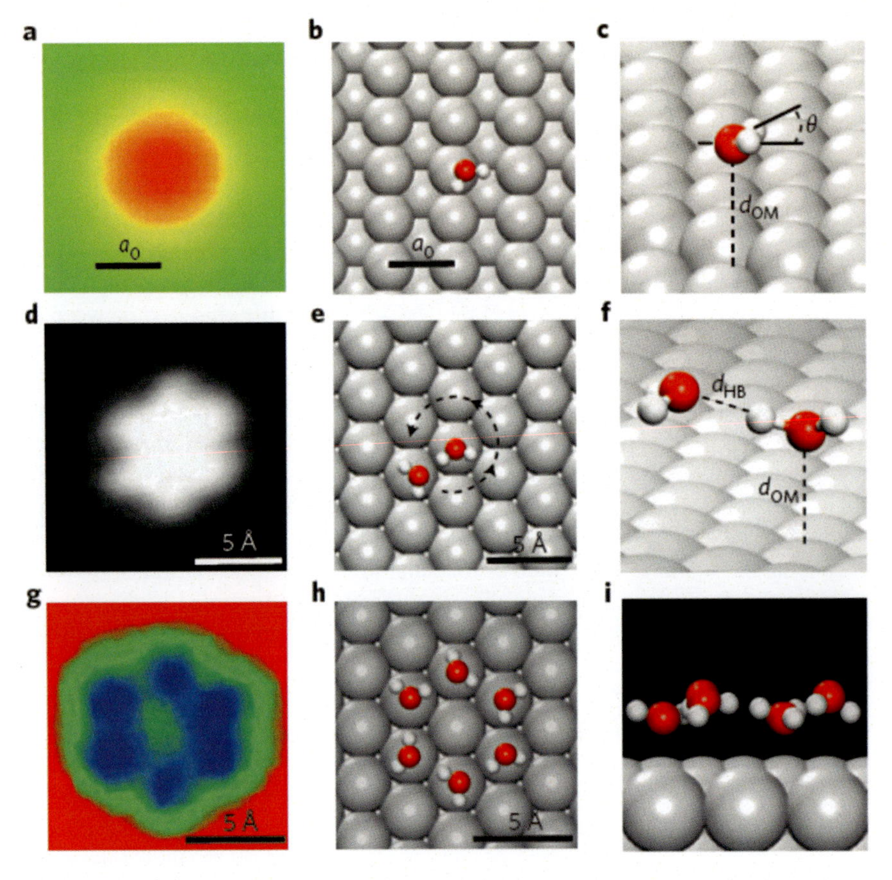

Fig. 6.3 STM images of **a** single water molecule on Cu(110); **d** water dimer on Pt(111); **g** water hexamer on Cu(111). **b**, **e**, **h** are top views of the corresponding atomic structure diagrams; **c**, **f**, **i** are side views [5]

Fig. 6.4. The figure shows that a water molecule at the edge of the hexamer can form a hydrogen bond with another water molecule to become heptamer, and the water molecule at the edge appears as a bright spot in the STM image because it is higher than that in the hexamer. Similarly, the octamer is composed of two water molecules adsorbed at the edge of the hexamer. The STM image of the water ninamer has three bright spots and appears roughly as a triple symmetric "triangle"/"trimer" structure, which suggests that great care should be taken when using STM images of water clusters for structure determination. Water clusters with larger size are also shown in figure. As shown more clearly in Fig. 6.5, the stable water hexamer or a large cluster formed by triple polymerization in a hexagonal manner can adsorb one or more water molecules to form bright spots.

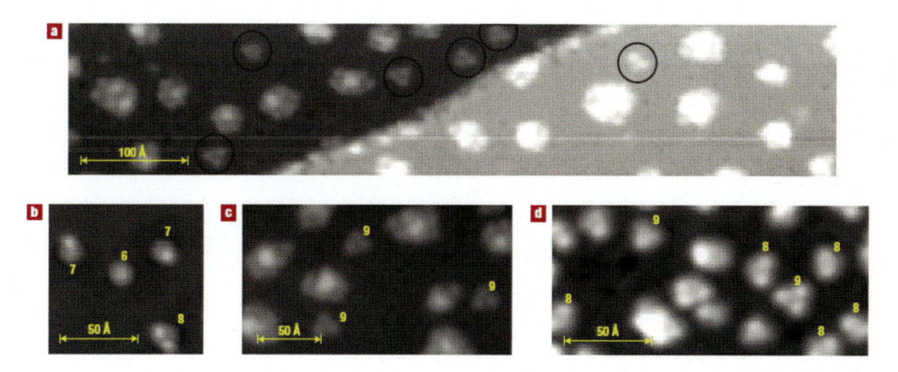

Fig. 6.4 STM image of a larger water cluster [6]. The numbers near the water cluster indicate the number of water molecules in that cluster. **b** shows D_2O/Ag(111), all others are H_2O/Cu(111)

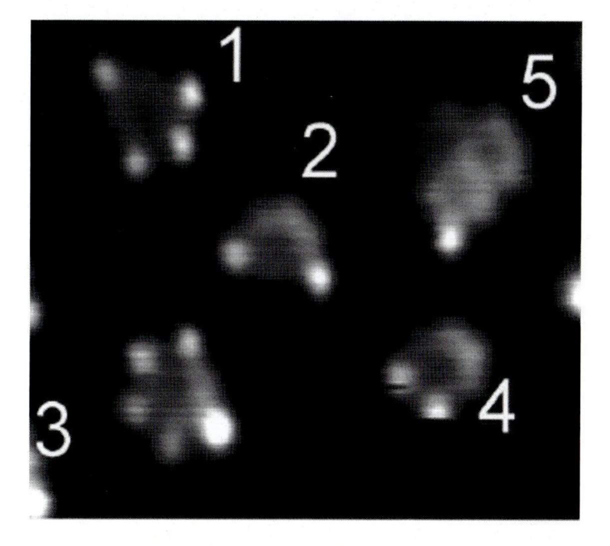

Fig. 6.5 Structure of water clusters on the surface of Ru(0001) [7]. Cluster #1 is a hexamer with four water molecules attached to it; #2 and #4 are hexamers with two water molecules attached to them; #3 is a hexamer with five water molecules attached to it; and cluster #5 shows one more water molecule attached to the edge of a tripolymer of three hexamers. To form the clusters, the temperature was first increased to 130 K, and then scanned at 50 K

Remarkably, the structure of gaseous clusters in free space corresponding to these larger clusters is unknown. Surface interactions fix the clusters of water and provide experimental means for studying them.

The cluster structure of water was also observed by Jiang et al. [4] on a thin layer of NaCl insulation. On such surface, water is most likely to form a tetrameric structure. It has a symmetrical quadruple structure, in which each water molecule provides and also accepts a hydrogen bond with neighboring water molecules, and thus is also a ring structure. Using STM tip, Ying Jiang et al. could manually manipulate individual water molecules to form a tetramer, and their manipulation process is shown in Fig. 6.6. Since the directionality of the hydrogen bonds in the tetramer could be counterclockwise or clockwise, it exhibits different chirality on the surface. Using high-resolution STM orbital imaging, Ying Jiang et al. were able to discriminate the chirality and thus were able to directly distinguish between the two tetramer structures with completely simplified energy, as shown in Fig. 6.7.

The decomposed adsorption structures of water molecules on the surface of oxides such as TiO_2, Fe_2O_3, such as OH, and H adsorption, and mixed adsorption formed with water molecules have also been much studied and will not be mentioned here.

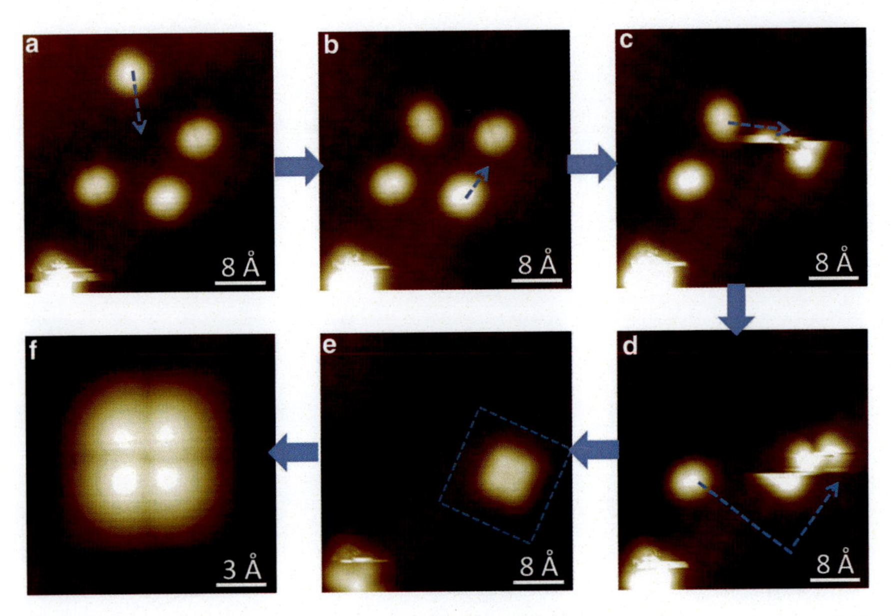

Fig. 6.6 Manipulation of four water molecules on a thin layer of NaCl on the surface of Au(111) to form a water tetramer

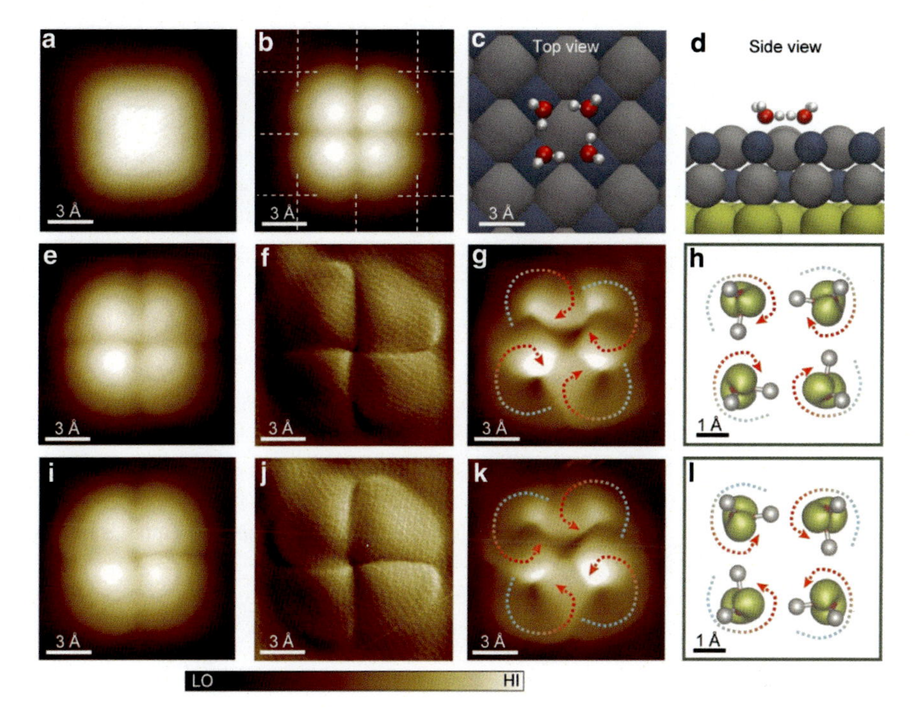

Fig. 6.7 Directional resolution of hydrogen bonding in the water tetramer. **a**, **b** are STM images of the water tetramer under two biases. **c**, **d** are atomic conformations. **e**, **i** are STM images of the two chiral tetramers. **f**, **j** are differential images. **g**, **k** are higher resolution STM images. **h**, **l** are calculated HOMO orbital distributions of the tetramer. **e–h** corresponds to the counterclockwise pointing hydrogen bonding conformation case. **i–l** corresponds to the clockwise pointing hydrogen bonding conformation case

6.2 Rosette and Chain Structure of Surface Water

More water molecules can present a petal-like structure by aggregating with each other. An example is shown in Fig. 6.8. In these structures, the water molecules inside the cluster are arranged in a hexagonal honeycomb structure, showing multiple interconnected hexagonal arrangements. This structure is similar to an ice bilayer in a bulk phase ice Ih, indicating a gradual transition of the water clusters to a bulk phase structure. Many bright spots or bright lines appear at the edges of these petal-like clusters, corresponding to the OH bonds of water molecules at the edges that are not hydrogen bonded. Due to the special orientation of their OH dangling-bonds, water molecules at the edges are higher than those inside, appearing as spots much brighter than the central area in the STM image [7].

These petal-like clusters can join together to form a ribbon structure when increasing coverage or undergoing high temperature annealing [8] (Fig. 6.9). Some

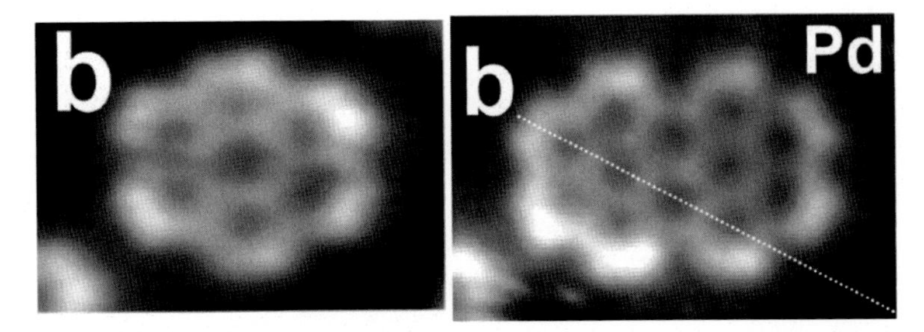

Fig. 6.8 Petal-like cluster structure formed by water molecules on the surface of Pd [7]

hexagonal star-shaped holes forming inside these ribbon structures, the width of which is about width of 1–3 water hexagons. These infinitely connected band structures are the transitional form between the small clusters of finite size and the complete two-dimensional film of infinite size forming by waters. Similar to the finite-sized petal-like structures, the middle part of these ribbon structures is also darker, while bright spots appear at the edges or around holes.

A detailed analysis shows that the darker parts of the water clusters and water bands correspond to water molecules lying flat on the surface of metal, while the bright spots at the edges correspond to those with adsorption conformation that the molecular plane perpendicular to the surface. Since the water clusters on the surface need to optimize the adsorption of water molecules to the surface and the hydrogen bonding between water molecules, the competition between these two interactions forces some of the water molecules in the clusters to adopt a perpendicular conformation that interacts more weakly with the surface. Once these vertically standing water molecules appear, they stop gathering to form larger structures due to the difficulties of formation of hydrogen bonds. Therefore, these vertically standing ones are often found at the edges of the clusters. Some characteristic arrangement structures of water

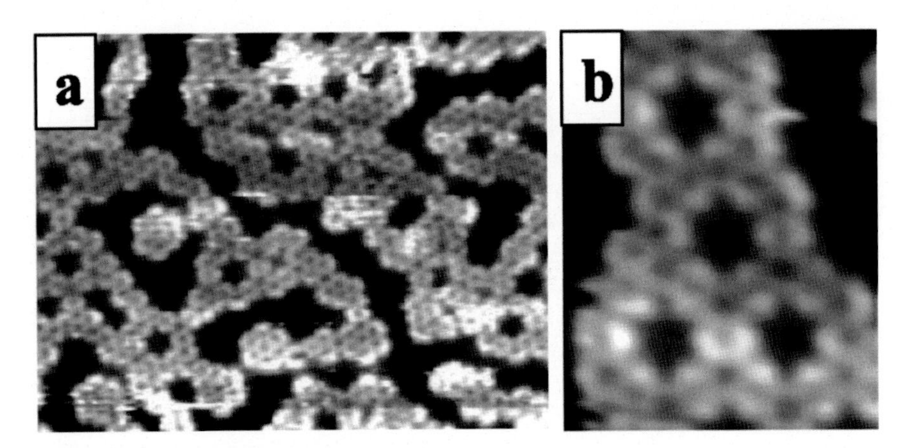

Fig. 6.9 Band structure formed by water molecules [8]

molecules are shown in Fig. 6.10. Ulteriorly, Salmeron et al. found [7] the variety in strength of the water-surface interactions and hydrogen bonding interactions on different surfaces, giving rise to different situations of competition between them. For example, on the Pd(111) surface, water-Pd adsorption is relatively weak and water molecules form large clusters with more hydrogen-bonds, while on the Ru(0001) surface, the water-Ru interaction is stronger and it is more likely to form isolated single molecules or small clusters.

Briefly mentioned here that, water molecules between zero-dimensional clusters and two-dimensional thin layers form other interesting structures on metal surfaces, such as one-dimensional chain-like structures. Here are a few well-known examples:

(1) Selective adsorption of water on the steps of Pt(111) surface formed a nearly unimolecular wide one-dimensional water chain [9]. Meng Sheng et al. found by calculation that the adsorption energy of water on the Pt step increases by 80 meV (about 1/4 of the adsorption energy at the platform) compared to that

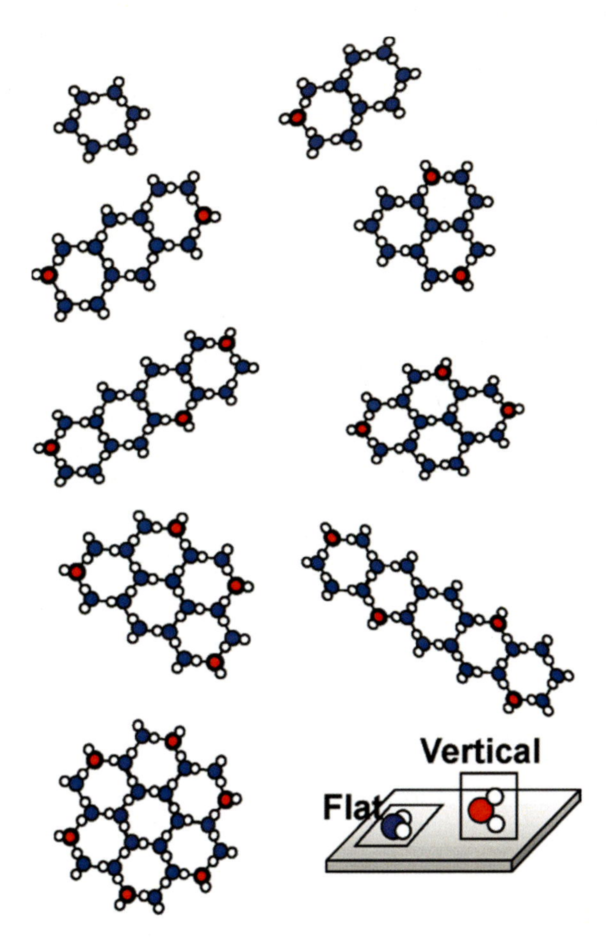

Fig. 6.10 Schematic diagram of the cluster structure of water molecules on the surface. Blue shows water molecules lying flat; red shows water molecules standing upright

at the platform, indicating that water tends to absorb on the steps to form chain structures [10].

(2) As early as 1995 Held and Menzel observed that H_2O forms a stripe-like structure on the surface of Ru(0001) based on the $\sqrt{3} \times \sqrt{3}R30°$ fundamental unit with a width of 6.5 periods of surface atom [11]. This one-dimensional stripe-like structure was not observed in the D_2O aqueous layer and thus exhibits isotopic effects. However, subsequent studies of this one-dimensional stripe-like structure on this surface are relatively rare.

(3) On the Cu(110) surface at 78 K, Yamada et al. observed the formation of many one-dimensional water chains [12] oriented along the surface [100] with a width of about 6 Å and a length of more than 1000 Å via STM. Detailed analysis showed that the water chains form a serrated structure with a period of 7.2 Å. They also found repulsive interactions between these chains, forming an almost equally spaced (on average about 50 Å) distribution. The STM images of Carrasco et al. confirm these findings; more importantly, by comparing the experimental images with those obtained from first-principle simulations, they found that the chain-like structure of these waters is not composed of six-membered rings of water molecules, but of interlaced five-membered rings [13], see Fig. 6.11. This is a surprising result that overturns the commonly held notion that water must form a hexagonal structure to be stable (e.g., the shape of snowflakes usually have hexagonal symmetry). In fact, we have long proposed to explain the adsorption of water on the surfaces of SiO_2 and salt in terms of quadrangular and octagonal ice structures. This also shows the diversity of the structure of water on the microscopic scale (Fig. 6.12).

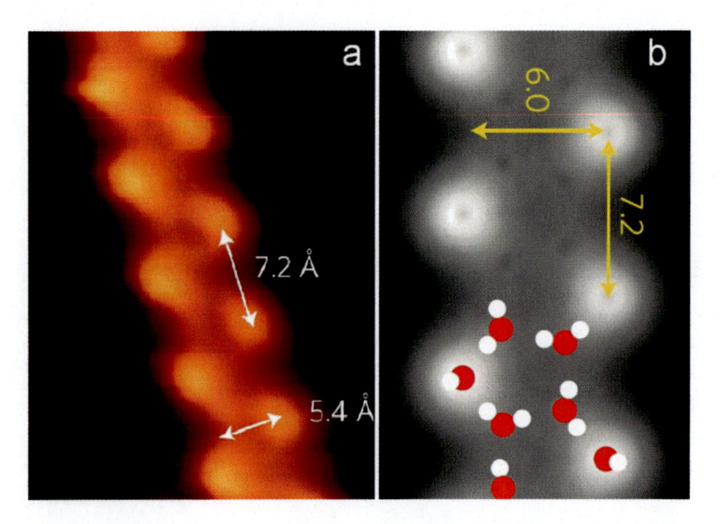

Fig. 6.11 **a** Experimental STM image and **b** theoretical simulation image and molecular structure of the chain-like structure consisting of water five-membered rings on the surface of Cu(110) [13]

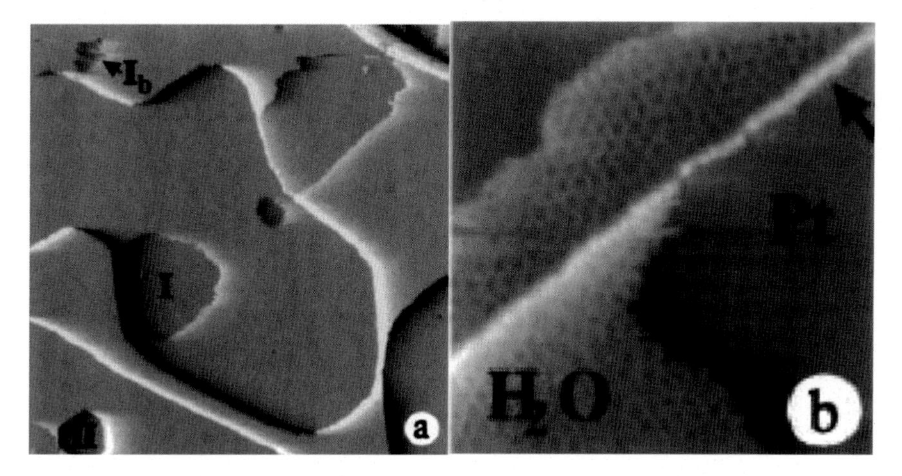

Fig. 6.12 **a** STM images of ice multilayers on the surface of Pt(111). **b** Image of ice monolayers on Pt(111). The arrows point to the one-dimensional ice chain at the edge of the step [9]

4) There are also a variety of one-dimensional chain-like structures on the Cu(110) surface, forming (2 × 1) and other reconfigurations. The direction of the chains is instead along the [110] direction, and these are usually due to the involvement of a large number of OH roots in the adsorption [14].

6.3 Monolayer and Multilayer Structure of Water on the Surface

At larger water coverage, water molecules form complete mono- and multi-layer structures on the surface [9], see Fig. 6.12. After more than 30 years of careful study, much has been learned about the structure of thin layers of surface water. For example, on the Pt(111) surface, low energy electron diffraction experiments (LEED) have found the formation of an ordered periodic structure $\sqrt{3} \times \sqrt{3}R30^\circ$ (abbreviated as $\sqrt{3}$), i.e., one period for every three metal atoms in the direction of 30º to the nearest surface layer atoms and each period adsorb two water molecules, corresponding to a coverage of 2/3 ML (1 ML = 1 Monolayer, corresponding to the coverage of one adsorbate per surface atom). In contrast to the lattice structure of the bulk-phase ice Ih (4.51 Å × 4.51 Å × 7.35 Å), it is widely believed that water molecules form a so-called ice bilayer (BL) in this two-dimensionally ordered system: water molecules are connected to three surrounding ones by hydrogen bonds, forming a folded hexagonal arrangement [3, 15–17], very similar to the structure in bulk ice (Ice Ih) [18]. However, this $\sqrt{3} \times \sqrt{3}R30^\circ$ pattern has only been observed in some small regions on the surface [3]. In addition, helium atom scattering (HAS) experiments have found that two other complex periodic structures can be grown at higher temperatures (130–140 K): $\sqrt{39} \times \sqrt{39}R16.1^\circ$ ($\sqrt{39}$) and $\sqrt{37} \times \sqrt{37}R25.3^\circ$

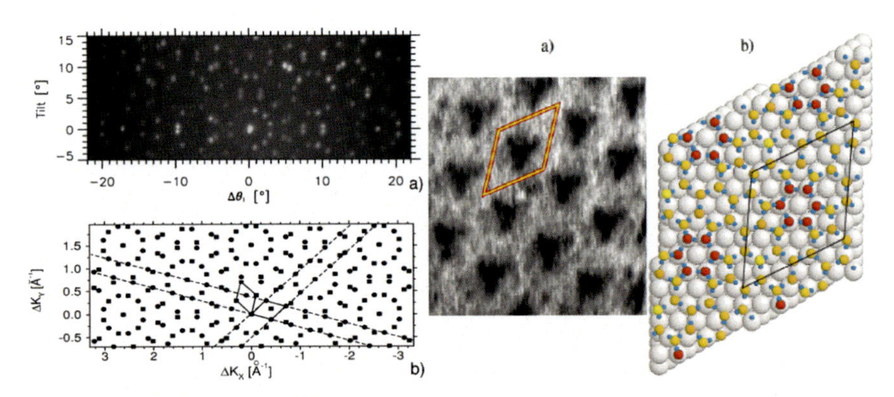

Fig. 6.13 He atomic scattering pattern, STM picture, and atomic structure model for the $\sqrt{39} \times \sqrt{39}R16.1°$ structure of water on the surface of Pt(111) [19]

($\sqrt{37}$) [16], see Fig. 6.13a. LEED experiments confirm that under electron beam irradiation or when growing more than five bilayers, the structure of $\sqrt{39}$ will reverse the orientation of the whole and becomes an island arranged by $\sqrt{3}$. This $\sqrt{3}$ multilayer forms a non-public thin film structure with the metal substrate [17].

Recently, Nie et al. used STM for direct real-space imaging observations of the $\sqrt{39}$ structure found in the HAS and LEED experiments [19]. They found that a triangular-like dark region appears at the center of each $\sqrt{39}$ protocell, see Fig. 6.13. After careful analysis and comparison with the theoretical model, they concluded that this dark region corresponds to the hexamer where all water molecules lie flat; this hexamer interacts most strongly with the surface and thus has the lowest height of O atoms and is the darkest in STM imaging. In addition, they found that in order to maximize hydrogen bonding within the water layer, several pairs of pentagon-heptagonal structures exist around this hexamer, similar to the Stone-Waals defects of the graphene structure, such that the total energy of the system is lowest.

Surprisingly, the $\sqrt{39}$ water (mono-)layer on the Pt(111) surface is hydrophobic. As shown in Fig. 6.14, By depositing Kr on the water layer on Pt(111) surface and measuring the spectrum of temperature programmed desorption (TPD) of Kr, Kay et al. found [20] that: depositing water on the Pt surface below 120 K, the water layer completely infiltrates the Pt surface; continuing to deposit water molecules on the water monolayer above 135 K, the water fails to infiltrate the monolayer. Experimentally, the desorption temperature of the inert gas Kr covering the water layer of different thicknesses is different: the desorption temperature of Kr on the bare Pt surface is highest, at about 65 K; the second highest is that on the water monolayer, at about 48 K; as the water layer becomes thicker, the surface adsorption potential energy to which Kr is subjected becomes smaller, and the characteristic desorption temperature decreases. At larger water coverage, no desorption peak around 65 K is observed indicating that Pt is completely covered by water. However, no matter how much the water coverage is, there is always a desorption peak around 48 K,

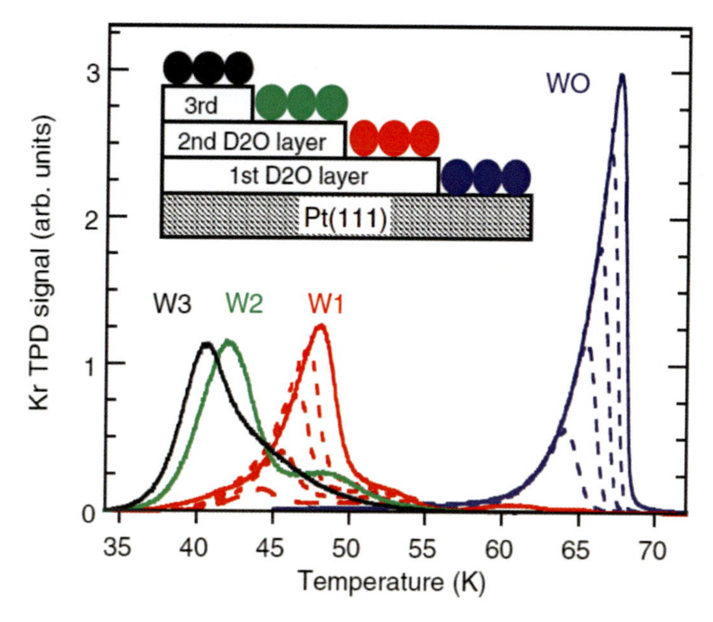

Fig. 6.14 Kr desorption spectrum of the ice layer formed on Pt(111) [20]

indicating constant presence of a bare water monolayer, or that the water monolayer cannot be completely covered and infiltrated by water.

On the Ru(0001) surface, it was measured early that the water layer forms a structure with $\sqrt{3}$ symmetry to the surface [18]. Later, measurements of LEED I-V experiment [21] showed that the water layer on the Ru surface is a bilayer extremely compressed in structure, with the distance between the two layers being only 0.10 ± 0.02 Å. The distances of the two layers of O atoms from the surface are 2.08 and 2.23 Å, respectively, see Fig. 6.15. In contrast, the interlayer distance of the ice bilayer in the bulk phase ice Ih is 0.9 Å, which is extremely different from that of the ice on Ru surface. This nearly planar surface bilayer ice structure caused much confusion until Feibelman proposed a model for a semi-decomposed water layer to explain it in 2002 [22]. In Feibelman's model, as shown in Fig. 6.16, half of the water molecules decompose into OH and H adsorbed on the surface, forming a structure with energy 0.2 eV/molecule less than the undecomposed water layer, which is possibly the structure in Held and Menzel's LEED experiment. Numerous experimental and theoretical studies later showed that the decomposed water layer on Ru(0001) is caused by the influence of the external environment, i.e., the undecomposed infiltrated water layer is excited to decompose under the irradiation of a large number of electrons or photons or at higher temperatures and is converted into a more stable semi-decomposed structure [23]. However, under general conditions, completely undecomposed infiltrated water layers still exist. So far, there is a lack of quantitative models for the structure of the undecomposed infiltrated water layer on Ru. Hodgson et al. suggested that there is a short chain structure consisting entirely of flat lying water molecules in this water layer, and that the connections and distribution between these short chain structures are rather disordered, so that the water layer is an average $\sqrt{3}$ structure in the horizontal direction, but may be a disordered dynamic structure in the height distribution of water molecule [24, 25].

There are water layers consist of H_2O and OH on the surface of Cu(110). They also exhibit an interconnected hexagonal honeycomb structure, as shown in Fig. 6.17. However, irregular distribution of bright spots appears in the STM diagram [26]. A careful study especially compared with the theoretical model indicates that a Bjerrum hydrogen bonding defect forms between two adjacent OH groups in this structure, i.e. the H of the two OH groups point towards each other and do not form a hydrogen bond. This defect itself is unstable and requires energy to form, but the presence of this defective structure makes the hydrogen bonding between the water molecules and the OH groups much stronger, and all the undecomposed water molecules lie flat on the surface and form a strong adsorption with the surface, so that the total energy of the system is instead lower than that without the hydrogen bonding defect. The surface water layer can have a large number of hydrogen bonding defects! This is also a surprising result.

Recently, Kay et al. found that water forms a two-layer structure on a graphene substrate deposited on a Pt surface [27]. Using Ar atomic desorption spectroscopy measurements, they found that unlike on the Pt surface where a single layer of water coverage is sufficient to completely cover the surface, on the graphene/Pt surface, two layers of water coverage are required, see Fig. 6.18. This indicates the special two-layer structure formed by water. Further measurements of the IR adsorption spectrum of this ice structure led them to infer that the water forms a flat two-layer structure where each water molecule within each layer forms three hydrogen bonds

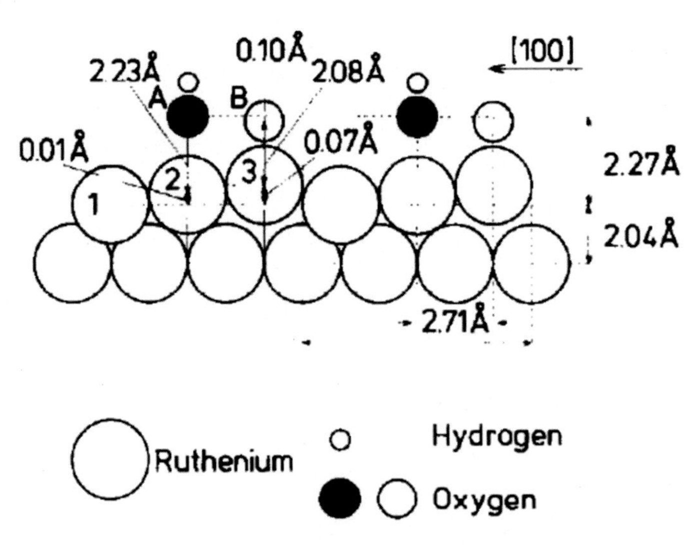

Fig. 6.15 Schematic representation of the atomic structure of the water layer on Ru(0001) surface obtained from the LEED I-V fit [21]

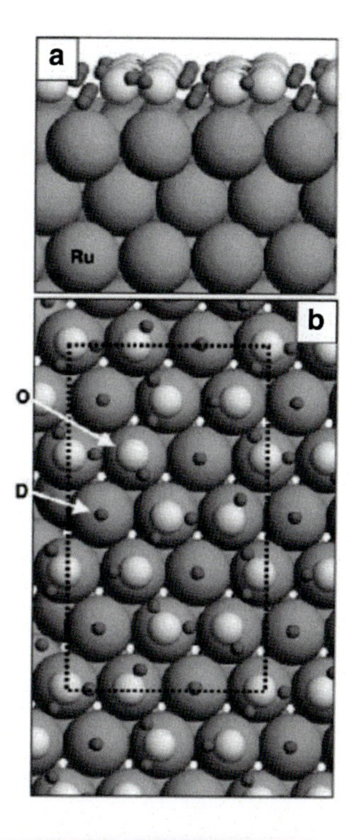

Fig. 6.16 Water semi-decomposition structure on the surface of Ru(0001) proposed by Feibelman [22]

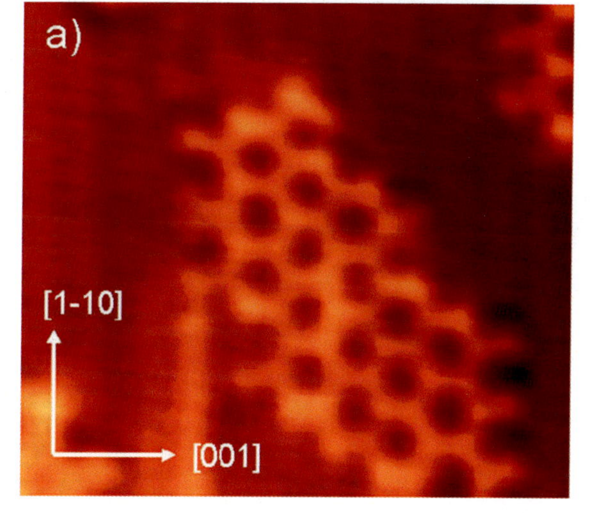

Fig. 6.17 **a** STM image of the structure of a small H_2O + OH water layer on the surface of Cu(110). **b** Atomic structure corresponding to the most stable H_2O + OH mixed water layer given by density generalized theory. Boxes are the original cells used for the calculations. The elliptical circles mark a Bjerrum hydrogen bonding defect [26]

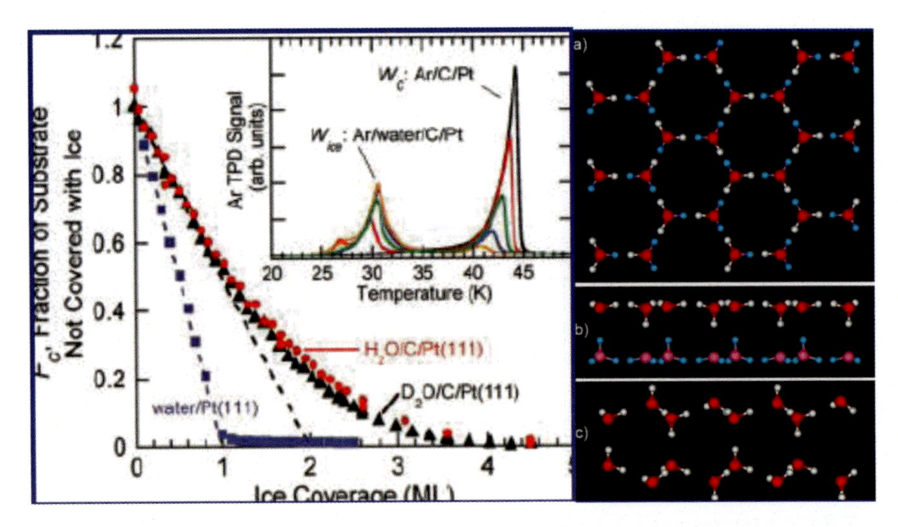

Fig. 6.18 Variation of the percentage of surface area exposed to water coverage on different substrates. The inset shows the Ar desorption spectrum. The structure of the water layer on the graphene substrate is deposited on the Pt surface. The top and side views of the atomic structure of the corresponding two ice layers are shown on the right. conventional ice layer structure [27]

with three surrounding near-neighbor molecules, and each water molecule in turn forms a vertically oriented hydrogen bond connecting the top and bottom layers, such that each water molecule forms four hydrogen bonds and is fully saturated with a very stable structure. Each water layer is very flat and has no hanging bonds, which is very different from the traditional two-layer structure similar to that of bulk-phase ice. Each water layer of the latter is a conventional "bilayer" with an OO thickness of 0.97 Å and a free OH pointing to the vacuum above and below the ice layer, respectively.

6.4 Structure of Surface Water at Ambient Conditions

The structure of surface water at ambient temperature and pressure is more important, but also very complex. The structure and role of water on general surfaces under normal conditions is often poorly understood due to a lack of suitable research tools.

Yuan-Yang Shen et al. measured the vibrational properties of ice and water on solid and liquid surfaces using surface and frequency generation spectroscopy to infer the structure of surface water [28]. For the quartz-ice interface, the measured water vibrational spectrum is shown in Fig. 6.19, with the main vibrational peaks concentrated at 3100 cm^{-1}, which corresponds to the vibrations of strong hydrogen bonds in bulk ice. There are some small bumps at 2900, 3500, 3700 cm^{-1}. At the air–liquid water interface, the shape of the vibrational peaks is quite different. There are three main vibrational peaks distributed at frequencies of 3100, 3400, 3700 cm^{-1};

they correspond to the stretching vibrational modes of OH with strong hydrogen bond, with weak hydrogen bond and without hydrogen bond formation in liquid water, respectively. The latter, in particular, has the highest intensity and the smallest width, indicating the presence of a large number of free OH suspension bonds at the surface of liquid water, roughly 25% of the total number of OHs at the surface. At the hexane-water and quartz-oil–water interfaces, the vibrations of water is similar to that at the water–air interface, but with different details. At the hexane-water interface, the OH vibrational frequency distribution of hydrogen bond formation is wider and more evenly distributed, and the two strengths of hydrogen bonds can no longer be distinguished. This suggests that unlike the air layer, the presence of the hexane layer results in a more uniform structure and polarity of the interfacial water. At the quartz-oil–water interface, not only the two strengths of hydrogen bonds no longer present, but the OH vibrational peaks forming the hydrogen bonds become sharper, similar to the vibrational spectrum in ice, except that the peaks are slightly redshifted, at around 3200 cm^{-1}. But the biggest difference from interfacial ice is that a large number of un-hydrogen-bonded OH suspension bonds exist on this surface, and their strength is greater than that of the water–air surface. By a similar experimental technique, Y. R. Shen et al. also measured the structure of water on metallic Pt(111) surfaces, Al_2O_3 surfaces, SiO_2 surfaces, and organic self-assembled film surfaces, and found that the presence of these surfaces often induced an ordered distribution of water in the range of one to tens of water layer thicknesses near the surface.

By Atomic Force Microscopy (AFM) measurements, J. Hu et al. found that an ice-like water structure forms on the mica surface even at room temperature, and named this structure as room temperature ice [29]. The water layers show different morphologies at different humidity levels. In particular, these water structures at room temperature show specific shapes with solid like sharp edges, especially these sharp edges show certain angles with the lattice direction of the mica surface, see Fig. 6.20. By counting, it is found that the values of these angles are basically 0°, 60°, or 120°, which coincides with the hexagonal lattice structure of solid ice Ih. So it is highly likely that the structure of this room temperature water is similar to that of solid ice. Optical measurements also indicate that the structure has ice-like vibrational properties. Until today, this ice structure that can exist at room temperature still attracts great interest, which has important implications for biotechnology, lubrication and environmental issues.

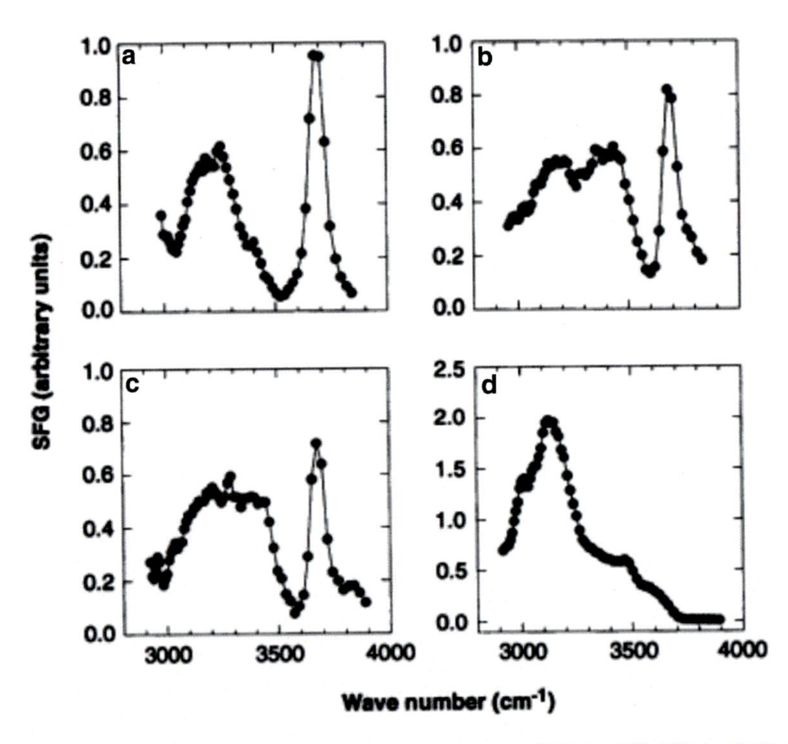

Fig. 6.19 Sum frequency vibrational spectra [28] from **a** quartz-OTS (i.e. $CH_3(CH_2)_{17}SiCl_3$) oil–water interface; **b** air–water interface; **c** hexane-water interface; **d** quartz-ice interface, respectively

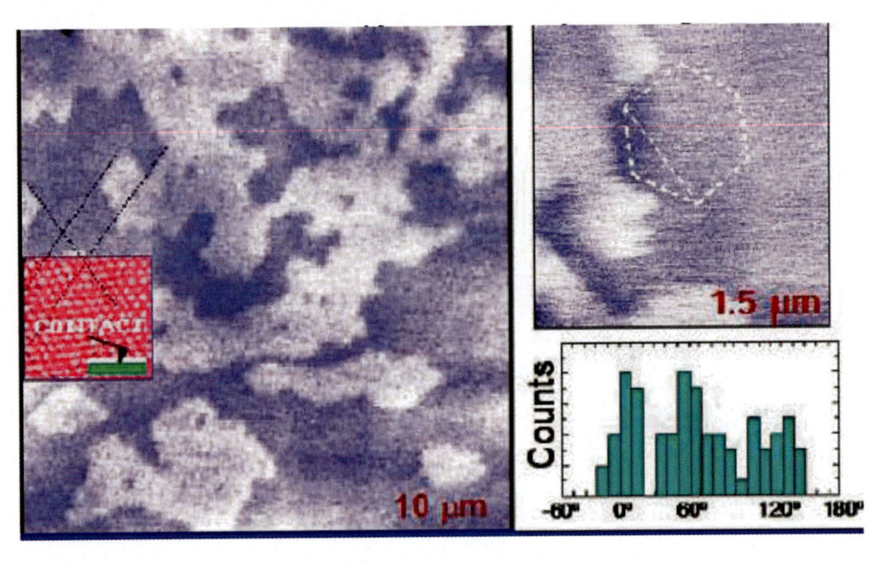

Fig. 6.20 AFM image of the water layer on a mica sheet at room temperature [29]. Top right, magnified image. Bottom right, statistical results of the angle between the edge line of the water block and the lattice direction of the mica surface

References

1. T. Mitsui, M.K. Rose, E. Fomin, D.F. Ogletree, M. Salmeron, Water diffusion and clustering on Pd(111). Science **297**, 1850 (2002)
2. V.A. Ranea, Water dimer diffusion on Pd{111} assisted by an H-bond donor-acceptor tunneling exchange. Phys. Rev. Lett. **92**, 136104 (2004)
3. K. Jacobi, K. Bedürftig, Y. Wang, G. Ertl, From monomers to ice—new vibrational characteristics of H_2O adsorbed on Pt(111). Surf. Sci. **472**, 9 (2001)
4. J. Guo, X.Z. Meng, J. Chen, J.B. Peng, J.M. Sheng, X.Z. Li, L.M. Xu, J.R. Shi, E.G. Wang, Y. Jiang, Real-space imaging of interfacial water with submolecular resolution. Nat. Mater. **13**, 184 (2014)
5. J. Carrasco, A. Hodgson, A. Michaelides, A molecular perspective of water at metal interfaces. Nat. Mater. **11**, 667 (2011)
6. A. Michaelides, K. Morgenstern, Ice nanoclusters at hydrophobic metal surfaces. Nat. Mater. **6**, 597 (2007)
7. M. Tatarkhanov, D.F. Ogletree, F. Rose, T. Mitsui, E. Fomin, S. Maier, M. Rose, J.I. Cerda, M. Salmeron, Metal- and hydrogen-bonding competition during water adsorption on Pd(111) and Ru(0001). J. Am. Chem. Soc. **131**, 18425 (2009)
8. J. Cerda, A. Michaelides, M.-L. Bocquet, P.J. Feibelman, T. Mitsui, M. Rose, E. Fomin, M. Salmeron, Novel water overlayer growth on Pd(111) characterized with scanning tunneling microscopy and density functional theory. Phys. Rev. Lett. **93**, 116101 (2004)
9. M. Morgenstern, T. Michely, G. Comsa, Anisotropy in the adsorption of H_2O at low coordination sites on Pt(111). Phys. Rev. Lett. **77**, 703 (1996)
10. S. Meng, E.G. Wang, S.W. Gao, Water adsorption on metal surfaces: a general picture from density functional theory calculations. Phys. Rev. B **69**, 195404 (2004)
11. G. Held, D. Menzel, Structural isotope effect in water bilayers adsorbed on Ru(001). Phys. Rev. Lett. **74**, 4221 (1995)
12. T. Yamada, S. Tamamori, H. Okuyama, T. Aruga, Anisotropic water chain growth on Cu(110) observed with scanning tunneling microscopy. Phys. Rev. Lett. **96**, 036105 (2006)
13. J. Carrasco, A. Michaelides, M. Forster, S. Haq, R. Raval, A. Hodgson, A one-dimensional ice structure built from pentagons. Nat. Mater. **8**, 427 (2009)
14. J. Lee, D.C. Sorescu, K.D. Jordan, J.T. Yates Jr., Hydroxyl chain formation on the Cu(110) surface: watching water dissociation. J. Phys. Chem. C **112**, 17672 (2008)
15. L.E. Firment, G.A. Somorjai, Low-energy electron diffraction studies of molecular crystals: the surface structures of vapor-grown ice and naphthalene. J. Chem. Phys. **63**, 1037 (1975)
16. A. Glebov, A.P. Graham, A. Menzel, J.P. Toennies, Orientational ordering of two-dimensional ice on Pt(111). J. Chem. Phys. **106**, 9382 (1997)
17. S. Haq, J. Harnett, A. Hodgson, Growth of thin crystalline ice films on Pt(111). Surf. Sci. **505**, 171 (2002)
18. D. Doering, T.E. Madey, The adsorption of water on clean and oxygen-dosed Ru(001). Surf. Sci. **123**, 305 (1982)
19. S. Nie, P.J. Feibelman, N.C. Bartelt, K. Thuermer, Pentagons and heptagons in the first water layer on Pt(111). Phys. Rev. Lett. **105**, 026102 (2010)
20. G.A. Kimme, N.G. Petrik, Z. Dohnalek, B.D. Kay, Crystalline ice growth on Pt(111): observation of a hydrophobic water monolayer. Phys. Rev. Lett. **95**, 166102 (2005)
21. G. Held, D. Menzel, The structure of the p($\sqrt{3}\times\sqrt{3}$)R30° bilayer of D_2O on Ru(001). Surf. Sci. **316**, 92 (1994)
22. P.J. Feibelman, Partial dissociation of water on Ru(0001). Science **295**, 99 (2002)
23. S. Meng, E.G. Wang, C. Frischkom, M. Wolf, S.W. Gao, Consistent picture for the wetting structure of water/Ru(0 0 0 1). Chem. Phys. Lett. **402**, 384 (2005)
24. S. Haq, C. Clay, G.R. Darling, G. Zimbitas, A. Hodgson, Growth of intact water ice on Ru(0001) between 140 and 160K: experiment and density-functional theory calculations. Phys. Rev. B **73**, 115414 (2006)

25. M. Gallagher, A. Omer, G.R. Darling, A. Hodgson, Order and disorder in the wetting layer on Ru(0001). Faraday Discuss. **141**, 231 (2009)
26. M. Forster, R. Raval, A. Hodgson, J. Carrasco, A. Michaelides, c(2x2) water-hydroxyl layer on Cu(110): a wetting layer stabilized by bjerrum defects. Phys. Rev. Lett. **106**, 046103 (2011)
27. G.A. Kimmel, J. Matthiesen, M. Baer, C.J. Mundy, N.G. Petrik, R.S. Smith, Z. Dohnalek, B.D. Kay, No confinement needed: observation of a metastable hydrophobic wetting two-layer ice on graphene. J. Am. Chem. Soc. **131**, 12838 (2009)
28. Q. Du, E. Freysz, Y.R. Shen, Surface vibrational spectroscopic studies of hydrogen bonding and hydrophobicity. Science **262**, 826 (1994)
29. J. Hu, X.D. Xiao, D.F. Ogletree, M. Salmeron, Imaging the condensation and evaporation of molecularly thin films of water with nanometer resolution. Science **268**, 267 (1995)

Chapter 7
Water Adsorption on Pt(111) Surfaces

The interaction between water and solid surfaces is widespread in nature and plays an important role in phenomena such as catalysis, electrochemistry, metal corrosion, rock weathering, permafrost formation, etc. Metal surfaces are often chosen as typical systems for water–solid surface interaction for two reasons: (1) they have the simplest atomic structure, and (2) metal-water contact has a wide range of technological applications, such as catalysis, fuel cells, automotive exhaust treatment, medical implants, etc. Over the last two decades, a large amount of experimental work has focused on the adsorption of water on metal surfaces, especially on the surfaces of several typical noble and transition metals, like Pt(111), Pd(111), Rh(111), Au(111), Ag(111), Cu(111), Ru(0001), etc. [1, 2]. Among them, Pt(111) is the most extensively studied surface experimentally. Since the Pt(111) surface is the most typical, and for the more primary purpose of being able to make a careful comparison with experimental data, the Pt(111) surface is chosen below as the starting point for a detailed microscopic image of the interaction of water with a solid surface.

Contemporary experiments of surface science are performed in ultra-high vacuum (UHV) chambers to obtain clean and contamination-free surfaces. With different intracavity air pressure, substrate temperature, and molecular beam injection, water can form a variety of different structures on the Pt(111) surface. Examples include adsorption structures present as individual molecules and small clusters [3–6], quasi-one-dimensional (1D) water chains along a step [7], regular ordered two-dimensional (2D) ice layers on a platform [8–12], several ice layers and even three-dimensional (3D) bulk ice [13–19].

The location, structure, and orientation of water molecules in these structures are very interesting and fundamental issues. For the adsorption of individual water molecules, early theoretical work [20, 21] debated whether to adsorb at the top site of the Pt(111) surface or at the vacancy site. Experiments by sum frequency generation spectroscopy (SFG) found that within one to thirty water layers at the Pt-H_2O interface, water molecules are uniformly oriented along the normal direction to form ferroelectrics [22]; and the orientation of water molecules can be controlled by adding a voltage to the surface [23]. Low energy electron diffraction (LEED)

S. Meng and E. Wang, *Water*,
https://doi.org/10.1007/978-981-99-1541-5_7

experiments found that water forms an ordered periodic structure $\sqrt{3} \times \sqrt{3}R30°$ (abbreviated as $\sqrt{3}$) at the surface [8, 9, 11, 13, 14]. It is usually assumed that water molecules form a so-called bilayer (BL) structure in this two-dimensional ordered system: it consists of an arrangement of folded hexagonal rings, very similar to the structure in bulk ice (Ice Ih) [24]. However, this $\sqrt{3} \times \sqrt{3}R30°$ pattern is only observed in a few small regions on the surface [9]. In addition, the helium atom scattering (HAS) experiments have found that two other complex periodic structures can be grown at higher temperatures (130–140 K): $\sqrt{39} \times \sqrt{39}R16.1°$ ($\sqrt{39}$) and $\sqrt{37} \times \sqrt{37}R25.3°$ ($\sqrt{37}$) [10]. LEED experiments have confirmed that under electron beam irradiation or more than five bilayers have been grown, the structure of $\sqrt{39}$ will reverse its orientation as a whole and becomes islands arranged in $\sqrt{3}$ pattern [11]. However, the microscopic mechanism of this transformation is not clear. Even for the simplest bilayer structures $\sqrt{3} \times \sqrt{3}R30°$, there is still a debate in the literature. Meng Sheng et al. found that two structures with different H pointing (H-up, H-down) are close in energy and have a small transition potential, both of which are possible in practice [25, 26]. Ogasawara et al. combined experiment and theory to conclude that half of the water molecules in this structure interact directly with the surface through the H atom, i.e., it is an H-down bilayer [27]. But Feibelman questioned this view by comparing the results of calculations about adsorption energies of both $\sqrt{3}$ and $\sqrt{39}$ structures, and he pointed out that the experimental water layer should be the structure of $\sqrt{39}$ with several H_3O^+, OH^- groups, and that the experiment should not be interpreted in terms of the proto-cell of $\sqrt{3}$ [28]. Recent experimental and theoretical work reveals the possible existence of a large number of water penta- and hepta-rings in the structure of the $\sqrt{39}$ aqueous layer on Pt(111). Further investigations are still needed for these issues.

Experimentally, LEED and HAS are commonly used to measure the structural period of the aqueous layer on metal surfaces. The kinetic properties of the water layer, i.e. the vibrational spectrum, are measured by high resolution electron energy loss spectroscopy (HREELS), infrared absorption spectroscopy (IRAS), SFG and HAS. Measurements of the electron binding states between water molecules and between water and surfaces are achieved by ultraviolet photoelectron spectroscopy (UPS), X-ray absorption spectroscopy (XAS), and X-ray emission spectroscopy (XES). Temperature-programmed desorption spectroscopy (TPD) is used to measure the thermodynamic stability of the water layer and the strength of the interaction with the surface. However, all of these are based on the macroscopic properties of a large number of water molecules and structural details specific to individual water molecules cannot be observed directly. Recently, methods have been developed to directly observe the morphology of water layers on metal surfaces, the dynamics of desorption, and even the diffusion of individual water molecules and small clusters on the surface [29] or the appearance of six-molecule clusters [30], etc., using scanning tunneling microscopy (STM) [7]. It greatly improves our ability of microscopic detection and to study the structure and dynamics of water on surfaces. Unfortunately, however, details about the structure and action of water, such as changes in bond lengths and bond angles, hydrogen bond pointing, water structure configuration, and even water molecule adsorption sites, are difficult to know due to limitations of resolution and the experimental technique itself. Moreover, the experimental results are difficult to interpret simply due to the influence of surface impurities and defects. First-principles calculations based on density generalized function theory can study the structure and dynamics of each water molecule, the composition and stability of the whole system, and even the general laws of water-surface interaction in detail from the perspective of electron bonding and energy, and are therefore very powerful and useful research tools. In particular, comparing theoretically calculated vibrational spectra with experimentally measured vibrational spectra can help us to identify the details of the composition and action of the experimental structure at the molecular level, in a way known as "vibrational identification" [25]. Theoretical tools and experimental data complement each other and together take us on a journey to a deeper understanding of water at the molecular level.

Via total energy calculations and molecular dynamics simulations based on first principles, we first investigated the adsorption of various water molecule structures on the Pt(111) surface. This includes the adsorption of single water molecules; dimers, trimers and hexamers of water molecules; one-dimensional water chains at the step.

The water bilayers of $\sqrt{3}$, $\sqrt{39}$, $\sqrt{37}$; the adsorption of multiple water bilayers such as 2–6. Their structures and energies were obtained and water was found to be essentially molecularly adsorbed on Pt(111), in agreement with experiments. The vibrational spectra of these adsorbed structures were also calculated and obtained in very good agreement with the experiments. This is elaborated in the sections of this chapter, respectively. Moreover, through vibrational identification, we found for the first time the existence of two hydrogen bonds of different strengths in the surface water bilayer structure, both of which can be identified by OH stretching vibrations.

7.1 Adsorption of Individual Water Molecules and Small Clusters of Water

7.1.1 Adsorption of Individual Water Molecules

The adsorption of a single water molecule on a surface represents the simplest water-surface interaction scenario, and it contains the most basic information about surface-water interaction. Early theoretical studies focused on this, but were mostly based on cluster model calculations [20, 21, 31] or model potential calculations [32], and thus were very incomplete. Besides, early work is still debating whether individual water molecules adsorb on the top site of Pt(111) surface or on the vacant site [21].

We used the Vienna ab initio computational simulation program VASP to perform first-principles calculations [33–35]. As shown in Fig. 7.1, a crystal plate consisting of four layers of atoms was used as a model for the Pt(111) surface, with each plate separated by a 13 Å thick vacuum (the lattice constant used in the calculations was

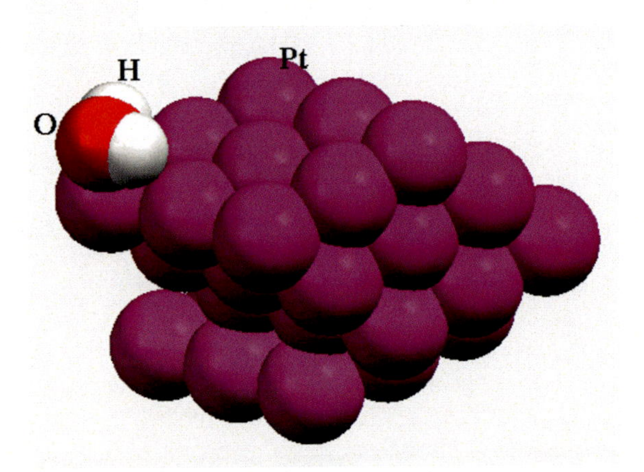

Fig. 7.1 Supercells (crystal slab model) used in the calculations

3.99 Å). Water molecules were placed on the surfaces of the crystal plates. A supercell of p(3 × 3) was chosen. For the surface Brillouin zone, the integration was carried out using the Monkhorst–Pack method [36] taking 3 × 3 × 1 k points. The energy cut-off of the plane wave is 300 eV. The electron energy level was broadened by a Gaussian method with a width of 0.2 eV [37]. This set of parameters ensures that the total energy converges to the level of 0.01 eV per atom. During the optimization of the structure, the water molecules and the uppermost Pt atoms are relaxed simultaneously. The calculation stops when the force acting on all relaxed atoms is less than 0.03 eV/Å. Vanderbilt's ultra-soft pseudopotential (USPP) [38] and Perdew and Wang's (PW91) universal gradient approximation to the exchange–correlation energy [39] were used in the calculations.

Structural optimization revealed that water molecules are most stably adsorbed at the top position (Table 7.1). The adsorption energy at the top site is 291 meV, about three times higher than that at the bridge and vacant sites; the O–Pt bond length is 2.43 Å, which is 0.7 Å shorter than at the bridge and vacant sites, also indicating that the top site has the strongest interaction with H_2O where the adsorption energy is defined as the energy of the substrate with a single gaseous water molecule and minus the total energy of the system.

$$E_{\text{ads}} = \left(E[\text{Metal}] + n \times E[(\text{H}_2\text{O})_{\text{gas}}] - E[(\text{H}_2\text{O})_n/\text{Metal}]\right)/n \tag{7.1}$$

n is the number of molecules in the supercell, here $n = 1$. The water molecules essentially lie flat on the top site, and the angle θ between the molecular plane and the surface is only 13° (Fig. 7.2). Experimentally determined that for water on Pd(100) and Cu(100) $\theta = 32, 33°$, the conformation is consistent with this [40]. Compared to the free water molecule ($d_{\text{OH}} = 0.972$ Å, $\angle\text{HOH} = 104.53°$, see Table

Table 7.1 Adsorption configurations and energies of single water molecules on Pt(111). Unless otherwise stated, the units of distance, energy, and angle are meV, Å, and degree, respectively

Number of layers	E_{cut} (eV)	Top site		Bridge site		Vacant site		d_{OH}	HOH	θ
		d_{Opt}	E_{ads}	d_{OPt}	E_{ads}	d_{OPt}	E_{ads}			
4	300	2.43	291	3.11	123	3.12	121	0.978	105.36	13
6	400	2.40	304	2.89	117	3.02	102	0.980	105.62	14

5.1), the adsorbed H_2O bond length increases ($d_{OH} = 0.978$ Å) and the bond angle widens ($\angle HOH = 105.36°$). In the bridge position, it has been proposed that the water molecules are standing vertically on the surface with H pointing upwards [41–43], and we found that the adsorption energy for this configuration is 40 meV lower than that lying flat in the bridge position, which is important when considering the diffusion of water molecules. All of these results can be explained from the point of view of electron transfer. H_2O couples to the 5d band on the Pt surface via a lone pair of electrons [44], leading to electron transfer from O to Pt. While on the Pt(111) surface, the top-site electron density is relatively small and the lone pair electron interaction with water is stronger due to charge smoothing effect, leading to that top-site adsorption is the most stable. Since the tetrahedral distribution of electrons, the water molecules lie flat also for optimal coupling of lone pair electrons. And the reduction of electrons in H_2O makes the sp^3 hybridization weaker and the bond length as well as bond angle increases. The results obtained with a crystal plate of six layers of Pt atoms and a higher plane wave energy cut-off ($E_{cut} = 400$ eV) are similar to the above, with little structural and energy variation (Table 7.1). So both sets of parameters are likely to be used later.

To gain further insight into the dynamic properties of adsorbed water molecules at the surface, Fig. 7.2 plots the total energy as a function of the O–Pt bond length, d_{OPt}, and the molecular angle to the plane, θ. The total energy follows d_{OPt} in the form of a Morse potential, which takes a minimum at 2.43 Å. $\theta = 90°$ corresponds to the water molecule configuration with H standing vertically and pointing upwards, while $\theta = -90°$ corresponds to that with H standing vertically and pointing downwards. The calculations have kept the O atom fixed in the equilibrium position while the polar angle direction rotates the entire H_2O. The molecular flip potentials corresponding to these two sites are 140 meV and 820 meV, respectively. The former is lower than the calculation by Michaelides et al. of 190 meV [45]. This is due to the smaller proto-cell 2×2 they used. If we consider that the O atoms can be made to move freely, the potential barrier for the upward flip of the H_2O molecule is still 140 meV and the potential barrier for the downward flip can be 194 meV, indicating that the water molecule is harder to flip downward. Furthermore, H_2O can be turned almost unimpeded along the azimuthal angle φ with a potential barrier of less than 2 meV.

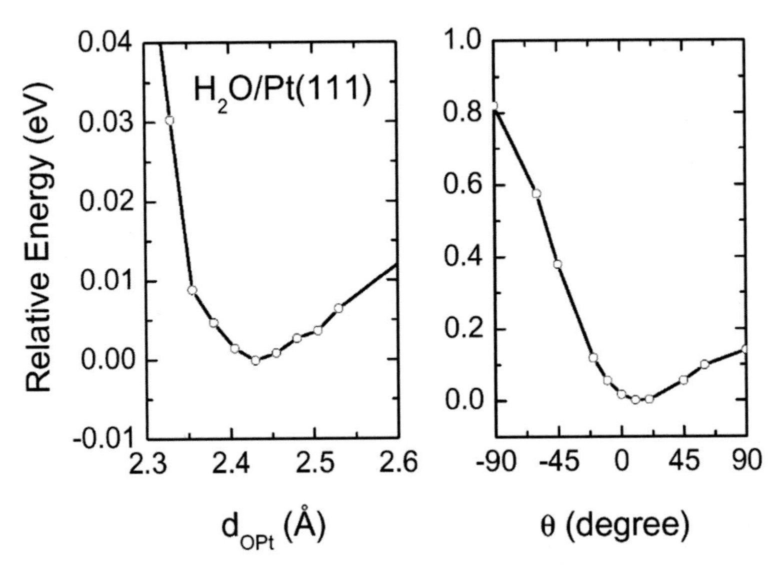

Fig. 7.2 Variation of the energy of single H_2O adsorption on the Pt(111) surface with the H_2O–Pt distance d_{OP} and the tilt angle θ

7.1.2 Adsorption of Small Clusters of Water Molecules

As in free space, water molecules form small clusters on the surface. The adsorption of small clusters is very interesting because it contains both water-surface interactions and hydrogen bonding between water molecules. Some interesting structures of small clusters of water on metal surfaces have also been observed experimentally [4, 6, 9, 29, 30].

Calculations are still performed using VASP. For dimeric (or bimolecular, $(H_2O)_2$) and trimeric molecules (trimolecular, $(H_2O)_3$) the $p(3 \times 3)$ supercell is still chosen; for hexameric molecules (hexamolecular, $(H_2O)_6$) a somewhat larger supercell, $2\sqrt{3} \times 2\sqrt{3}$, is chosen to avoid the influence of periodic boundaries. The obtained optimal structural configuration is shown in Fig. 7.3, and the specific structural parameters are listed in Table 7.2 (specific to each H_2O molecule).

In general, the structure of the small clusters of water on the Pt(111) surface is similar to the free case [46, 47]. Each water molecule adsorbs on the top site and lies as flat as possible to allow the strongest interaction between the metal and the water [48, 49]. For the case of $(H_2O)_2$ adsorption, both H_2O molecules adsorb at the top site and form two O–Pt bonds with the surface, with another hydrogen bond between them (Fig. 7.3b). The adsorption configuration of the proton donor is very similar to the single-molecule adsorption configuration, with an O-Pt bond length of 2.26 Å and an angle $\theta = 25.1°$ with the surface. The acceptor molecule is farther from the surface ($d_{OPt} = 3.05$ Å) and the H is directed downward toward the surface at an angle of $\theta = -41.8°$ with the surface. Taking this configuration, the angle made by the hydrogen bond to the plane of the acceptor molecule is about 123°, similar to that

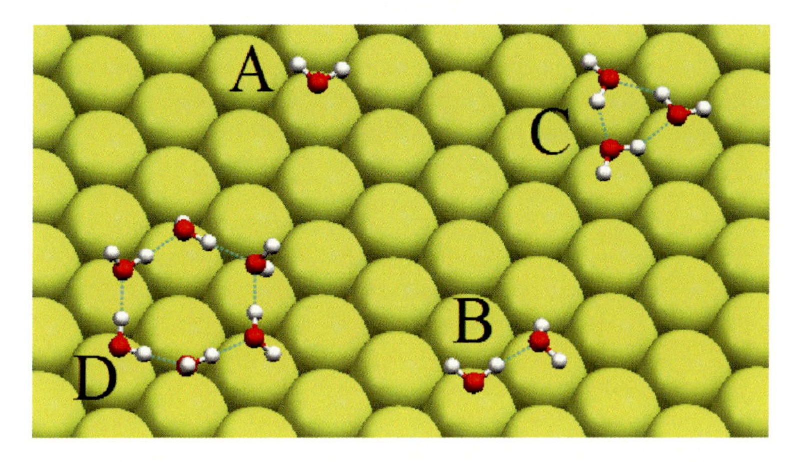

Fig. 7.3 Adsorption configurations of single H_2O molecules and small clusters on metal surfaces

Table 7.2 Adsorption configurations and energies of single water molecules and small clusters on Pt(111). Units are the same as in Table 7.1

	E_{ads}	d_{Opt}	θ	d_{OH1}	d_{OH2}	HOH	d_{OO}
H_2O	304	2.40	13.8	0.980	0.980	105.62	–
$(H_2O)_2$	433	2.26	25.1	0.978	1.012	106.72	2.70
		3.05	−41.8	0.981	0.982	103.52	–
$(H_2O)_3$	359	2.76	3.5	0.975	0.985	107.75	2.78
		2.76	3.5	0.975	0.985	107.86	2.80
		2.76	3.1	0.974	0.985	107.71	2.79
$(H_2O)_6$	520	2.32	31.1	0.997	1.001	106.22	2.99
		3.38	32.9	0.974	0.991	104.49	2.80
		2.77	−1.8	0.978	0.990	107.25	2.89
		3.35	0.3	0.975	0.988	106.88	3.01
		2.77	3.7	0.979	0.987	107.14	2.80
		3.39	32.3	0.974	0.991	104.83	2.88

of a free dimer molecule ($\beta = 127°$, Table 4.2 USPP). Except that the OH bond that does not form a hydrogen bond in the donor water molecule is lying face down on the surface. The distance between two oxygen atoms is 2.70 Å, much smaller than the O–O spacing of the free $(H_2O)_2$, 2.86 Å; the OH (1.012 Å) to which the hydrogen bond is attached also grows (0.985 Å in the free dimer molecule). This suggests that the hydrogen bonds are strengthened due to adsorption. Based on Table 7.2, one would expect the hydrogen bond energy to be between (E_{ads} [bimolecule]—E_{ads} [monomolecule]) $\times$ 2 = (433 − 304) $\times$ 2 = 258 meV and (E_{ads}[bimolecule] $\times$ 2 − E_{ads} [monomolecule]) = (433 $\times$ 2 − 304) = 562 meV in the adsorbed bimolecule. greater than the calculated hydrogen bond energy of the free bimolecule, 250 meV (experimental value of 236 ± 30 meV [50]). The hybridization of the lone pair of electrons of the underlying oxygen atom with the 5d band electrons of Pt leads to the transfer of electrons and thus lead to an increase in the instantaneous electric dipole moment, which may account for the enhanced hydrogen bonding.

The adsorbed trimolecules and hexamolecules remain in a ring-like structure, similar to the free case. The $(H_2O)_3$ water molecules in the structure essentially lie flat on Pt(111) ($\theta \sim 3.5°$). The $(H_2O)_6$ structure forms a folded hexagonal ring arrangement with three H_2O molecules interacting directly with the surface (d_{Opt} = 2.32, 2.77, 2.77 Å) and another three further away from the surface (d_{Opt} = 3.4 Å), connected to the former by hydrogen bond. This is the basic unit that forms the bulk ice (Ice Ih, Ic). One water molecule is as close to the surface as a single water molecule adsorbs and provides two protons to form hydrogen bonds; it is critical to stabilize the adsorption of the entire six-molecule cluster. The average distance of OO in $(H_2O)_6$ is somewhat larger than in $(H_2O)_2$ and $(H_2O)_3$ due to the increase of numbers of H_2O. In addition to the hexagonal ring structure, the free $(H_2O)_6$ also has triangular columnar and cage structures [46, 47]. Calculations show that the adsorption energy of columnar $(H_2O)_6$ is only 321 meV, which is 200 meV lower than that of the ring, and thus difficult to exist on the surface.

Among these small clusters, $(H_2O)_6$ has the largest adsorption energy (520 meV), and is the most stable; $(H_2O)_3$ has the smallest adsorption energy (359 meV); and $(H_2O)_2$ is in the middle (433 meV). This mainly reflects the difference in the number of Pt-O bonds and hydrogen bonds they form. The small adsorption energy of $(H_2O)_3$ is due to the fact that the triangular arrangement of hydrogen bonds does not allow optimal coupling of the water molecule lone pair of electrons to H. Consistent with the theory, the most observed in STM experiments are six-molecule $(H_2O)_6$ [29, 30]. Mitsui et al. also observed the formation of dimer and trimer molecules by molecular diffusion [29]. Although Morgenstern and Nieminen pointed out that the STM image of $(H_2O)_6$ on Ag(111) is an equilateral triangle formed by the three H_2O molecules on the upper side [30]. However, our calculations suggest that a better match to the figure should be the shape of an equilateral triangle, with the two upper H_2O molecules adjacent to that H_2O similar to the single molecule are connected into a shorter side length. For example, on the Pt(111) surface this isosceles triangle should be (4.7, 4.9, 4.9 Å).

7.2 One-Dimensional Water Chains at Surface Steps

The 1D chain-like structure of water is of great importance because water molecules in living organisms are linked in chains across cell membranes [51], which is one of the most fundamental and important processes for the movement of life. Current simulations of 1D water chains are usually studied in tubular radial confinement, such as in carbon nanotubes [52–54]. However, such structures do not exist in practice and have not been implemented in experiments. A more realistic 1D structure, a 1D water chain on a Pt(111) surface step, has been observed by STM experiments [7].

The step that supports the quasi-1D water chain in the experiment is the <110>/{100} step. We also take the same type of steps for our calculations. A 15-layer Pt(322) surface is used to represent this type of step. The proto-cell size is shown in Fig. 7.4. To form a hydrogen-bonded chain along the step, one OH of the water molecule points along the step and the other OH points either vertically inward or vertically outward to the step, creating the two simplest adsorption configurations: H-in (Fig. 7.4, 1) and H-out (Fig. 7.4, 2). If the H_2O of these two conformations are interleaved, a third 1D water chain structure is formed: the sawtooth structure H-mix (Fig. 7.4). The results of calculations for these three chain structures are listed in Table 7.3. For comparison, the adsorption of individual water molecules on the steps (H-in and H-out) has also been calculated. The results for individual molecules and 1D chains on the platform are shown in the last row.

The H_2O single molecule adsorbs strongly on the steps with adsorption energies of 449 meV (H-in) and 426 meV (H-out). Among the three aqueous chains, only the serrated H_2O chain is stable and it has an adsorption energy of 480 meV. The 1D chains of H-in and H-out have adsorption energies of 431 meV and 385 meV,

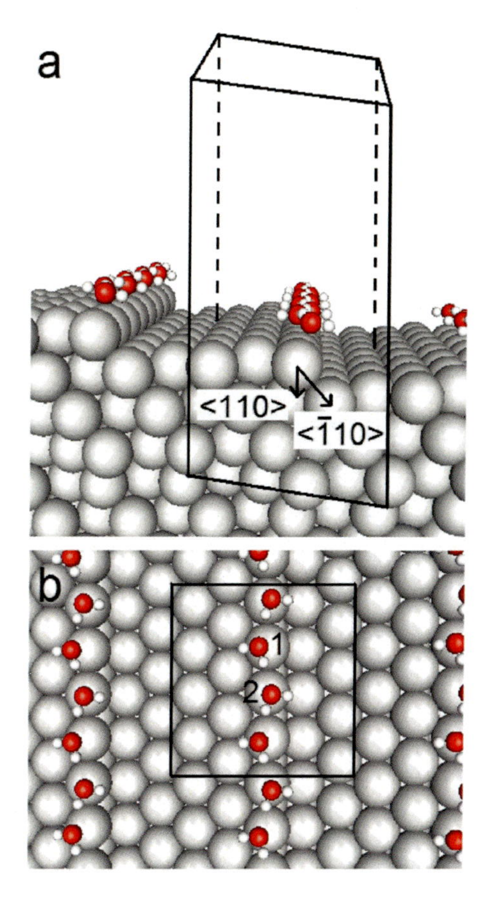

Fig. 7.4 One-dimensional H_2O chain on a Pt(111) surface step. **a** side view, **b** top view. The black line shows the proto-cell used for the calculation

Table 7.3 Adsorption of single water molecules and one-dimensional chains on the Pt(111) step. Units are the same as in Table 7.1

	Single H_2O		1D chain	
	d_{Opt}	E_{ads}	d_{OPt}	E_{ads}
H-in	2.22	449	2.42	431
H-out	2.25	426	2.48	385
H-mix	–	–	2.45	480
On terrace	2.43	291	2.62,272	246

respectively, which are unstable compared to the single molecule adsorption. The hydrogen bonding energy can be deduced to be about 85 meV in the serrated chain of H-mix. Such chain is the most stable because it makes the strongest intermolecular hydrogen bonding and dipole moment interactions. Compared to the platform, the single molecule adsorption on the step is stronger by ~ 150 meV, while the one-dimensional chain is more stable by ~ 230 meV which suggests a stronger interaction of water with the substrate at the step. It also explains why the 1D water chain was not observed on the platform in the experiment. This all stems from the more extended electron distribution and lower density on the steps. These are also consistent with the experimental observations.

7.3 Double and Multilayer Adsorption of Water

At higher coverage, the water forms bilayer (BL) and multilayer structures. Bilayer structures and the first few bilayers are the most interesting, as they mark the initial stages of bulk ice formation. Several types of layer structures have been observed experimentally. $\sqrt{3} \times \sqrt{3}R30°$ ($\sqrt{3}$) biayer ice is the simplest and the best known one [8, 9, 11]. Water molecules are arranged in folded hexagonal rings in this two-dimensional structure, forming a structure very similar to that of bulk ice (Ice Ih) [24, 25]. In addition, two other complex periodic structures were found that can be grown experimentally at higher temperatures (130–140 K): $\sqrt{39} \times \sqrt{39}R16.1°$ ($\sqrt{39}$) and $\sqrt{37} \times \sqrt{37}R25.3°$ ($\sqrt{37}$) [10]. LEED experiments confirmed that at higher energy electron irradiation or when growing more than five bilayers, the structure of $\sqrt{39}$ reverses its orientation of the whole and becomes an island of $\sqrt{3}$ arrangement (see Fig. 7.5) [11]. The details of these structures are still debated in the literature.

Figure 7.6 shows the bilayer structure of $\sqrt{3}$. Each supercell the of $\sqrt{3} \times \sqrt{3}R30°$ supercell has two water molecules. The lower H_2O molecule is directly bonded to the surface, while the higher H_2O is attached to the adjacent H_2O molecule below by hydrogen bonding, forming a bilayer junction structure. Three of the four OHs in the supercell form hydrogen bonds, leaving one either pointing along the surface normal or perpendicular towards the surface, called the H-up (Fig. 7.6a) and H-down bilayers (Fig. 7.6b), respectively. The perpendicular distances Z_{OO} of the OO in the

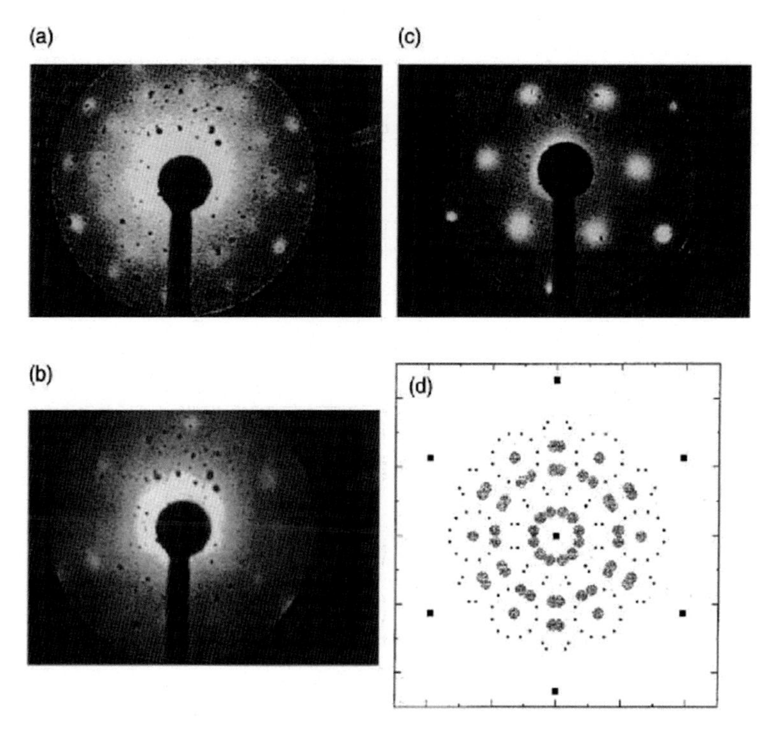

Fig. 7.5 LEED patterns (**a**, **b**, **d**) for the ($\sqrt{39}$) phase on the Pt(111) surface $\sqrt{39} \times \sqrt{39}R16.1°$. Turn into $\sqrt{3} \times \sqrt{3}R30°$ ($\sqrt{3}$) islands (**c**) after electron irradiation [11]

normal direction in the H-up and H-down bilayers are 0.63 Å and 0.35 Å, respectively (Table 7.4). Compared to the distance in the bulk ice Ih ($Z_{OO} = 0.97$ Å), both structures are compressed due to adsorption. They are almost parsimonious in energy, 522(H-up) and 534 meV (H-down), respectively (see Table 7.4). Both can be used as possible configurations for the bilayer structure at the Pt(111) surface. As a control, the bilayer structure found by Feibelman to be partially decomposed on the Ru(0001) surface [55] was also calculated (last column of Table 7.4). Its adsorption energy is 291 meV, which is much lower than the adsorption energy of the molecular state bilayer structure. Thus, the decomposed structure is not dominant in the structure of Pt(111) surface $\sqrt{3}$. The bilayer structure in the molecular state was originally proposed on the basis of high-resolution electron energy loss spectroscopy (HREELS) [15] and ultraviolet photoemission spectroscopy (UPS) [16] measurements and was recently confirmed by X-ray absorption spectroscopy (XAS) [27]. The results obtained support these experiments.

Since the two configurations are similar in structure and comparable in energy, it is interesting to examine their transformation relations. Using the nudged elastic band method (NEB) [56] it was calculated that there is a potential barrier from the H-up configuration to the H-down configuration with a height of 76 meV. The minimum energy path (MEP) between them is shown in Fig. 7.7. The configuration of the

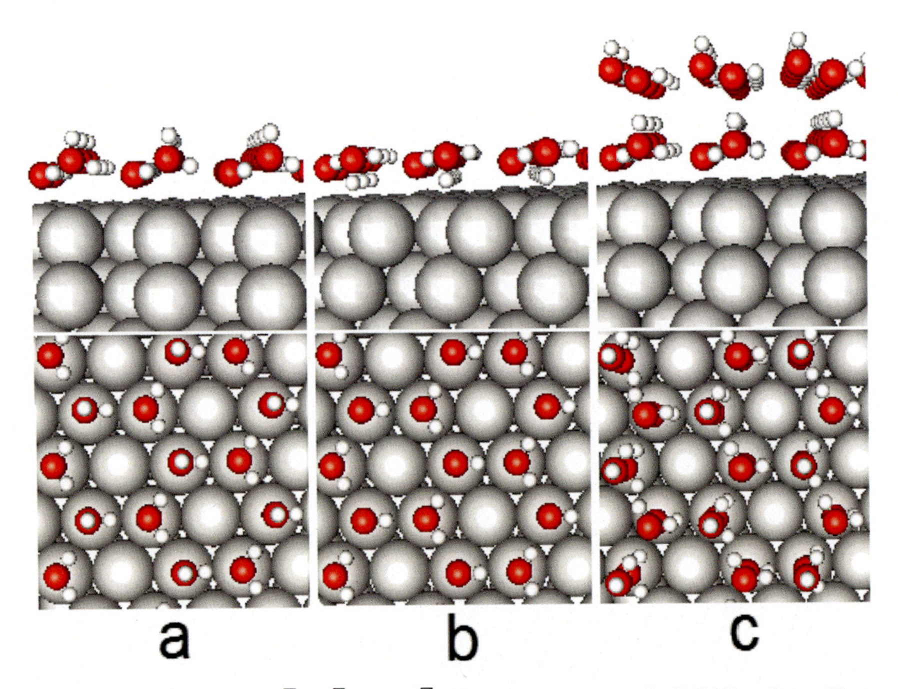

Fig. 7.6 Two-dimensional $\sqrt{3} \times \sqrt{3}R30°$ ($\sqrt{3}$) bilayer ice structure on the Pt(111) surface. **a** H-up, **b** H-down, **c** 2 bilayers

Table 7.4 Structure and energy of the water bilayer adsorbed on the surface of Pt(111). Calculations for high energy cutoff are in parentheses. Units are the same as in Table 7.1

	Z_{OO}	Z_{OPt1}	Z_{OPt2}	E_{ads}
H-up bilayer	0.63 (0.64)	2.70 (2.69)	3.37 (3.38)	522 (505)
H-down bilayer	0.35 (0.33)	2.68 (2.71)	3.14 (3.15)	534 (527)
Partially decomposed structure	0.06	2.12	2.23	291

transition state (saddle point) is also drawn in this figure. The whole path is mainly determined by the rotation of the above water molecules in the molecular plane. The saddle point (saddle point) is reached when the free OH turns from an angle of 0° to an angle of 33° normal to the surface. The potential of 76 meV is much smaller than the corresponding calculation by Michealides et al. on Ru(0001) [57], 300 meV; however, it is comparable to the corresponding structural transition potential in the free dimer molecule, 78 meV (the energy difference between structures 1 and 9 in Fig. 2 of literature [58]). This suggests that Pt(111) has little effect on this conversion. That is because the upper H_2O molecule in the bilayer ice is largely unaffected by the surface.

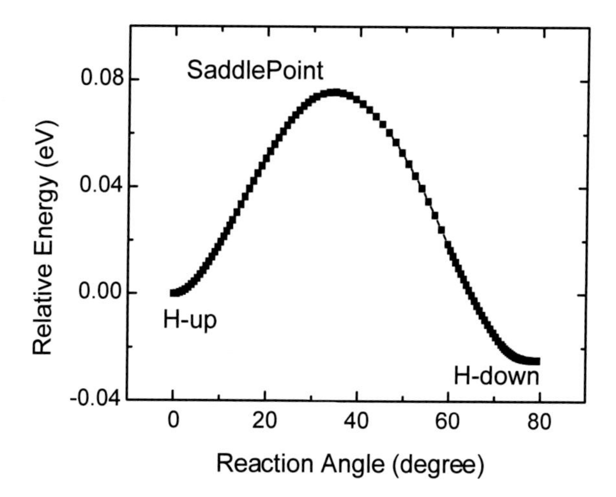

Fig. 7.7 Double-layer structural transition potentials for H-up and H-down

We also studied two periodic structures of $\sqrt{39}$ and $\sqrt{37}$ [26]. Three layers of Pt(111) facets were taken as substrates, with 32 and 26 H_2O molecules in the supercell, respectively. They do not differ much from $\sqrt{3}$ overall and are also folded hexagonal ring structures (Fig. 7.8). Compared to bulk ice, the lattices of the three 2D structures are expanded by 7.2% ($\sqrt{3}$), 4.3% ($\sqrt{37}$), or compressed by 3.3% ($\sqrt{39}$), respectively. The atoms in the $\sqrt{39}$ structure are more disordered due to the compression and orientation of ice supercell. The density distribution of O and H atoms along the surface normal (z) is also plotted on the right side of Fig. 7.8, where the gray line corresponds to the $\sqrt{3}$ structure and is used for comparison. As seen in the figure, the peaks of the atomic density distribution in the $\sqrt{39}$ structure are very broad, with some atoms far from the surface. The nearest O atom is only 2.10 Å from the surface, while the farthest reaches 4.4 Å. This ice layer is therefore 2.3 Å thick. This structure is far from the flat plate structure proposed recently by Ogasawara et al. where the vertical distance between the upper water molecules and the lower water molecules is only 0.25 Å [27]. The atomic distribution in the $\sqrt{37}$ structure is also more disordered than $\sqrt{3}$, but more regular than $\sqrt{39}$. The largest O-Pt distance is 3.58 Å, which is similar to that in $\sqrt{3}$: 3.37 Å (H-up) or 3.14 Å (H-down). This is because both $\sqrt{3}$ and $\sqrt{37}$ are stretched structures. A feature that also differs from them is that a small amount of H_2O molecules are decomposed in the $\sqrt{39}$ ice phase [28].

To investigate the dependence of ice structure and energy on the degree of coverage, we calculated the adsorption of two bilayers (Fig. 7.6c) to six bilayers of ice on the Pt(111) surface in the $\sqrt{3}$ period (Table 7.5). The OO distance between two neighboring bilayers ranges from 2.75 to 2.83 Å. The distance between the bottom H_2O molecule and the surface of the bottommost bilayer decreases as the coverage increases. In the structures with one to six bilayers, the distances are 2.69, 2.63, 2.56, 2.49, 2.52, and 2.47 Å. In contrast, the distance of the upper H_2O molecule from the surface is always maintained at 3.25 ± 0.02 Å. This indicates that in the bilayer

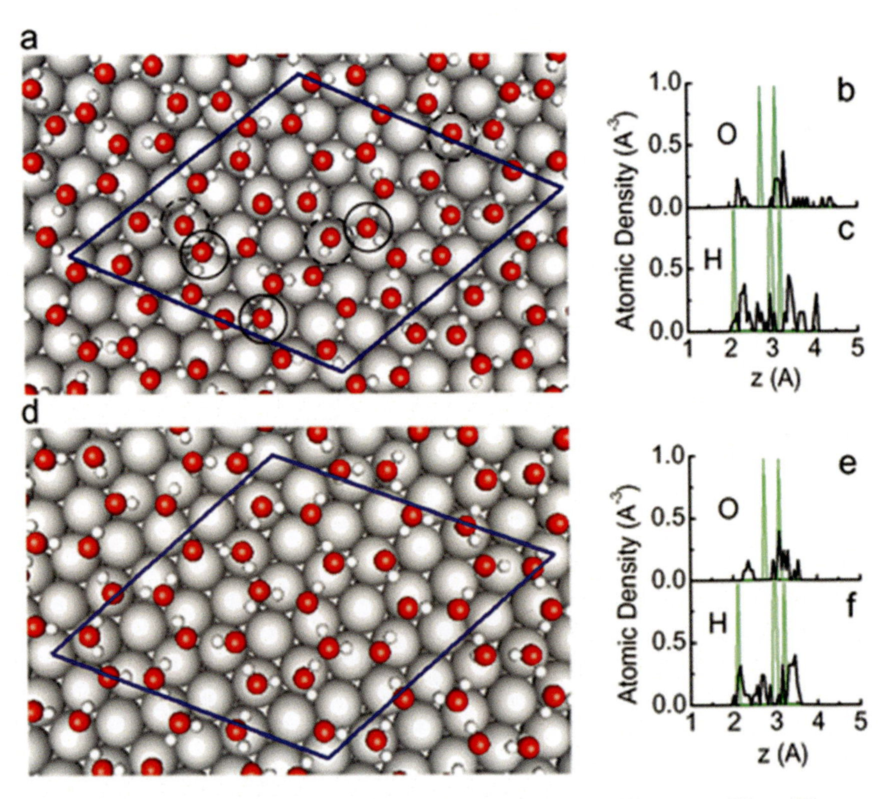

Fig. 7.8 Ice structures on the Pt(111) surface $\sqrt{39} \times \sqrt{39}R16.1°$ ($\sqrt{39}$, **a**) and $\sqrt{37} \times \sqrt{37}R25.3°$ ($\sqrt{37}$, **d**). The density distribution of atoms along the surface normal direction is shown on the right. The solid and dashed circles indicate the positions of H_2O and OH, respectively

and multilayer structures, only the bottommost water molecule is directly bonded to the surface, while the upper H_2O molecule is almost unaffected. It indicates that water-metal interactions are rather localized.

The energies of the $\sqrt{3}$, $\sqrt{37}$ and $\sqrt{39}$ ice layers are compared in Fig. 7.9. The figure shows that at one bilayer $\sqrt{39}$ has a somewhat larger adsorption energy (80 meV) than $\sqrt{3}$, at 615 and 534 meV, respectively, while the energy of $\sqrt{37}$ is in the middle, at 597 meV. With increasing coverage, at 3BL, the $\sqrt{3}$ phase becomes more stable than the other two phases. This is consistent with the experimental findings. The ice layer of $\sqrt{39}$ in the experiment transforms into the structure of $\sqrt{3}$ at about 5BL [11]. This is due to the fact that the structures of $\sqrt{39}$ and $\sqrt{37}$ are more irregular in the z-direction, which is not conducive to the formation of multilayer structures.

Table 7.5 Adsorption energies (E_{ads}) and hydrogen bonding energies (E_{HB}) of various adsorption structures on Pt(111). The number of molecules (n) and the number of bonding (N) in the supercell are also listed. The units are the same as in Table 7.1

Type	Supercell	n	E_{ads}	N_{H_2O-Pt}	N_{HB}	E_{HB}
H_2O	3×3	1	304	1	0	–
$(H_2O)_2$	3×3	2	433	2	1	258
$(H_2O)_3$	3×3	3	359	3	3	55
$(H_2O)_6$	$2\sqrt{3} \times 2\sqrt{3}$	6	520	3	6	368
1 BL	$\sqrt{3} \times \sqrt{3}$	2	505/527	1	3	235
2 BL	$\sqrt{3} \times \sqrt{3}$	4	564	1	7	312
3 BL	$\sqrt{3} \times \sqrt{3}$	6	579	1	11	303
4 BL	$\sqrt{3} \times \sqrt{3}$	8	588	1	15	307
5 BL	$\sqrt{3} \times \sqrt{3}$	10	593	1	19	307
6 BL	$\sqrt{3} \times \sqrt{3}$	12	601	1	23	320
1 BL	$\sqrt{37} \times \sqrt{37}$	26	597	13	39	297
1 BL	$\sqrt{39} \times \sqrt{39}$	32	615	16	48	309
2 BL	$\sqrt{39} \times \sqrt{39}$	64	582	16	112	275
3 BL	$\sqrt{39} \times \sqrt{39}$	96	572	16	176	276

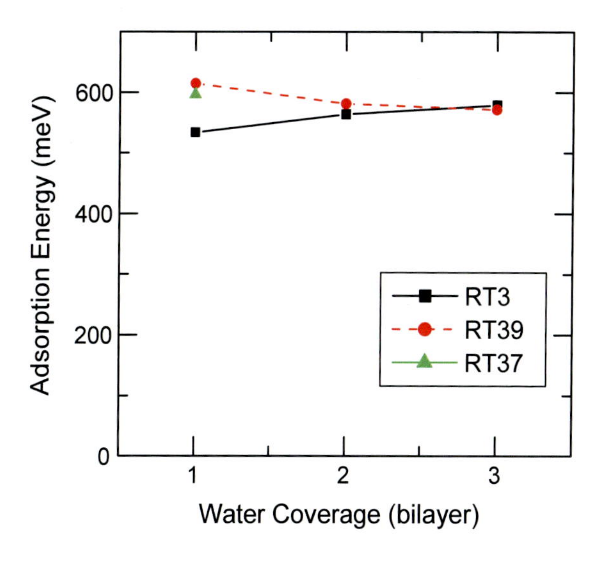

Fig. 7.9 Energy comparison of $\sqrt{3}$, $\sqrt{37}$, $\sqrt{39}$ structures

7.4 Vibrational Identification of Water Structures on Surfaces

To further confirm these structures, vibrational spectra were calculated for single molecules, bimolecular and bilayer structures (Table 7.6). These vibrational spectra were derived by Fourier transforming of the velocity autocorrelation functions obtained from ab initio molecular dynamics simulations. A microcanonical system synthesis was adopted with a total simulation time of 2 ps, a time step of 0.5 fs, and a temperature of about 90 K [25, 49]. The shear motion frequency of HOH during the adsorption of single H_2O molecules is low, 190 meV. It is due to the transfer of electrons from O to the Pt surface. In agreement with previous analysis [59]. The reduced OH vibrational frequency for bimolecular adsorption is due to the effect of hydrogen bonding.

Both H-up and H-down bilayer structures were simulated. The vibration spectrum (see Fig. 7.10) has a sharp peak at 4(6) meV on the left; there are five other vibration modes at 18(16), 32(34), 53(57), 69(69), and 87(91) meV. (The numbers in parentheses represent the H-down case). In the HREEL spectra, similar vibrational peaks have been observed at 16.5, 33, 54, 65 and 84 meV (Fig. 7.11). They are considered to be the Pt–OH_2 vertical vibrations of the top H_2O molecule (32 meV) and the bottom water molecule (16 meV) (Fig. 7.12), as well as the suppressed rotational and oscillatory modes (54, 69, 87 meV). Energy of the Pt-OH_2 vibrational mode is usually considered to be around 68 meV [15, 17]. The recent experiments of Jacobi et al. [9] indicate that these modes should correspond to strong energy loss peaks at 16.5 and 33 meV. Molecular dynamics simulations support this new view. The suppressed advective mode of the bilayer structure was observed to be 5.85 meV in the helium atom scattering experiment (HAS) [6], while the sharp peak at 4(6) meV is close to this measurement. The higher energy vibrational modes are associated with motion of internal molecular, i.e., the HOH bending mode at 198 meV and the OH stretching vibrational mode in the range of 380 to 470 meV. Based on Fig. 7.10 and Table 7.6, it is clear that for the bilayer structure, the vibration frequencies derived from the calculations agree well with the experimental data for both the H-up and H-down configurations.

The OH relative vibrational modes on the right side of the vibrational spectrum is more interesting. These modes are sensitive to intermolecular interactions, especially

Table 7.6 Vibrational energies (in meV) for the H_2O/Pt(111) structure

	$T_\parallel$	T_2	T_3	L_3	L_4	L_5	δ_a	δ_b	δ_{HOH}	$\nu_{O\text{-}Hb}$	$\nu_{O\text{-}H}$
H_2O	4	16		40	61	89	113	121	190		440
$(H_2O)_2$	8	20	32	44	65	85	105	133	198	347	432,452
H-up	4	18	32	53	69	87	107	119	198	388,432	467
H-down	6	16	34	57	69	91	111	119	196,202	384,424	438
Experiment	5.85	16.5	33	54	65	84	115	129	201	424	455

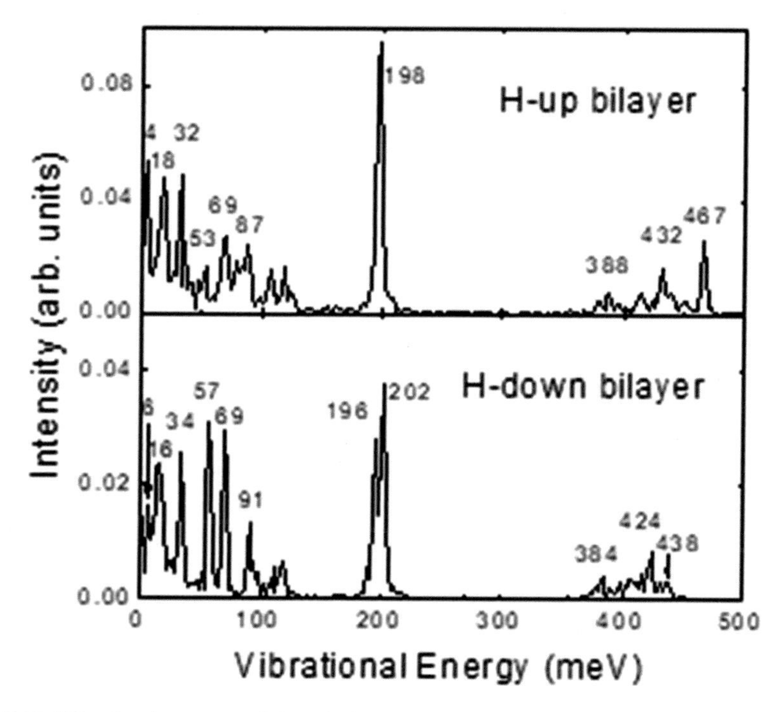

Fig. 7.10 Vibrational spectrum of the calculated bilayer structure

to the formation of hydrogen bonds. In Fig. 7.10, the three modes located at 388 (384), 432 (424) and 467 (438) meV are distinguishable. Compared to the OH relative vibrational frequencies of 454 meV (symmetric) and 466 meV (asymmetric) [60] in the gas phase, it can be argued that the high frequency mode in Fig. 7.10 is the OH relative vibrational mode without the formation of hydrogen bonds, while the two lower frequency modes located at 388(384) and 432(424) meV correspond to the OH stretching vibrations within the bilayer structure. The latter clearly appears redshifted due to the formation of hydrogen bonds. These modes directly reflect the conformation of the bilayer structure. As shown in Fig. 7.13, the water molecule in the top layer contributes one hydrogen atom and forms a hydrogen bond with a neighboring water molecule (black line in Fig. 7.13), while the water molecule in the bottom layer contributes two hydrogen atoms and forms two hydrogen bonds with neighboring water molecules (gray line). Thus, the one hydrogen bond in the top layer is much stronger, resulting in a larger redshift, while the two hydrogen bonds in the bottom layer are relatively weak and thus have a smaller redshift.

The above explanation can be confirmed by the trajectories of the OH stretching vibrations. Figure 7.14 gives the amplitudes of all four OH bonds as a function of time in the bilayer structure of H-up. The above graph gives the case of free OH bonds and OH bonds that form strong hydrogen bonds. The free OH bond has the shortest bond length and the highest vibrational frequency, while the OH bond forming a strong

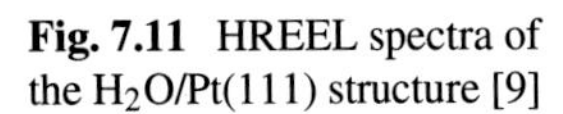

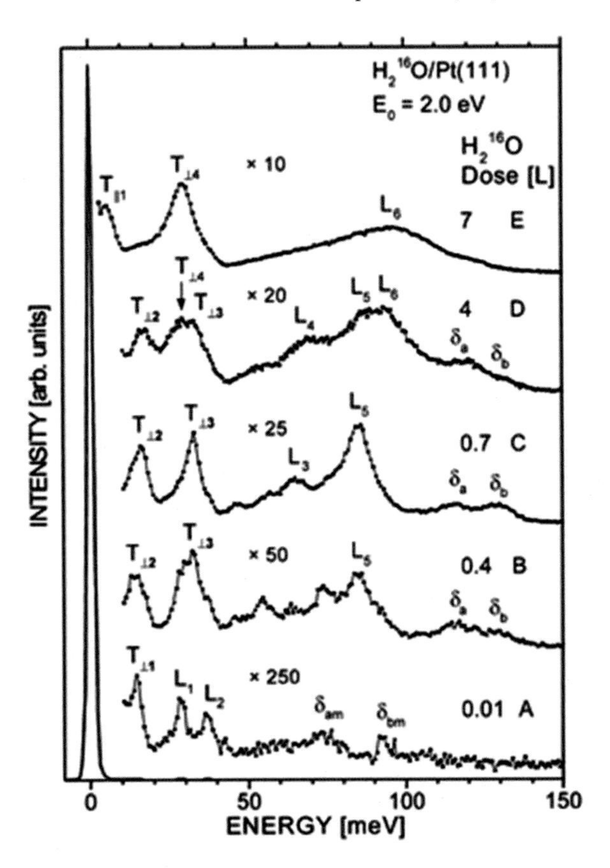

Fig. 7.11 HREEL spectra of the H_2O/Pt(111) structure [9]

hydrogen bond has the lowest vibrational frequency, but the longest bond length. The following diagram gives two scenarios of OH bonds that form weak hydrogen bonds, which are intermediate between the two scenarios in the above diagram. The average bond lengths of the vibrations are 0.973, 0.987 and 1.000 Å for the OH bonds that are free, with weak hydrogen bond and with strong hydrogen bond, respectively. Fourier transform of the curves in Fig. 7.14 gives frequencies very close to 467, 432 and 388 meV, which further confirms the correctness of the vibrational modes identified above. Figures 7.10 and 7.14 clearly show the sensitivity of OH vibrations to hydrogen bond formation in the bilayer structure. We believe that this approach can be a promising and universal method for identifying the structure of hydrogen bonding networks.

In contrast to the experiment, the vibrational modes located at 424 and 455 meV have been observed in the HREELS spectrum [9]. They correspond to the two modes on the right-hand side of Fig. 7.10. Other experiments have also identified a peak in the 360–400 meV range [1, 2]. However, this peak was first considered to be an OH-Pt vibration towards the surface [4]. The peak forming strong hydrogen bonds cannot be distinguished in the HREELS spectrum obtained from H_2O adsorbed

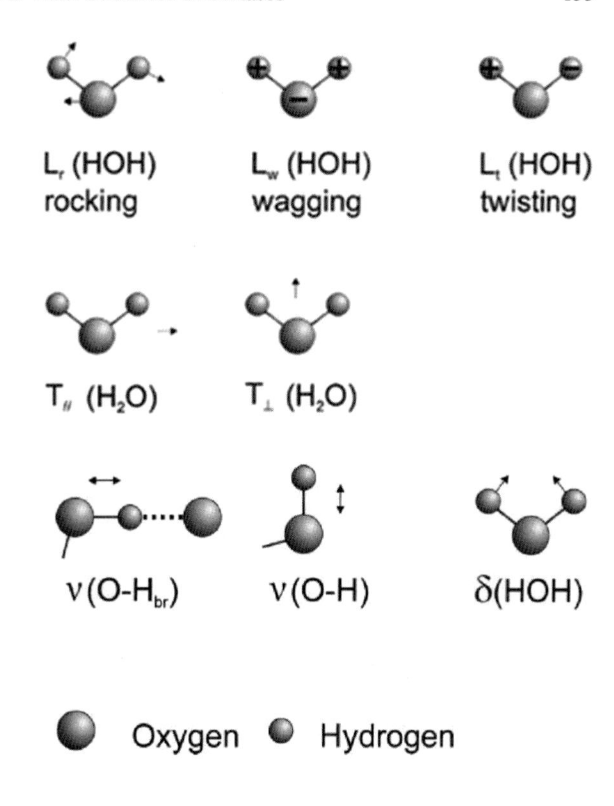

Fig. 7.12 Various modes of vibration of water on the surface of Pt(111) [9]

on the surface of Pt(111) [9, 15, 17], but appears in the IR reflectance absorption spectrum obtained from D_2O adsorbed on the surface of Pt(111) [4]. In the latter case, a broader peak is seen at 2200^{-1} cm. Multiplied by an isotope factor of 1.35, this peak corresponds to an energy of about 3000 cm^{-1}(370 meV) for H_2O This value is comparable to the energy of 384(388) meV for the OH vibrational mode that forms strong hydrogen bonds. Our findings were also recently confirmed by Denzler et al. in the experiment of D_2O/Ru(0001) [61]. The frequencies of the OH stretching vibrational modes in Fig. 7.14 are likewise well comparable to the OH stretching vibrational mode frequencies in ice Ih, 390 and 403 meV [1]. In this way, all vibrational characteristic modes of the bilayer structure have been identified.

The physical roots of both types of hydrogen bonding, as confirmed by the vibrational spectra, lie in the properties of bonding at the surface. The redistribution of charge due to the formation of hydrogen bonds is given in Fig. 7.15 for the free water bilayer (Fig. 7.15a), the adsorbed bilayer (Fig. 7.15b), and the water bilayer structure (Fig. 7.15c–f). The horizontal axis is through the O–H…O bond, while the vertical axis is along the direction of surface normal. For the free bimolecule (Fig. 7.15a), the formation of hydrogen bonds results in the transfer of electrons from the proton donor (dashed line in the left molecule) to the acceptor in the bonding region. This redistribution of charge is more pronounced in the adsorbed bilayer (Fig. 7.15b), indicating that hydrogen bonding is strengthened in the adsorbed state, as revealed

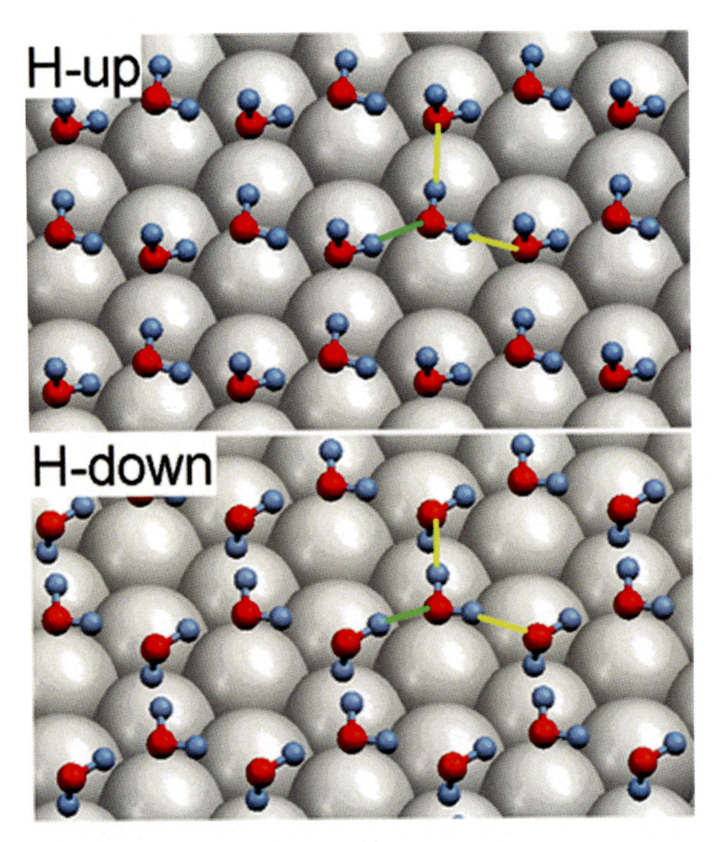

Fig. 7.13 $\sqrt{3}$ bilayer structure of water on the surface of Pt(111). The black and gray lines indicate strong and weak hydrogen bonds, respectively

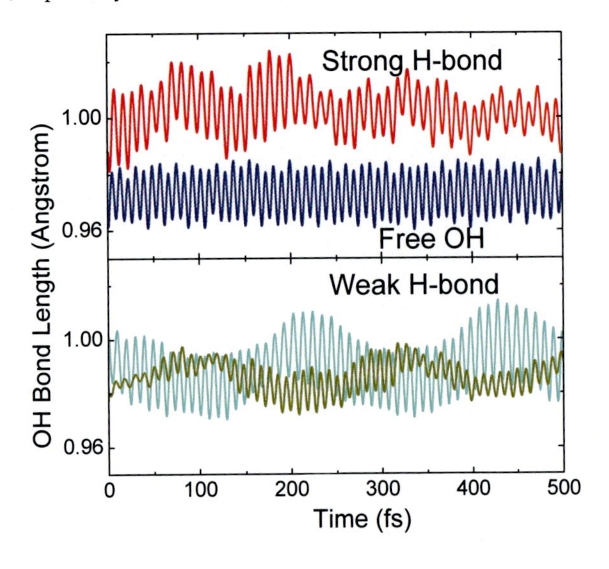

Fig. 7.14 OH bond length versus time

by total energy calculations. For strong hydrogen bonding in the bilayer structure (see Fig. 7.15c for the H-up case and Fig. 7.15d for the H-down case), the charge redistribution is very similar to that of the adsorbed bimolecule. However, for the weak hydrogen bonding of the underlying molecule (Fig. 7.15e, f), the redistribution of charge is not so significant. This difference reaffirms the finding above that there are two types of hydrogen bonds in the bilayer structure: a strong hydrogen bond and two weak hydrogen bonds. This strengthening of hydrogen bonding due to adsorption is dramatically different from the usual picture of the interaction between adsorbates. According to the Pauling principle, the usual picture is that the interaction between adsorbates weakens after the adsorbate has bonded to other atoms on the surface. Whether the previous conclusions hold only for water, or whether they also hold generally for other hydrogen-bonded systems, and what the origin of this anomalous behavior is, remains to be further investigated.

In summary, this chapter investigates the adsorption of single water molecules, small clusters of water molecules, one-dimensional water chains at the steps, water bilayers, and multiple water bilayers such as 2–6 on Pt(111). Structural optimization

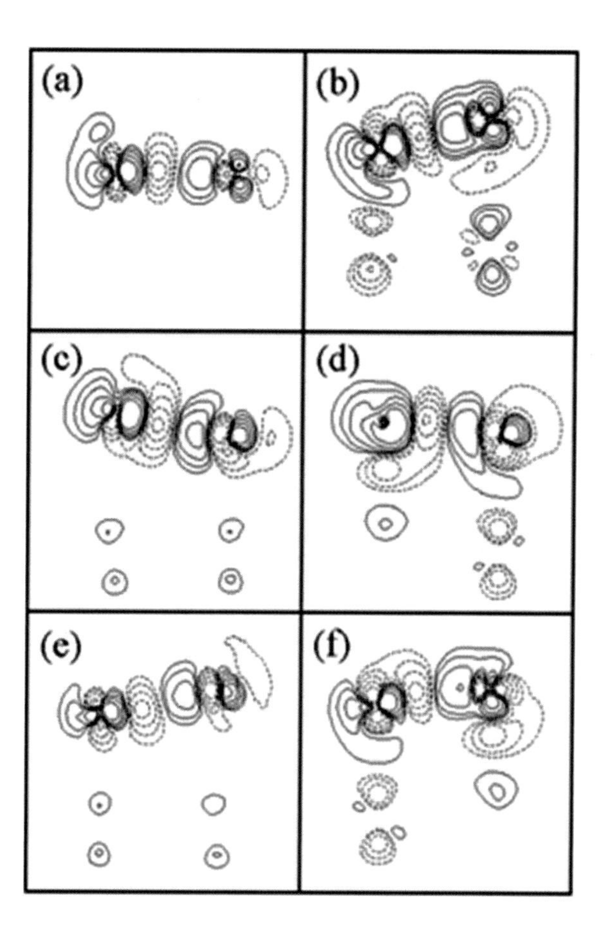

Fig. 7.15 Charge transfer density diagram for hydrogen bonding: **a** free bimolecule for water; **b** bimolecule adsorbed on the surface of Pt(111), **c** strong hydrogen bond for the bilayer structure of H-up, **d** strong hydrogen bond for the bilayer structure of H-down, **e** weak hydrogen bond for the bilayer structure of H-up, **f** weak hydrogen bond for the bilayer structure of H-down. The density of electron transfer is defined as $\Delta\rho = \rho[(H_2O)_2] - \rho[H_2O_{(1)}] - \rho[H_2O_{(2)}]$ for free bilayers; for other cases, $\Delta\rho = \rho[2(H_2O)/Pt] - \rho[H_2O_{(1)}/Pt] - \rho[H_2O_{(2)}/Pt] + \rho[Pt]$. The medium density line in the figure is taken as $\Delta\rho = \pm 0.005 \times 2^n e/Å^3$, where n = 0, 1, 2, 3, 4. The solid and dashed lines correspond to the cases $\Delta\rho > 0$ and $\Delta\rho < 0$, respectively

reveals that water molecules adsorb most stably at the top position, where the angle between the molecular plane and the surface is only 13°. In general, the structure of small clusters of water on the surface of Pt(111) is similar to the free case. Each water molecule adsorbs on the top site and lies as flat as possible to allow the strongest interaction between the metal and the water. H_2O adsorbs strongly on the steps and forms stable serrated one-dimensional H_2O chains. These are due to the coupling of H_2O to the 5d band on the Pt surface via lone pairs of electrons, leading to electron transfer from O to Pt.

At higher coverages, water forms bilayers and multilayers. The $\sqrt{3} \times \sqrt{3}R30°$ ($\sqrt{3}$) phase has two energetically almost concise structures: the H-up and H-down. The potential barrier between them is only 76 meV. The structures of $\sqrt{39} \times \sqrt{39}R16.1°$ ($\sqrt{39}$) and $\sqrt{37} \times \sqrt{37}R25.3°$ ($\sqrt{37}$) are overall not very different from $\sqrt{3}$ and are also folded hexagonal ring structures. However, the atoms in the $\sqrt{39}$, $\sqrt{37}$ structures are more disordered in the direction of surface normal. The distance between the bottom H_2O molecule of the lowermost bilayer and the surface gradually decreases as the coverage increases. It indicates that in the bilayer and multilayer structures, only the bottommost water molecules are directly bonded to the surface, while the upper H_2O molecules are almost unaffected. This suggests that water-metal interactions are rather localized. At 3BL, the $\sqrt{39}$ phase transforms to a $\sqrt{3}$ phase, in agreement with experimental findings.

We found that water is essentially molecularly adsorbed on Pt(111), which is consistent with the experiment. We also calculated the vibrational spectra of various adsorption structures and obtained peak positions that corresponded to the experiments. In order to compensate for the lack of experimental knowledge of the molecular structure and details of the interaction, we proposed a "vibrational identification" approach. Applying this approach, we have discovered for the first time that there are two hydrogen bonds of different strengths in the apparent water bilayer structure: the top water molecule contributes one hydrogen atom, which forms a hydrogen bond with the neighboring water molecule and is therefore stronger, while the bottom water molecule contributes two hydrogen atoms and forms two hydrogen bonds with neighboring water molecules and are therefore weaker. Both of them be identified by OH stretching vibrations. This method can be a promising universal method for identifying the structure of hydrogen bond networks.

References

1. P.A. Thiel, T.E. Madey, The interaction of water with solid surfaces: fundamental aspects. Surf. Sci. Rep. **7**, 211 (1987)
2. M.A. Henderson, The interaction of water with solid surfaces: fundamental aspects revisited. Surf. Sci. Rep. **46**, 1 (2002)
3. H. Ogasawara, J. Yoshinobu, M. Kawai, Water adsorption on Pt(111): from isolated molecule to three-dimensional cluster. Chem. Phys. Lett. **231**, 188 (1994)
4. H. Ogasawara, J. Yoshinobu, M. Kawai, Clustering behavior of water (D2O) on Pt(111). J. Chem. Phys. **111**, 7003 (1999)

5. M. Nakamura, Y. Shingaya, M. Ito, The vibrational spectra of water cluster molecules on Pt(111) surface at 20 K. Chem. Phys. Lett. **309**, 123 (1999)
6. A.L. Glebov, A.P. Graham, A. Menzel, Vibrational spectroscopy of water molecules on Pt(111) at submonolayer coverages. Surf. Sci. **427**, 22 (1999)
7. M. Morgenstern, T. Michely, G. Comsa, Anisotropy in the adsorption of H_2O at low coordination sites on Pt(111). Phys. Rev. Lett. **77**, 703 (1996)
8. M. Morgenstern, J. Muller, T. Michely, G. Comsa, The ice bilayer on Pt(111): nucleation, structure and melting. Z. Phys. Chem. **198**, 43 (1997)
9. K. Jacobi, K. Bedürftig, Y. Wang, G. Ertl, From monomers to ice—new vibrational characteristics of H_2O adsorbed on Pt(111). Surf. Sci. **472**, 9 (2001)
10. A. Glebov, A.P. Graham, A. Menzel, J.P. Toennies, Orientational ordering of two-dimensional ice on Pt(111). J. Chem. Phys. **106**, 9382 (1997)
11. S. Haq, J. Harnett, A. Hodgson, Growth of thin crystalline ice films on Pt(111). Surf. Sci. **505**, 171 (2002)
12. S.K. Jo, J. Kiss, J.A. Polanco, J.M. White, Identification of second layer adsorbates: water and chloroethane on Pt(111). Surf. Sci. **253**, 233 (1991)
13. L.E. Firment, G.A. Somorjai, Low-energy electron diffraction studies of molecular crystals: the surface structures of vapor-grown ice and naphthalene. J. Chem. Phys. **63**, 1037 (1975)
14. L.E. Firment, G.A. Somorjai, Low energy electron diffraction studies of the surfaces of molecular crystals (ice, ammonia, naphthalene, benzene). Surf. Sci. **84**, 275 (1979)
15. B.A. Sexton, Vibrational spectra of water chemisorbed on platinum (111). Surf. Sci. **94**, 435 (1980)
16. G.B. Fisher, J.L. Gland, The interaction of water with the Pt(111) surface. Surf. Sci. **94**, 446 (1980)
17. F.T. Wagner, T.E. Moylan, A comparison between water adsorbed on Rh(111) and Pt(111), with and without predosed oxygen. Surf. Sci. **191**, 121 (1987)
18. W. Ranke, Low temperature adsorption and condensation of O_2, H_2O, and NO on Pt(111), studied by core level and valence band photoemission. Surf. Sci. **209**, 57 (1989)
19. N. Materer, U. Starke, A. Barbieri, M.A. Van Hove, G.A. Somorjai, G.-J. Kroes, C. Minot, Molecular surface structure of ice(0001): dynamical low-energy electron diffraction, total-energy calculations and molecular dynamics simulation. Surf. Sci. **381**, 190 (1997)
20. M.A. Leban, A.T. Hubbard, Quantum mechanical description of electrode reactions. II. Treatment of compact layer structure at platinum electrodes by means of the extended Huckel molecular orbital method. Pt (111) surfaces. J. Electroanal. Chem. **74**, 253 (1976)
21. A.B. Anderson, Reactions and structures of water on clean and oxygen covered Pt(111) and Fe(100). Surf. Sci. **105**, 159 (1981)
22. X. Su, L. Lianos, Y.R. Shen, G.A. Somorjai, Surface-induced ferroelectric ice on Pt(111). Phys. Rev. Lett. **80**, 1533 (1998)
23. M.S. Yeganeh, S.M. Dougal, H.S. Pink, Vibrational spectroscopy of water at liquid/solid interfaces: crossing the isoelectric point of a solid surface. Phys. Rev. Lett. **83**, 1179 (1999)
24. D. Doering, T.E. Madey, The adsorption of water on clean and oxygen-dosed Ru(001). Surf. Sci. **123**, 305 (1982)
25. S. Meng, L.F. Xu, E.G. Wang, S.W. Gao, Vibrational recognition of hydrogen-bonded water networks on a metal surface. Phys. Rev. Lett. **89**, 176104 (2002)
26. S. Meng, L.F. Xu, E.G. Wang, S.W. Gao, Vibrational recognition of hydrogen-bonded water networks on a metal surface—reply. Phys. Rev. Lett. **91**, 059602 (2003)
27. H. Ogasawara, B. Brena, D. Nordlund, M. Nyberg, A. Pelmenschikov, L.G.M. Pettersson, A. Nilsson, Structure and bonding of water on Pt(111). Phys. Rev. Lett. **89**, 276102 (2003)
28. P.J. Feibelman, Comment on "vibrational recognition of hydrogen-bonded water networks on a metal surface." Phys. Rev. Lett. **91**, 059601 (2003)
29. T. Mitsui, M.K. Rose, E. Fomin, D.F. Ogletree, M. Salmeron, Water diffusion and clustering on Pd(111). Science **297**, 1850 (2002)
30. K. Morgenstern, J. Nieminen, Intermolecular bond length of ice on Ag(111). Phys. Rev. Lett. **88**, 066102 (2002)

31. G. Estiu, S.A. Maluendes, E.A. Castro, A.J. Arvia, Theoretical study of the interaction of a single water molecule with Pt(111) and Pt(100) clusters: Influence of the applied potential. J. Phys. Chem. **92**, 2512 (1988)
32. K. Raghavan, K. Foster, K. Motakabbir, M. Berkowitz, Structure and dynamics of water at the Pt(111) interface: molecular dynamics study. J. Chem. Phys. **94**, 2110 (1991)
33. G. Kresse, J. Hafner, Ab-initio molecular-dynamics simulation of the liquid-metal amorphous-semi -conductor transition in Germanium. Phys. Rev. B **49**, 14251 (1994)
34. G. Kresse, J. Furthmüller, Efficiency of ab-initio total energy calculations for metals and semiconductors using a plane-wave basis set. Comput. Mat. Sci. **6**, 15 (1996)
35. G. Kresse, J. Furthmüller, Efficient iterative schemes for ab initio total-energy calculations using a plane-wave basis set. Phys. Rev. B. **54**, 11169 (1996)
36. H.J. Monkhorst, J.D. Pack, Special points for Brillouin-zone integrations. Phys. Rev. B **13**, 5188 (1976)
37. M. Methfessel, A.T. Paxton, High-precision sampling for Brillouin-zone integration in metals. Phys. Rev. B **40**, 3616 (1989)
38. D. Vanderbilt, Soft self-consistent pseudopotentials in a generalized eigenvalue formalism. Phys. Rev. B **41**, 7892 (1990)
39. J.P. Perdew, Y. Wang, Accurate and simple analytic representation of the electron-gas correlation energy. Phys. Rev. B **45**, 13244 (1992)
40. S. Andersson, C. Nyberg, C.G. Tengstål, Adsorption of water monomers on Cu(100) and Pd(100) at low temperatures. Chem. Phys. Lett. **104**, 305 (1984)
41. J.E. Müller, J. Harris, Cluster study of the interaction of a water molecule with an aluminum surface. Phys. Rev. Lett. **53**, 2493 (1984)
42. M.W. Ribarsky, W.D. Luedtke, U. Landman, Molecular-orbital self-consistent-field cluster model of H_2O adsorption on copper. Phys. Rev. B **32**, 1430 (1985)
43. S. Izvekov, G.A. Voth, Ab initio molecular dynamics simulation of the Ag(111)-water interface. J. Chem. Phy. **115**, 7196 (2001)
44. H.P. Bonzel, G. Pirug, J.E. Müller, Reversible H_2O adsorption on Pt(111)+K: work-function changes and molecular orientation. Phys. Rev. Lett. **58**, 2138 (1987)
45. A. Michaelides, V.A. Ranea, P.L. de Andres, D.A. King, General model for water monomer adsorption on close-packed transition and Nobel metal surfaces. Phys. Rev. Lett. **90**, 216102 (2003)
46. K. Liu, J.D. Cruzan, R.J. Saykally, Water clusters. Science **271**, 929 (1996)
47. J.K. Gregory, D.C. Clary, K. Liu, M.G. Brown, R.J. Saykally, The water dipole moment in water clusters. Science **275**, 814 (1997)
48. S. Meng, E.G. Wang, S.W. Gao, A molecular picture of hydrophilic and hydrophobic interactions from ab initio density functional theory calculations. J. Chem. Phys. **119**, 7617 (2003)
49. S. Meng, E.G. Wang, S.W. Gao, Water adsorption on metal surfaces: a general picture from density functional theory calculations. Phys. Rev. B **69**, 195404 (2004)
50. L.A. Curtiss, D.L. Frurip, M. Blander, Studies of molecular association in H_2O and D2O vapors by measurement of thermal conductivity. J. Chem. Phys. **71**, 2703 (1979)
51. D. Zahn, J. Brickmann, Molecular dynamics study of water pores in a phospholipid bilayer. Chem. Phys. Lett. **352**, 441 (2002)
52. G. Hummer, J.C. Rasaiah, J.P. Noworyta, Water conduction through the hydrophobic channel of a carbon nanotube. Nature **414**, 188 (2001)
53. C. Dellago, M.M. Naor, G. Hummer, Proton transport through water-filled carbon nanotubes. Phys. Rev. Lett. **90**, 105902 (2003)
54. D.J. Mann, M.D. Halls, Water alignment and proton conduction inside carbon nanotubes. Phys. Rev. Lett. **90**, 195503 (2003)
55. P.J. Feibelman, Partial dissociation of water on Ru(0001). Science **295**, 99 (2002)
56. H. Jonsson, G. Mills, K.W. Jacobsen, Nudged elestic band method for finding minimum energy paths of transitions, in *Classical and Quantum Dynamics in Condensed Phase Simulations*, ed. by B.J. Berne, G. Ciccotti, D.F. Coker (World Scientific, 1998)

57. A. Michaelides, A. Alavi, D.A. King, Different surface chemistries of water on Ru{0001}: from monomer adsorption to partially dissociated bilayers. J. Am. Chem. Soc. **125**, 2746 (2003)
58. B.J. Smith, D.J. Swanton, J.A. Pople, H.F. Schaefer III., L. Radom, Transition structure for the interchange of hydrogen atoms within the water dimer. J. Chem. Phys. **92**, 1240 (1990)
59. R.H. Hauge, J.W. Kauffman, J.L. Margrave, Infrared matrix-isolation studies of the interactions and reactions of Group 3A metal atoms with water. J. Am. Chem. Soc. **102**, 6005 (1980)
60. F. Sim, A. St-Amant, I. Papai, D.R. Salahub, Gaussian density functional calculations on hydrogen-bonded systems. J. Am. Chem. Soc. **114**, 4391 (1992)
61. D.N. Denzler, C. Hess, R. Dudek, S. Wagner, C. Frischkorn, M. Wolf, G. Ertl, Interfacial structure of water on Ru(0001) investigated by vibrational spectroscopy. Chem. Phys. Lett. **376**, 618 (2003)

Chapter 8
Water Adsorption on Metal Surfaces

After meticulously studying the various structures, dynamics, and trends of water on Pt(111) surfaces, we consider the nature and general laws of interactions that occur between water and general metal surfaces, especially transition and noble metals surfaces. These studies also help us to understand the interactions between water and general solid surfaces, such as oxides.

Two basic factors determine the structure and stability of adsorbed water: (a) the bonding form between water and metal surface, and (b) the strength of hydrogen bonds between water molecules. Usually, the two are coupled to each other and are thus difficult to distinguish exactly. On most metal surfaces, the two effects are of comparable magnitude. The competition between these two results in a rich and diverse range of adsorption structures and modes of motion on the surface. Over the past two decades, due to a wide range of technological applications in catalysis, fuel cells, automotive exhaust treatment, medical implants, etc., a large amount of experimental work in ultra-high vacuum (UHV) has focused on the adsorption of water on densely packed surfaces of several typical noble and transition metals, like Pt(111), Pd(111), Rh(111), Au(111), Ag(111), Cu(111), Ru(0001), etc. [1, 2]. Summarizing the experimental results, one finds that the growth of ice layers on metal surfaces is a slightly different from bonding law (BFP rule) in bulk ice, and is subject to the BFP rule at the surface [3].

(1) Water is bonded to the metal surface through the lone pair of electrons of O atom.
(2) Even in two-dimensional clusters or incomplete ice layers, the water molecules still maintain a tetrahedral binding configuration.
(3) Each water molecule forms at least two bonds (both O-bonding to the surface and hydrogen bonds between water molecules).
(4) All free lone pairs of electrons are confined in a direction essentially perpendicular to the metal substrate.

It is usually assumed that, unlike decomposed adsorption on oxide surfaces, water exists as a bilayer structure in the molecular state on clean metal surfaces unless the

S. Meng and E. Wang, *Water*,
https://doi.org/10.1007/978-981-99-1541-5_8

surface is contaminated with free O atoms or other contaminants [1]. However, recent studies by Feibelman on water adsorption on Ru(0001) surfaces [4, 5] have suggested that half of the water molecules in the bilayer structure are partially decomposed, i.e., the OH bonds in them that do not form hydrogen bonds with other molecules are broken. This semi-decomposed structure is 0.2 eV/molecule lower in energy than the bilayer structure in the molecular state. What about the other surfaces? Even on Ru(0001), the vibrational spectrum of D_2O/Ru(0001) was recently measured by Denzler et al. using a sum-frequency generation (SFG) approach and the measurements were consistent with the molecular adsorption state, whereupon the authors rejected Feibelman's proposal that half of the water molecules decompose and adsorb [6]. The contradiction between theory and experiment still needs more research to be resolved. Scanning tunneling microscopy (STM) experiments also revealed interesting phenomena when observing the diffusion of individual water molecules and small cluster molecules on the surface of Pd(111): dimeric and trimeric molecules diffuse 3–4 orders of magnitude faster than single molecules [7].

This chapter attempts to reveal the unified law of water-metal surface interactions from the perspective of first-principles calculations and to explain the phenomena in experiments. Based on the view of atomic and electronic interactions, we first investigate the adsorption of individual water molecules and water bilayers on the surfaces of Ru, Pd, Au, etc. and show that water-substrate interactions are essentially caused by the coupling of O lone pairs of electrons to the d-bands of these metals. Typically, this causes a transfer of electrons from O to the substrate atom, allowing this water molecule to provide H and form a hydrogen bonding reinforcement. Comparing the water-metal interaction (O-M) and hydrogen bonding in the water layer on these surfaces, we microscopically explain the difference between hydrophilicity and hydrophobicity and obtain a surface infiltration order consistent with the experiment. Finally, we investigate various dynamic phenomena such as vibration, transformation, decomposition, proton transport, and diffusion of water structures on metal surfaces.

8.1 Adsorption of Water on Ru, Pd, Au, and Other Metal Surfaces

We have studied the adsorption of water on Ru(0001), Rh(111), Pd(111), Au(111). These surfaces differ from Pt(111) surfaces in terms of (1) different elemental chemical activities and (2) different lattice constants and thus different matches to the bulk ice period, affecting the hydrogen bond strength. The former mainly determines the strength of the water-metal interaction, especially in the case of single H_2O molecule adsorption. And it affects the hydrogen bonding in the adsorption structure together with the lattice matching effect. So we have studied two typical adsorption scenarios: the adsorption of single water molecules and the adsorption of water bilayers. This allows us to clearly understand the water-surface interaction and hydrogen bonding in the adsorption structure.

8.1.1 Adsorption of Individual Water Molecules

As with water on Pt(111), we start our study with individual molecular adsorption. The $p(3 \times 3)$ protocells were taken and the calculations were performed using the Vienna ab initio simulation program VASP. The theoretical lattice constants used in the calculations are shown in Table 8.1, and they agree well with the experiment. The parameters used in the calculations are essentially the same as in Sect. 7.1: the planar energy cutoff is taken to be 300 eV, the vacuum thickness is about 13 Å, and the water molecules are placed on one side of the crystal plate.

The conformation of single molecules adsorbed on these surfaces is essentially the same as on Pt(111) [8] (Fig. 8.1). H_2O molecules choose to adsorb in the top position, at angles of 6°–24° to the surface. The adsorption usually results in an increase in OH bond length and a broadening of the HOH bond angle. This suggests a charge transfer from the lone pair of electron orbitals of O to the metal surface, in agreement with the analysis on the Pt surface. The specific structures and energies are listed in Table 8.2.

The table shows that water has the highest adsorption energy and the strongest effect on Ru and Rh surfaces; the second strongest on Pd and Pt surfaces; and the weakest on Au surfaces. This indicates that the coupling strength of water to metal is in the order of Ru > Rh > Pd > Pt > Au from the largest to the smallest. Interestingly,

Table 8.1 Lattice constants (Å) for some hcp(Ru) and fcc(Rh, Pd, Pt, Ag, Au) metallic materials

	Ru	Rh	Pd	Pt	Ag	Au
Theoretical	2.72	3.83	3.96	3.99	4.17	4.18
Experimental	2.71	3.81	3.89	3.92	4.09	4.08

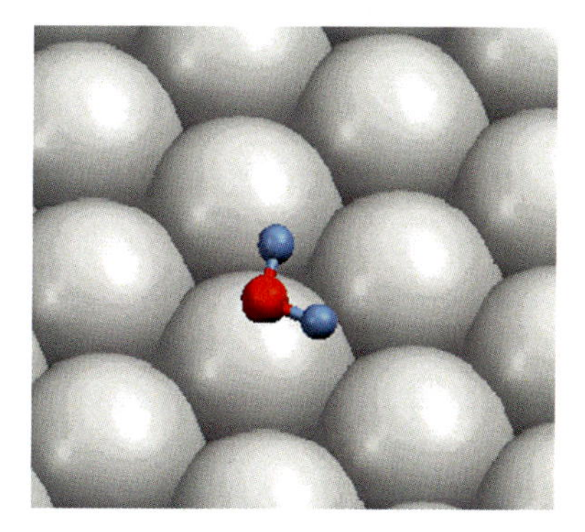

Fig. 8.1 Typical configuration of a single water molecule adsorbed on a densely arranged metal surface

Table 8.2 Adsorption configurations and energies of single water molecules on metal surfaces. The units of distance, energy, and angle are meV, Å, degree(°), respectively

Surfaces	Layers	Top site		Bridge		Vacancy		d_{OH}	HOH	θ
		d_{OM}	E_{ads}	d_{OM}	E_{ads}	d_{OM}	E_{ads}			
Ru(0001)	5	2.28	409	2.55	92	2.56	67	0.981	105.66	16
Rh(111)	4	2.32	408	2.57	126	2.70	121	0.978	105.95	24
Pd(111)	4	2.42	304	2.74	146	2.77	130	0.977	105.63	20
Pt(111)	4	2.43	291	3.11	123	3.12	121	0.978	105.36	13
Au(111)	7	2.67	105	2.80	32	2.80	25	0.977	105.04	6

the water-metal atomic distance (d_{OM}), the polar angle θ, and the HOH bond angle also show the same order, reinforcing our conclusion. This order of bonding strength matches well with the chemical activity of the elements. In the periodic table, Ru, Rh, Pd belong to the same fifth row of elements and are arranged adjacent from left to right; Pt and Au are also adjacent from left to right in the sixth row, and Pt and Pd belong to the same column. So their order of activity is also Ru > Rh > Pd > Pt > Au. Recently Michaelides et al. have meticulously studied the adsorption of individual water molecules on metal surfaces [9]. Our results are essentially in agreement with them, but with slight differences in details. For example, they obtained adsorption energies in the order Rh > Ru > Pd > Pt > Au. But in fact, the difference in adsorption energies of water molecules on Rh and Ru, and on Pd and Pt is very small. Also, they obtained the H_2O molecules without a certain order in the polar angle θ. This may originate from the large calculation error when the polar angle is small and the small size of the original cell (2×2) they used.

8.1.2 *Adsorption of Water Bilayers*

We have also calculated the structure of $\sqrt{3} \times \sqrt{3}R\ 30°$ water bilayers (bilayer ice) on Ru, Rh, Pd, Pt, and Au surfaces. This structure has been confirmed experimentally many times [1, 2]. The simultaneous presence of water-surface interactions, and hydrogen bonding between water molecules in it, provides us with a convenient way to study the competition and mixing of these two interactions.

The structural parameters and energies obtained from the calculations are listed in Table 8.3. Calculations were made for both H-up and H-down bilayer configurations. Similar to the situation on Pt(111), the bilayer ice on these surfaces is also a network structure with folded hexagonal rings arranged. The adsorption heights z_{OM1} of the lower water molecules increase gradually in the order Ru, Rh, Pd, Pt, Au; while the heights of the upper water molecules remain essentially horizontal, i.e., at 3.40 Å in the H-up structure and 3.20 Å in the H-down structure. So the OO vertical distances (z_{OO}) also decrease in this order. These relationships are more clearly shown in Fig. 8.2. Plotted in the figure are the patterns of these structural parameters (z_{OM1},

z_{OM2}, z_{OO}) in the H-up bilayer as a function of elemental order (the H-down bilayer structure is similar to this). These general structural laws once again indicate that the interaction between metal and water is rather localized and essentially restricted to the bottom water molecule, while the upper molecule is almost unaffected.

This structural regularity of the double layer ice on the surface of metals is consistent with the d-electron energy band occupation pattern of these metal elements. According to the periodic table of elements, the order of d-band filling i.e. Ru < Rh < Pd = Pt < Au (Fig. 8.3). It is known that the amount of d-band filling affects the adsorption of atoms such as O and H, and even the general activity of the metal surface. Figures 8.2 and 8.3 clearly show that the water-surface interaction is directly related to the d-band occupation of surface elements. We believe that the increase

Table 8.3 Structure and energy of water bilayers adsorbed on tightly disposed metal surfaces

Surfaces	Bilayers	z_{OO} (Å)	z_{OM1} (Å)	z_{OM2} (Å)	E_{ads} (meV/H_2O)
Ru(0001)	H-up	0.86	2.46	3.42	531
	H-down	0.42	2.69	3.22	533
	Half-disso	0.05	2.09	2.16	766
Rh(111)	H-up	0.79	2.50	3.40	562
	H-down	0.42	2.52	3.12	544
	Half-disso	0.04	2.09	2.16	468
Pd(111)	H-up	0.60	2.78	3.45	530
	H-down	0.36	2.66	3.18	546
	Half-disso	0.07	2.09	2.20	89
Pt(111)	H-up	0.63	2.70	3.37	522
	H-down	0.35	2.68	3.14	534
	Half-disso	0.06	2.12	2.23	291
Au(111)	H-up	0.46	2.90	3.38	437
	H-down	0.29	2.85	3.25	454
	Half-disso	0.14	2.20	2.43	-472

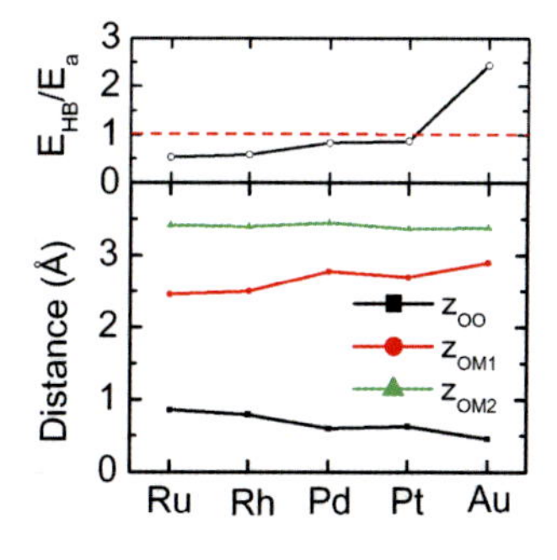

Fig. 8.2 Comparison of structural parameters and infiltration capacity of water bilayer (H-up) on metal surfaces

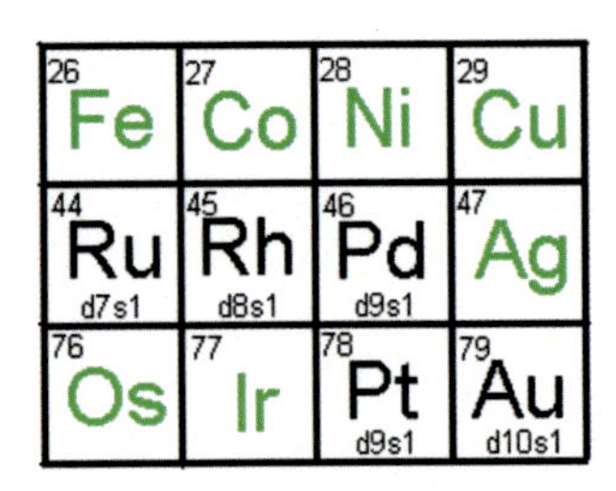

Fig. 8.3 Position and d-electron number of Ru, Rh, Pd, Pt, Au and other metals in the periodic table

in the surface lattice constant from Ru (2.75 Å) to Au (2.95 Å) has a much smaller effect on the water-surface interaction.

In addition to the $\sqrt{3} \times \sqrt{3}R\ 30°$ aqueous bilayers, we also calculated the adsorption of half of the water molecules decomposed by the structure on these surfaces, which is proposed by Feibelman. Typically, the two O-M adsorption bond lengths are around 2.10 and 2.20 Å, respectively (except on Au), which is much smaller than the O-M bond length in the bilayer adsorption of the molecular state. It is a very flat structure ($z_{OO} \sim 0.05$ Å). However, we find it to be energetically stable only on the Ru(0001) surface, while on the Au(111) surface the adsorption energy is negative and the water layer repels from the surface.

As for what the actual water bilayer structure looks like, recent experiments at Pt(111) [10] and Ru(0001) [6] seem to favor the H-down bilayer structure. Calculations on Pt(111) are consistent with this, although the H-down structure is only a dozen meV lower in energy. Interestingly, our calculations show that the H-up bilayer ice structure on Rh(111) would be more stable than the H-down, which has yet to be confirmed experimentally.

8.2 Water Adsorption on Open Metal Surfaces

8.2.1 Adsorption of Individual Water Molecules on the Cu(110) Surface

We have systematically studied the adsorption of water on Cu(110), Ag(110), Au(110), Pd(110), and Pt(110) [11, 12]. The adsorption of individual water molecules on the surface of Cu(110) is first considered. We chose a $c(2 \times 2)$ protocell and used the Vienna ab initio simulation program VASP to perform the calculations. The parameters used for the calculations were: the lattice constant was chosen to be the experimental value of 3.6149 Å, the planar energy cutoff was taken to be 400 eV, the vacuum thickness was over 23 Å, and the water molecules were placed on only one side of the crystal plate.

The main adsorption sites for single water molecules on the open system Cu(110) surface contain the top, bridge, and vacant sites (Fig. 8.4). The preference of water

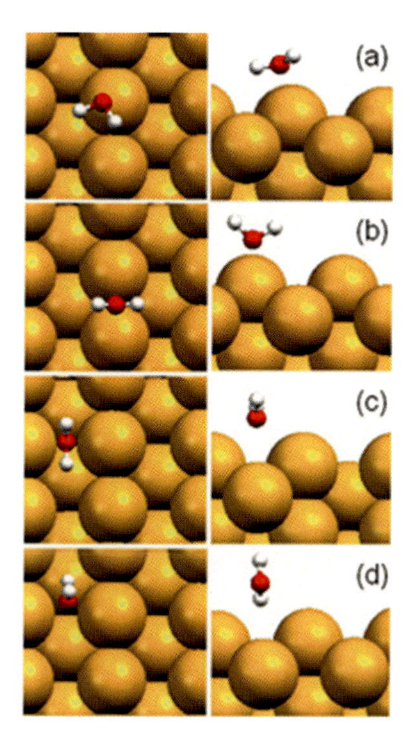

Fig. 8.4 Typical adsorption configuration of single water molecule on the surface of open system Cu (110): **a** top position, **b** bridge position along [110], **c** bridge position along [001], **d** vacancy. Red, white, and orange spheres represent O, H, and Cu atoms respectively

molecules to lie flat and adsorb at the top site relative to other adsorption configurations is mainly dependent on the fact that this configuration possesses the shortest Cu–O spacing and the strongest Cu–O bonding interactions. Calculations show that more charge is transferred from O to Cu atoms, which weakens the OH bonding of water. The selective top-site adsorption of water molecules is also consistent with the results obtained on other metal surfaces such as Pt(111), Pd(111), Rh(111), Ru(0001), Cu(100), Cu(111), and Cu(211).

It is experimentally observed that water molecules readily diffuse on metal surfaces even at very low temperatures. We calculate that the diffusion potential barriers of a single water molecule along the [110] and [001] directions are 0.12 eV and 0.23 eV, respectively, on the Cu(110) surface. This makes it easier for H_2O molecules to diffuse along the [110] direction. The asymmetric potential energy surface and the adsorption position of water molecules determine their diffusion behavior in different directions. Also, the smaller diffusion potential barrier facilitates the formation of water clusters.

Based on the structure of the top-site adsorption of H_2O molecules, Fig. 8.5 depicts the decomposition paths of individual water molecules and the corresponding decomposition potentials. The single water molecule gradually moves from the top position (primary state) to the bridge position (transition state). Accompanying this process, the position of the O atom approaches the bridge position in the [110] direction, and the length of one OH bond increases to 1.55 Å. From Fig. 8.5, the

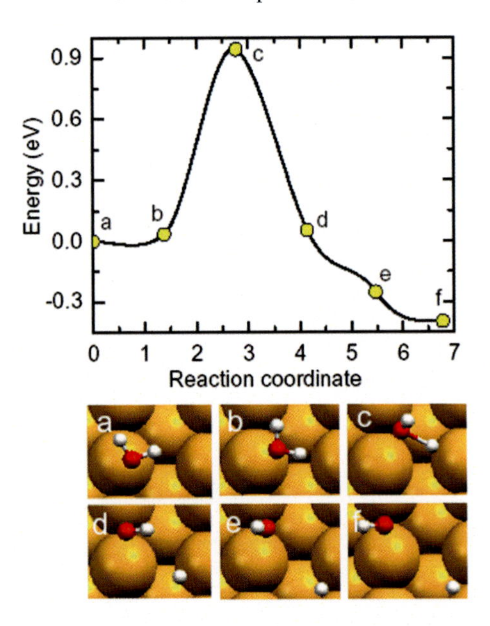

Fig. 8.5 Decomposition potential of a single water molecule on Cu(110) and several representative configurations on the decomposition path

decomposition potential barrier of a single water molecule is 0.94 eV. A single water molecule is well able to decompose on the Cu(110) surface relative to the adsorption and desorption energies of 0.38 eV. After the decomposition of a single water molecule, the OH bond occupies the bridge site on [110], and the disconnected H atom resides in the gully between the Cu atoms, and the energy of the whole system is then greatly reduced. Moreover, the decomposition of individual water is an exothermic process.

8.2.2 Adsorption of Water Bilayers

We also calculated several typical adsorption structures (Fig. 8.6) for the water bilayer (i.e., 1:1 ratio of surface Cu atoms to H_2O molecules) on the Cu(110) surface: molecular adsorption (Fig. 8.6a–d) and semi-decomposition adsorption (Fig. 8.6e). The calculated adsorption energy of 0.554 eV/H_2O for the hydrogen-down (H-down) structure (Fig. 8.6a) is much larger than that of 0.514 eV/H_2O for the hydrogen-up (H-up) structure (Fig. 8.6b), indicating that the former is more stable than the latter. This result is consistent with the X-ray absorption experiments that observed the formation of H-down water bilayers on the surface of Pt(111). On Cu(110), the chain-like water bilayer structure (Fig. 8.6c), which has adsorption energy only 6 meV lower than that of H-down (Fig. 8.6a), is also more stable, while on Ru(0001), the chain-like water bilayer structure is the most stable, with an adsorption energy 0.13 eV larger than that of the H-down structure [13]. We found that different H positions and pointing lead to some energy differences (<18 meV), but not enough

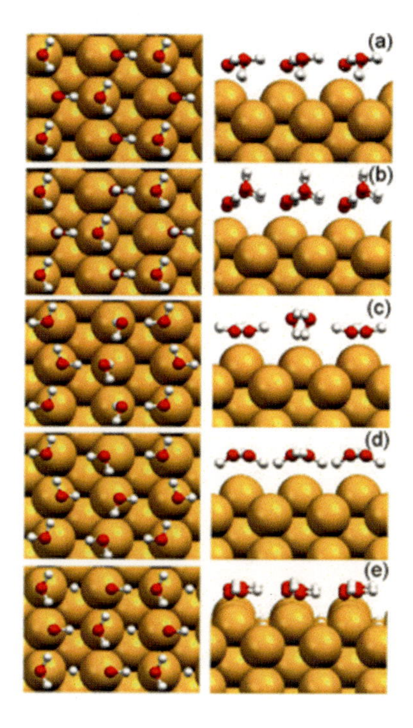

Fig. 8.6 Atomic structure of water bilayer adsorbed on Cu (110) surface. Complete adsorption: **a** H-down, **b** H-up, **c** chain, **d** quasi square configuration; and semi decomposition adsorption **e**

to change the stability of the H-down and chain-like water structures on Cu(110). The square-like water structure (Fig. 8.6d) is the most unstable (adsorption energy of 0.442 eV/H_2O) relative to the first three fully adsorbed configurations (Fig. 8.6a–c). Some of the other conformations are also unstable and are eventually transformed into the H-down water structure by structural optimization. Thus, the stability of the aqueous bilayer structure for intact adsorption onto Cu(110) is: H-down > chain structure > H-up > square-like structure.

However, once the OH bonds that do not form H bonds are broken, the water layer will form a stable semi-decomposed structure on the Cu(110) surface (Fig. 8.6e), which is in agreement with the previous Feibelman proposal that water will semi-decompose on Ru(0001). On Cu(110), the adsorption energy of the semi-decomposed structure reaches 0.632 eV/H_2O. The energy stability is very close compared to the intact non-decomposed adsorption configuration [14]. The above results are all calculated based on a small protocell of c(2 $\times$ 2); the transient structure of each protocell is not necessarily identical in the actual system. If a larger protocell is taken for calculation, Michaelides et al. found that the structure of water on the Cu(110) surface with H_2O: OH = 2:1 ratio and certain hydrogen bond defects (Bjerrum defects) is the most stable [15].

From the charge density diagrams of both H-down and semi decomposed structures (Fig. 8.7), the interaction between water molecule and Cu(110) surface occurs mainly through the coupling of the lone pair of electrons of O and the p_z orbital of the Cu atom, leading to the accumulation of 1π states and the loss of the d_{z2} state of

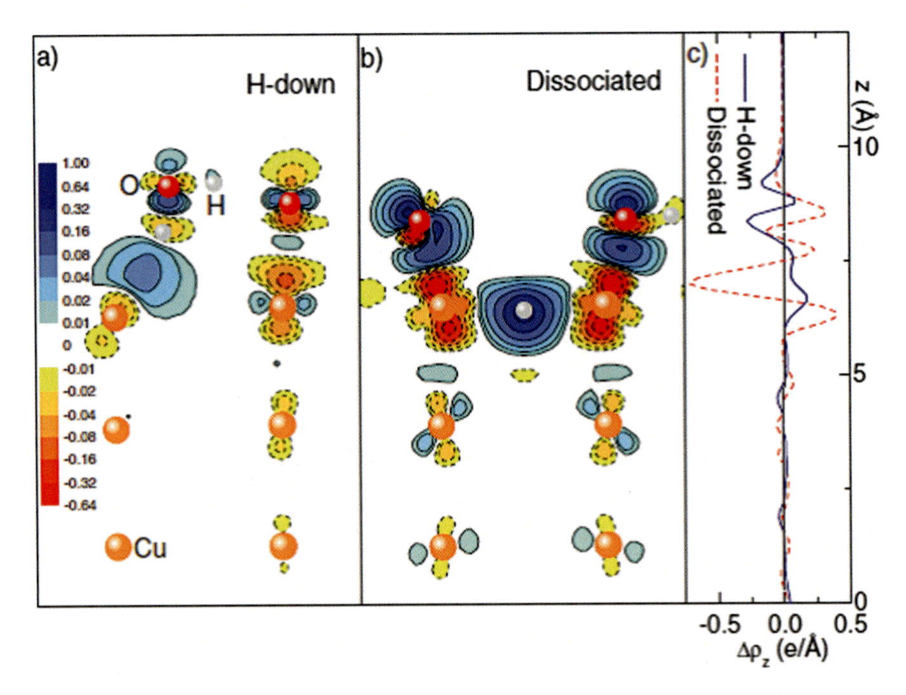

Fig. 8.7 Plot of charge density redistribution due to adsorption of water structures. Black and gray indicate regions of increasing and decreasing charge density, respectively; the units of the isodensity lines are e/Å^3. **a** Hydrogen downward water layer structure; **b** decomposed water layer structure. **c** One-dimensional charge density changes in the direction normal to the surface

Cu. Integrating the charge density along the z-axis in the surface normal direction, we find [11]: for the H-down water bilayer, 0.17 e/protocell of charge is transferred from H_2O to Cu(110); for the semi decomposed structure, 0.14 e/protocell of charge is transferred from Cu to the H_2O molecule.

Considering other precious metal (110) surfaces, including: Ag, Au, Au(1 × 2), Pd, Pt, Pt(1 × 2), the order of adsorption of individual water molecules at the top site is Pt > Pt(1 × 2) > Pd > Cu > Ag > Au > Au(1 × 2) (Fig. 8.8a). In general, the adsorption energies are all very close to each other. The order of the adsorption energies of the homologous elements Cu, Ag, and Au is also consistent with the chemical reactivity of their respective surfaces. Once the OH bond of the water molecule is broken, OH + H interacts with the Ag and Au surfaces in a mutually repulsive manner, implying that both surfaces are more resistant to water corrosion. Although individual water molecules can be broken down on Pt, the structure of Pt(1 × 2) is relatively more stable. Thus, only the Cu(110) surface favors the decomposition of individual water molecules. For the adsorption of water bilayers, we compared the more stable H-down intact adsorption structure with the semi-decomposed structure on these precious metal (110) surfaces (Fig. 8.8b). For the H-down structures, the order of adsorption of the water bilayers is Pt > Pd > Pt(1 × 2) > Cu > Ag > Au > Au(1 × 2). After the semi-decomposition of the water layer, the H_2O molecules and the decomposed

OH are almost in the same layer, indicating that the decomposed water layer is flat. Compared with the configuration of complete adsorption, the adsorption energy on all precious metal surfaces except the Cu surface is reduced, and the general trend of the adsorption energy [12] is Cu > Pt > Pd > Pt(1 × 2) > Ag > Au > Au(1 × 2). Especially on Au surfaces, the decomposed water layer is very unstable. Considering that most of the metal surfaces, such as Na and Fe, are susceptible to water adsorption, and that the water double layer has medium-sized binding energy for both intact and semi-decomposed adsorption on Cu(110), this reveals that the Cu(110) surface is the dividing line for water layer decomposition.

In summary, a single water molecule selects the top site of Cu(110) for adsorption and can readily diffuse on the surface, but is difficult to decompose. And for the water bilayer, it prefers the complete adsorption conformation of H-down. However, once the water molecule is decomposed, it will form a more stable semi-decomposed structure on Cu(110). Comparing with the results on other precious metal (110) surfaces (Ag, Au, Pd, Pt), we find that Cu(110) is the dividing line between complete and decomposed adsorption of the water bilayer.

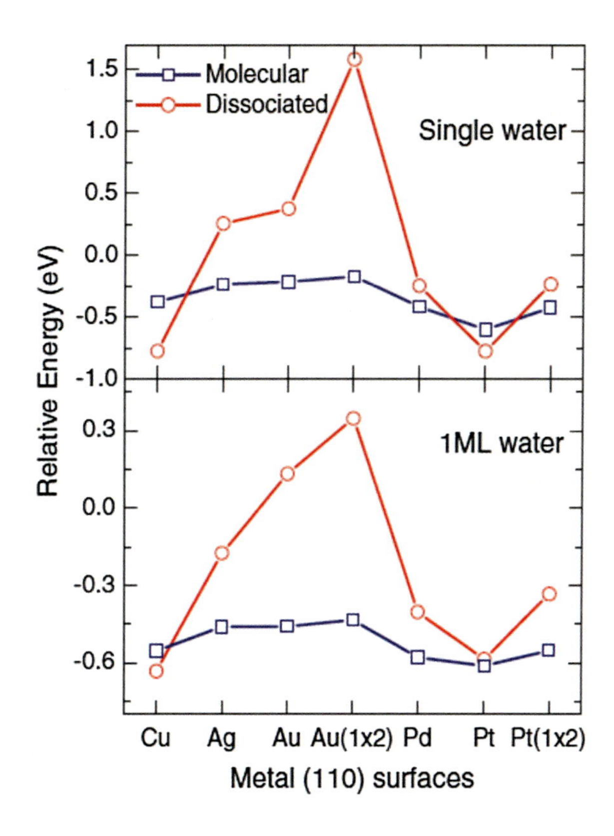

Fig. 8.8 Relative energies corresponding to intact and semi-decomposed adsorption of water on the (110) open surfaces of different metals. **a** Individual water molecules. **b** Water layers

8.3 The Nature of Water-Metal Substrate Interactions: An Electronic Prospective

Through the above studies of water adsorption on Ru(0001), Rh(111), Pd(111), Pt(111), Au(111) surfaces, we have summarized the nature of interaction between water and transition and precious metal surfaces. Water molecules are in contact with the surface by electronic coupling. The adsorption structure clearly shows that the strength of the interaction, the conformation are closely related with the occupation of the electronic states on the surface. A unified and unambiguous electronic picture of the water-surface interaction has not been given in the literature. We have tried to study the nature and laws of water interaction with these metallic surfaces from the electronic interactions, which are also revealing for understanding the interaction of water with general surfaces.

Figure 8.9 illustrates the deformation charge density distribution induced by adsorption in the single H_2O molecule (a), $(H_2O)_2$ (b), H-up (c), and H-down (d) bilayer structures on Pt(111). The horizontal axis is the surface [110] direction and also essentially along the direction of one OH group. The vertical axis is along the normal direction of the surface. The figure shows that in all cases the distribution of polarized charges near the Pt atom shows d_{xz} and d_{z^2} characteristics. This suggests that the d electronic bands on the surface of Pt(111), particularly the d_{xz} and d_{z^2} bands, are involved in the H_2O-Pt interaction, resulting in a transfer of about 0.02 electrons from the water molecule to the surface. The expansion of the HOH bond angle and the elongation of the OH bond length observed earlier are both due to the reduced electrons on the O atom. This image is consistent with previous studies of the adsorption of individual water molecules onto the surface.

The water molecule has four main occupied valence electron orbitals $2a_1$, $1b_2$, $3a_1$, $1b_1$, and two unoccupied orbitals $4a_1$, $2b_2$ (see Fig. 2.5). They determine the interactions of water with the outside world. A detailed analysis shows that the orbitals mixed with the metal atoms are mainly $3a_1$, $1b_1$, which are also the main components (together with $2a_1$) that make up the lone pair of electron orbitals. The $3a_1$, $1b_1$ density of states and the charge distribution of the corresponding orbitals for a single H_2O molecule adsorbed on Pt(111) are given in Fig. 8.10. The spatial distribution of charges on each orbital when a single molecule is adsorbed on Ru(0001) is also similar (Fig. 8.11). This suggests that the image of water molecules interacting with metal substrates via lone pairs of electrons ($3a_1$, $1b_1$ orbitals) of O is universal and exists in a variety of water-surface systems.

Figure 8.9 also shows that the interaction of water with the surface is rather localized. In the case of bimolecular adsorption (b) and even bilayer adsorption (c, d), the polarized charge is mainly distributed between the underlying water molecules and the substrate atoms in direct contact with them. There is little coupling between the upper H_2O molecule and the surface atoms in the bilayer structure. Figure 8.12a draws the electronic density of states [on Ru(0001)] for the two water molecules in the bilayer structure (H-up). The $1b_1$ energy level of the lower water molecule is significantly lower than that of the upper molecule. It also indicates that only the

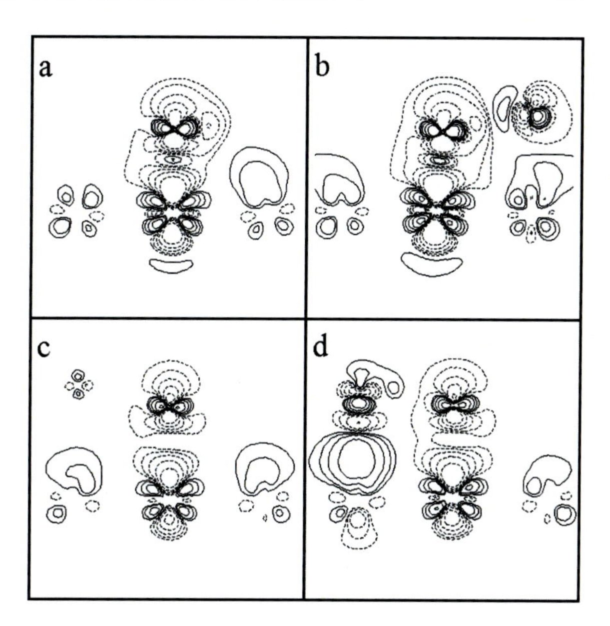

Fig. 8.9 Charge-transfer density diagram for the interaction of a metal surface (Pt) with water. **a** single water molecule, **b** bimolecule, **c** H-up bilayer, **d** H-down bilayer structure. The difference electron density is defined as $\Delta\rho = \rho[(H_2O)_n/Pt] - \rho[(H_2O)_n] - \rho[Pt]$. n is the number of H_2O molecules in the protocell. The isodensity line in the figure is taken as $\Delta\rho = \pm 0.005 \times 2^k e/Å^3$, where $k = 0, 1, 2, 3, 4$. The solid and dashed lines correspond to the cases $\Delta\rho > 0$ and $\Delta\rho < 0$, respectively

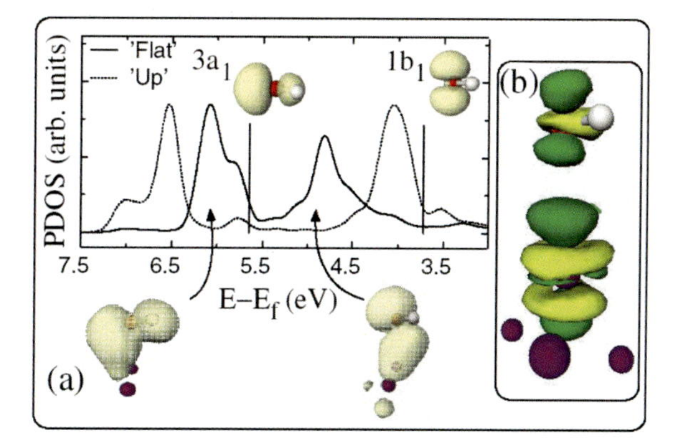

Fig. 8.10 Spatial distribution of electronic density of states and orbitals for adsorption of single H_2O molecules on Pt(111) in equilibrium configuration (Flat) and upright configuration (Up) [9]

lower molecule is more strongly coupled to the surface. Whereas in the half-dissolved (Half-disso.) structure since both molecules (H_2O and OH) are in direct contact with the surface, they both have a stronger mixing with the surface (Fig. 8.12b). These guide us to the following conclusion: **the water structure on the metal surface forms chemical bonds through the lone pair of electrons of O and the substrate electrons, especially the surface state electrons; this water-surface interaction is rather localized, concentrating mainly on the underlying water molecules that are directly bonded to the surface.** This picture of water-surface interactions is universally applicable, although the strength of the water-surface interaction varies from one surface to another.

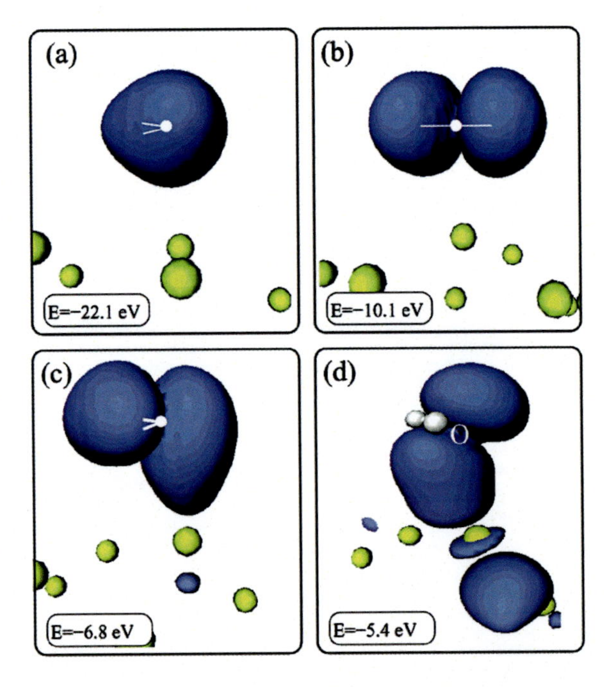

Fig. 8.11 Spatial distribution of each electronic orbital upon adsorption of a single H_2O molecule on Ru(0001) [16]

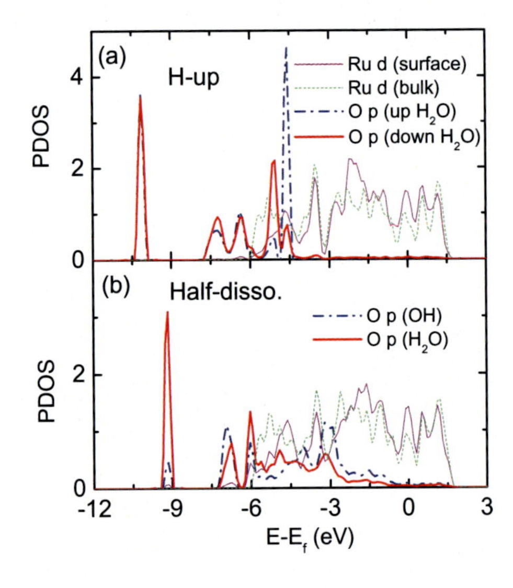

Fig. 8.12 Density of electronic states of **a** H_2O H-up bilayer and **b** semi decomposition structure projected onto the d orbital of surface and bulk Ru atom and the p orbital of O on Ru (0001)

The formation of chemical bonds between water and the surface is often accompanied by charge transfer. The variation of the work function caused by water adsorption on the Pt(111) surface is shown in Fig. 8.13. Depending on the protocell we used, single H_2O molecules, $(H_2O)_2$ and bilayer adsorption correspond to 1/9, 2/9, and 2/3 ML coverage, respectively (1 ML corresponds to one water molecule adsorbed on

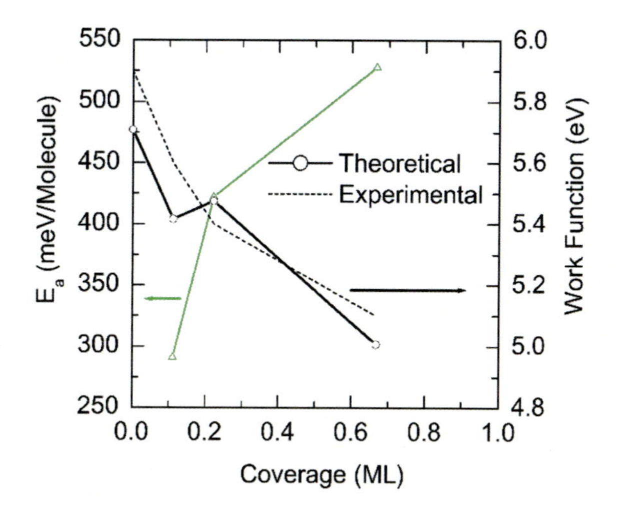

Fig. 8.13 Variation of Pt(111) work function with coverage due to water sorption

each surface atom). Where the bilayer structure results from averaging over both H-up and H-down configurations. For comparison, the adsorption energy is also plotted in the figure. The adsorption of water causes the surface work function to decrease from 5.8 eV at the bare surface to 5.0 eV at the bilayer. A monotonically decreasing of the work function with increasing water coverage was also found experimentally and it decreased by 0.7–0.8 eV at one bilayer [17]. The decrease in the work function directly implies the transfer of electrons from the water to the surface, which is consistent with the charge difference density distribution in Fig. 8.9.

8.4 Surface-Induced Hydrogen Bond Enhancement?

It is very interesting to study how hydrogen bonding is affected by surface adsorption. According to Pauling's principle, it is commonly assumed that the formation of hydrogen bonds between water molecules weakens the interaction of individual molecules with the substrate; similarly, bonding to a surface causes hydrogen bonds between water molecules to weaken. But this image has never been carefully examined in a hydrogen-bonded adsorption system. In principle, hydrogen bonding at the surface and the interaction of water with the surface are confounded and difficult to separate strictly. But a qualitative image of hydrogen bonds at the surface is still interesting and necessary. By carefully considering the energy changes and charge transfer in the adsorption of water structures on the densely packed surfaces of transition and noble metals, we conclude that the adsorption of water at the surface usually results in enhanced hydrogen bonds.

The primary question is what factors affect the strength of the hydrogen bonds at the surface? The two most important factors are (1) the underlying water molecule–metal surface coupling and (2) the formation of hydrogen bonds between the water molecules at the surface and the surrounding water molecules, where the number and configuration of the hydrogen bonds cause automatic modulation of the strength of hydrogen bonds. Both of these are responsible for the formation of stronger hydrogen bonds at the surface.

The simplest hydrogen bonding system on the surface is the double water molecule $(H_2O)_2$. Calculations of $(H_2O)_2$ adsorption on surfaces have shown that the hydrogen bonds in adsorbed $(H_2O)_2$ are much stronger than those in free $(H_2O)_2$. As in Table 8.2, the adsorption energy of a single molecule on Pt(111) is 304 meV, while the bimolecular adsorption energy is 433 meV/H_2O. As in Sect. 7.2, the hydrogen bonding energy in $(H_2O)_2$ can be estimated to be between 258 and 562 meV. Careful calculations (the adsorption energies of a single H_2O molecule fixed on the corresponding adsorption configuration in the $(H_2O)_2$ molecule are 282 and 134 meV, respectively) indicate that this hydrogen bond energy is roughly $433 \times 2 - 282 - 134 = 450$ meV, much stronger than in free (H_2O_2) (250 meV)! The deformation charge density distribution (Fig. 7.15b) also shows a stronger redistribution of charge than that in the free case (Fig. 7.15a).

This enhanced hydrogen bonding due to adsorption originates from the charge transfer between water and the surface. As shown in Fig. 8.9b, the lone pair of electrons of O couples to the surface d-band, resulting in the transfer of electrons from O to Pt. This is immediately followed by O capturing more electrons from H through OH bonding to compensate for the loss of lone pair of electrons, which leads to (1) an increase in the electropositivity of H and (2) a weakening of the binding of O to H. Both of these lead to stronger hydrogen bonding of that H atom to the adjacent water molecule. This is how surface adsorption leads to enhanced hydrogen bonding. Similarly, we can speculate that if the water molecule receiving the hydrogen bond couples to the surface, i.e., forms a strong O–M bond, this hydrogen bond will not be enhanced, but rather weakened. Calculations for such an $(H_2O)_2$ configuration show that the adsorption energy is 230 meV/H_2O and the hydrogen bonding is about 40–160 meV, which is weaker than the free case. But the most stable conformation naturally selected for surface adsorption tends to be the one with increased hydrogen bonding, as in Fig. 7.3b.

This image holds for small clusters and bilayer and multilayer structures of water as well. In the six-molecule $(H_2O)_6$ adsorbed configuration, the lowest water molecule is the most strongly coupled to the surface, and the two hydrogen bonds formed by its supply of H are enhanced by adsorption. Similarly, in bilayer and multilayer structures the hydrogen bonds formed by the lower water molecule providing hydrogen are enhanced by adsorption.

However, adsorption also automatically adjusts the hydrogen bond strength in small clusters and lamellar structures at the surface due to environmental changes caused by adsorption. For example, the strength of hydrogen bonds in the water structure is adjusted because of the presence of some suspended (i.e., not hydrogen-bonded) OH groups. In general, if only one OH forms a hydrogen bond in an H_2O and the other OH is suspended, that hydrogen is stronger than the two hydrogen bonds formed by providing two H's. Similarly, an H_2O in which it accepts two H to form hydrogen bond is stronger than by accepting only one H. In short, a water molecule (1) provides the less H to form hydrogen bonds, and (2) the more hydrogen bonds it accepts, the stronger those hydrogen bonds it forms by providing H. Again, this can be explained in terms of the charge transfer caused by hydrogen bonding. The H that forms a hydrogen bond gains electrons and the O loses electrons. The less H the water molecule provides, the few electrons the O in the molecule gains through the H and the less H is bound to it; the more hydrogen bonds it receives, the more electrons the O loses through the hydrogen bonds and the less H is bound to that molecule. This all makes the hydrogen bond formed by that molecule providing H stronger. This is most evident in the water bilayer structure and the uppermost bilayer of the multilayer structure. In the bilayer, the upper H_2O molecule provides one H and accepts two hydrogen bonds; the lower H_2O molecule provides two H and accepts one hydrogen bond. So the one hydrogen bond formed by the upper H_2O molecule providing H is stronger than both hydrogen bonds formed by the lower H_2O molecule providing H. They are namely the strong and weak hydrogen bonds in the bilayer structure, respectively. The difference in their strength can also be seen distinctly in the charge transfer diagram 5.16 and the vibrational spectrum diagram 5.11.

Similarly, the bond length (OH...O distance), the bond angle (∠OH...O), and the angle between the two molecular planes have a significant effect on the absolute strength of the hydrogen bond when the water structure automatically regulates the hydrogen bond. Roughly speaking, the closer the bond length is to 2.6–2.7 Å, the closer the bond angle is to ~180°, and the closer the molecular planes are to vertical, the stronger the hydrogen bond is.

We find that the disorder of H pointing at the surface has little effect on the hydrogen bond strength. In the case of $\sqrt{3} \times \sqrt{3}R\ 30°$ two bilayer structures grown on the surface, the second bilayer has two types of hydrogen bond pointing relative to the first bilayer underneath. This is shown in Fig. 8.14. The hydrogen bonds formed between the upper and lower bilayers in this way have different H pointing of the host H_2O molecule (on top), and that hydrogen bond is two different H pointing hydrogen bonds in the two structures. The adsorption energies of these two structures differ by 2 meV/H_2O and the OH vibrational frequencies involved in the hydrogen bonds differ by 2 meV, so the two hydrogen bonds can be considered to be at the same strength. This is different from the view of Li et al. that the hydrogen bonds of these two configurations in the ice of the bulk material are of different strengths, perhaps

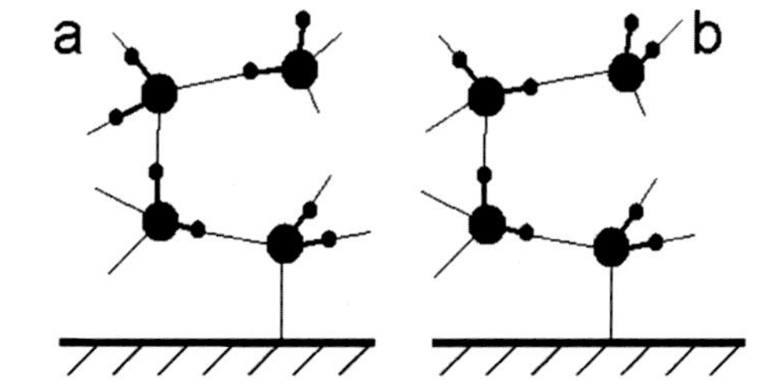

Fig. 8.14 Two $\sqrt{3} \times \sqrt{3}R\ 30°$ water bilayers grow on the surface, and the second bilayer forms two different H pointers **a**, **b**. The thin lines indicate O-M bonds or hydrogen bonds

indicating that the origin of the two peak positions (27 meV and 36 meV) in the neutron scattering experiment requires a new interpretation [18].

8.5 Microscopic Criteria for Surface Wetting

Hydrophilicity and hydrophobicity are one of the most common everyday phenomena. When water comes in contact with a surface, some surfaces adhere to water and are called hydrophilic; others repel water and are called hydrophobic surfaces. Both hydrophilicity and hydrophobicity play important roles in life, production, and life activities such as lubrication, cleaning, DNA synthesis, and protein folding, to name a few. Macroscopically, the difference in hydrophilicity can be expressed in terms of the infiltration angle θ [19], which is the angle made by the tangent surface of a droplet to the surface (Fig. 8.15). $\theta < 90°$ is a hydrophilic surface and $\theta > 90°$ is a hydrophobic surface. However, this image is macroscopic and suitable only for liquids. How can we understand the difference between surface hydrophilicity and hydrophobicity on a molecular scale?

Recent low-temperature desorption experiments [20, 21] have given us useful insights. The dynamics of desorption of thin ice layers and clusters on several surfaces were carefully measured in the experiments, and comparing their different performances revealed that the order of hydrophilicity of the surfaces was Pt(111) > graphite > Au(111) (Fig. 8.16). In particular, Pt(111) is a hydrophilic surface, while Au(111) is a hydrophobic surface. This reminds us that hydrophilic phenomena at surfaces can be studied microscopically from energetic, in particular by considering the involved atomic and electronic interactions.

First, we took two surfaces, Pt(111) and Au(111), for a comparative study. Table 8.4 lists the adsorption and hydrogen bonding energies for various structures on Pt, Au. Where the water structure and energy on Pt(111) surface is the same as in Table 7.5. The hydrogen bonding energy is defined as follows.

$$E_{\mathrm{HB}} = \begin{cases} \left(E_{\mathrm{ads}} \times n - E_{\mathrm{ads}}[\mathrm{monomer}] \times N_{\mathrm{M-H_2O}}\right)/N_{\mathrm{HB}} & \text{for clusters and 1BL} \\ \left(E_{\mathrm{ads}}[m\ \mathrm{BL}] \times 2m - E_{\mathrm{ads}}[(m-1)\ \mathrm{BL}] \times 2(m-1)\right)/4 & \text{for } m \text{ BL}, m > 1 \end{cases} \tag{8.1}$$

where E_{ads} is the adsorption energy of each structure and $N_{\mathrm{M\text{-}H2O}}$, N_{HB} is the number of surface-water bonding and hydrogen bonding in the protocell, respectively.

Fig. 8.15 Schematic diagram of hydrophilic hydrophobic interaction and infiltration angle

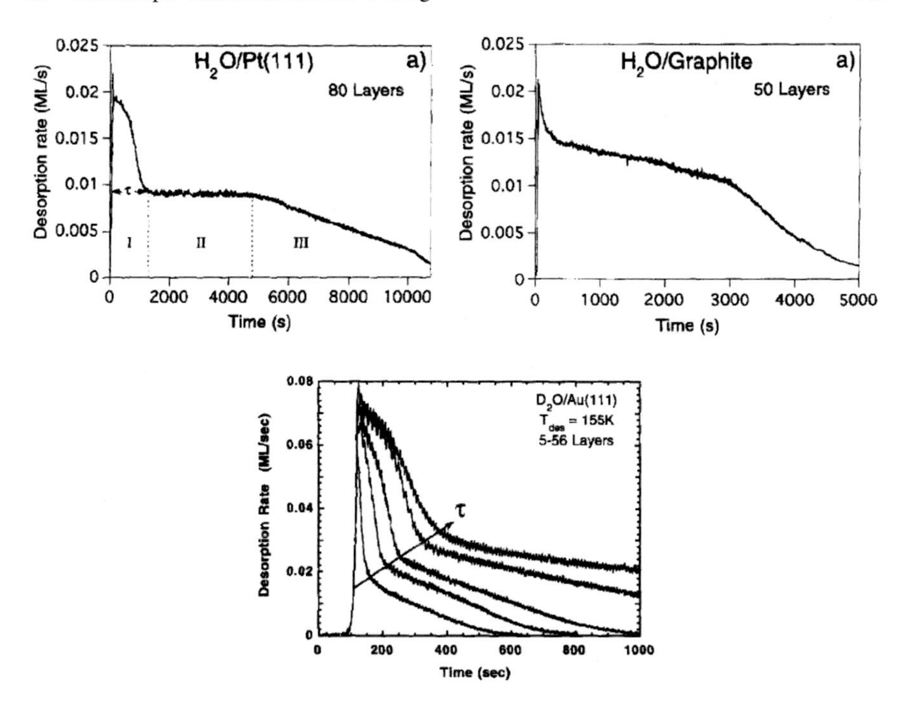

Fig. 8.16 Relationship between the rate and time of ice desorption on the surfaces of Pt, graphite and Au at low temperature, showing the order of hydrophilicity: Pt > graphite > Au. Collected from the literature [20] and [21]

Table 8.4 Comparison of adsorption energy (E_{ads}) and hydrogen bonding energy (E_{HB}) for various adsorption structures on Pt(111) and Au(111)

Type	Primitive cell ce	n	E_{ads}(Pt)	E_{ads}(Au)	N_{H2O-M}	N_{HB}	E_{HB}(Pt)	E_{HB}(Au)
H_2O	3×3	1	304	105	1	0	–	–
$(H_2O)_2$	3×3	2	433	259	2	1	258	308
$(H_2O)_3$	3×3	3	359	283	3	3	55	178
$(H_2O)_6$	$2\sqrt{3} \times 2\sqrt{3}$	6	520	402	3	6	368	350
1 BL	$\sqrt{3} \times \sqrt{3}$	2	505/527	437/454	1	3	235	256
2 BL	$\sqrt{3} \times \sqrt{3}$	4	564	489	1	7	312	271
3 BL	$\sqrt{3} \times \sqrt{3}$	6	579	508	1	11	303	271
4 BL	$\sqrt{3} \times \sqrt{3}$	8	588	520	1	15	307	279
5 BL	$\sqrt{3} \times \sqrt{3}$	10	593	532	1	19	307	290
6 BL	$\sqrt{3} \times \sqrt{3}$	12	601	545	1	23	320	305

The number of molecules (n) and the number of bonding (N) in the primary cell are also listed together. Energy units: meV

From the table: (1) The adsorption energy of a single molecule on Pt(111) (304 meV), is almost three times higher than that on Au(111) (105 meV). This indicates that the adsorption of Pt surface with water is stronger than that of Au. In addition, the adsorption energy on the Pt surface is 50 to 200 meV higher at various coverages in the table. (2) The adsorption energy varies considerably in small cluster structures due to large changes in structure and coordination number. However, the adsorption energy grows smoothly in the thin layer structures from 1 to 6 BL.

From the electronic images, these differences in energy arise from differences in electronic structure. Figure 8.17 plots the total and deformation charge density distributions for individual water molecules adsorbed on Pt(111) (a, b) and Au(111) (c, d). The horizontal axis is the [110] direction and the vertical axis is the surface normal direction. The polarized charges on Pt(111) show the characteristics of d_{xz} and d_z^2 (Fig. 8.17b), but on the Au(111) surface is the characteristic of $s + p_z$ (Fig. 8.17d). Moreover, the charge transfer due to the adsorption of water on the Pt(111) surface is much stronger than on Au(111). This is because the 5d electrons of Au are already occupied and located 3–10 eV below the Fermi energy level and are not involved in the surface interaction; whereas there is an ample d-band near the Fermi energy level on Pt(111). Thus, the basic difference between Pt(111) and Au(111) when water is adsorbed on their surfaces is the involvement or non-involvement of d electrons in the adsorption. The same image holds for small clusters and thin layers of water because, as mentioned earlier, the water-surface interaction is mainly limited to the interaction of the underlying molecule with the metal, similar to the single-molecule adsorption scenario. Figure 8.18 further shows the two-dimensional planar distribution of deformation charge (a, b) and the one-dimensional distribution of deformation charge density (c) along the normal for the five bilayers during both types of surface adsorption. We see that in addition to the predominantly localized effect of charge transfer, there is a long-range polarization effect in the thin layer. This long-range polarization comes mainly from the influence of the surface potential field.

Let us now discuss surface hydrophilicity and hydrophobicity from energy [22]. In general, the hydrophilic and hydrophobic phenomenon that manifests itself on the macroscopic scale through structure and geometry depends on two energetic parameters on the microscopic scale: the strength of the coupling between water and the surface, and the strength of the hydrogen bonding between the water molecules. These two parameters determine the structural configuration and stability of water at the interface, i.e.: is it more water-metal bonds or more hydrogen bonds that are formed to obtain the most stable structure? Water molecules are more likely to be adsorbed on clean surfaces where the strength of the coupling with water is greater than the strength of the hydrogen bonds, rather than on the ice layer at the surface, which exhibits hydrophilic properties. Conversely, surfaces with weaker coupling to water exhibit hydrophobic properties. Unfortunately, however, the two interactions are mixed and both contribute to the adsorption energy and cannot be strictly distinguished. Moreover, the changes in adsorption energy in Table 8.4 mainly reflect changes in the number of bonding and coordination numbers, rather than directly reflecting the strength of the two interactions, so they cannot be used to directly

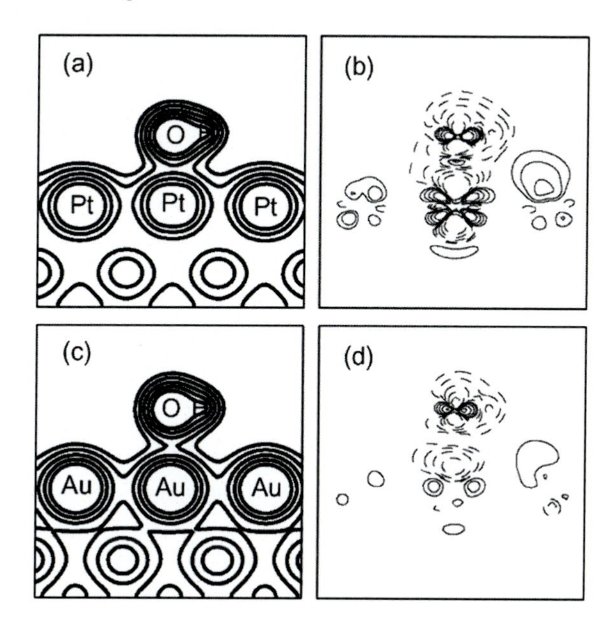

Fig. 8.17 Total and differential charge densities of single water molecules adsorbed on Pt **a**, **b** and Au **c**, **d**. The differential electron density is defined as $\Delta\rho = \rho[H_2O/Metal] - \rho[H_2O] - \rho[Metal]$. The isodensity line in the figure is taken as $\rho = 0.1 \times 2^k e/Å^3$, $\Delta\rho = \pm 0.005 \times 2^k e/Å^3$, where $k = 0, 1, 2, 3, 4$. The solid and dashed lines correspond to the cases $\Delta\rho > 0$ and $\Delta\rho < 0$, respectively

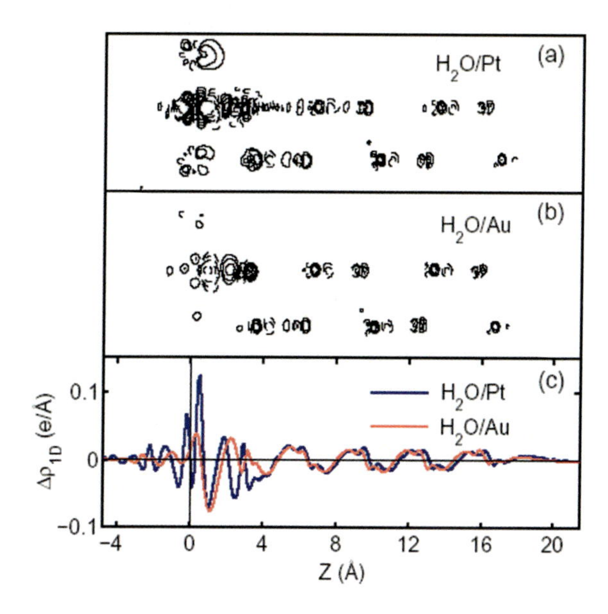

Fig. 8.18 Polarized charge distributions upon adsorption of five ice bilayer films on Pt **a** and Au **b**. The one-dimensional deformation charge density along the normal is shown in **c**. The isodensity line in the figure is taken in the same way as in Fig. 8.17

describe surface hydrophilic and hydrophobicity. In order to distinguish the relative magnitudes of the two interactions, appropriate simplifications and approximations are necessary. We used Eq. (8.1) to estimate the strength of hydrogen bonds in the surface adsorption structure, where the adsorption energy of a single water molecule is used as a universal parameter representing the strength of the water-metal bond.

This is because it describes the strength of water-surface coupling well even in clusters and bilayer structures. [This can be confirmed by the following step: if we keep the position of one H_2O molecule in the bilayer structure constant while removing the other molecule, the two adsorption energies obtained add up to 270 meV (H-up bilayer) and 340 meV (H-down bilayer), which is close to the adsorption energy of one H_2O molecule (304 meV). The same conclusion holds for $(H_2O)_2$]. The hydrogen bonds defined in Eq. (8.1) reflect the average strength of hydrogen bonds in an adsorbed water structure. In particular, in a multilayer structure, it reflects the strength of hydrogen bonding in the uppermost bilayer.

The hydrogen bond energy (E_{HB}) from Table 8.4 is plotted as a function of coverage in Fig. 8.19. The coverage is defined as the ratio of the number of H_2O molecules in the protocell to the number of atoms on the surface. 1 ML implies 1×1 surface coverage on Pt(111) or Au(111), and the value is multiplied by 3/2 to be the coverage in the bilayer (BL) and by 3 to be the coverage in water layer. The coverage of small clusters is qualitative, and it stems from the fact that we use periodic protocells in our calculations to describe these zero-coverage regimes. However, these do not affect the physical discussion that follows. As seen in the figure, the hydrogen bond strength varies considerably in the small clusters, which is due to the drastic changes in the structure of the small clusters. The hydrogen bonding energy increases steadily from the first bilayer to six bilayers to 320 meV (Pt) and 305 meV (Au). This is basically consistent with the hydrogen bond strength in bulk ice (315 meV), implying that the influence of the metal surface is minimal at this point. Comparing Figs. 8.18 and 8.19, we find that the hydrogen bond energy is closely related to the electronic structure at the interface. The rather localized charge transfer on Pt causes a jump in the hydrogen bond energy E_{HB} at a coverage equal to 4/3 ML (2 BL), and a flat region immediately thereafter (4/3–4 ML). But on Au(111) because charge transfer is rare, long-range polarization becomes more important, resulting in a slow increase in E_{HB} in the range 2/3 to 4 ML. Even at a coverage of 4 ML (6 BL), the hydrogen bonding energy of the water layer on Au is still lower than that of the water layer on Pt. This indicates that even for this thickness (~20 Å) of the thin ice layer, the role of the surface is still not completely negligible. Also, the lattice mismatch between the ice layer and the surface can affect this difference. The ice (Ice Ih) surface water molecule nearest-neighbor distance is 2.61 Å, while the Pt(111) surface lattice constant is 2.82 Å (8% difference) and the Au(111) lattice constant is 2.95 Å (13%). The better lattice match on Pt may lead to a larger hydrogen bond energy.

We define a quantity that reflects the hydrophilic and hydrophobic ability of the surface, $\omega = E_{HB}/E_{ads}$, as the ratio of the hydrogen bonding energy to the (single molecule) adsorption energy. As mentioned earlier, the latter better represents the strength of surface binding to water molecules in aqueous structures. Qualitatively, $\omega = 1$ is a rough dividing line between the hydrophilic and hydrophobic regions. We see that although ω is oscillatory in small clusters, the curve corresponding to H_2O/Au (dashed line) still lies in the region where $\omega \gg 1$, indicating that Au(111) is hydrophobic. However, the curve corresponding to H_2O/Pt (solid line) lies in the region where $\omega \leq 1$, indicating that Pt(111) is hydrophilic. This is consistent with the

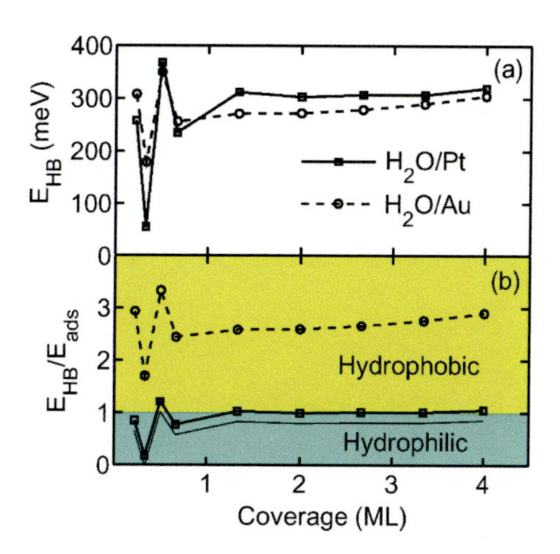

Fig. 8.19 Variation of **a** hydrogen bond energy (E_{HB}) and **b** infiltration capacity (E_{HB}/E_{ads}) with water coverage

experimental understanding. If one considers the zero-point energy correction, i.e., subtracting the zero-point energy of 60 meV for one hydrogen bond in the ice, the curve corresponding to Pt (thin solid line) falls exactly in the region of $\omega < 1$. (The zero-point energy correction does not change the character of the curve $\omega \gg 1$ for H_2O/Au.) The large gap between these two curves indicates a significant difference in the hydrophilicity of the two surfaces.

Extending this study, we also calculated ω in a bilayer ice for Ru(0001), Rh(111), and Pd(111) surfaces, as shown in the upper panel of Fig. 8.2. The figure shows $\omega_{Ru\leq}$ $\omega_{Rh} < \omega_{Pd} \leq \omega_{Pt} \leq \omega_{Au}$, indicating that the order of surface infiltration capacity is Ru > Rh > Pd > Pt > Au. Ru, Rh, Pd, Pt surfaces are hydrophilic surfaces, while Au is a hydrophobic surface. Such a difference stems mainly from the difference in the adsorption energy of single molecules on different surfaces, since the variation of hydrogen bonding energy is limited. Such results suggest a general conclusion: microscopically, the surface wetting ability (hydrophilicity) is determined by the competition between water-surface adsorption and hydrogen bonding, as illustrated in Fig. 8.20, where the surface behaves hydrophilically when the water-surface interaction is stronger than the hydrogen bonding, and the stronger the water-surface interaction the more hydrophilic the surface is; conversely, the surface is hydrophobic when the water-surface interaction is weaker than the hydrogen bonding. In our example, water-surface adsorption on surfaces such as Pt and Rh wins the competition and is a hydrophilic surface, while hydrogen bonding between water molecules on Au(111) wins and exhibits hydrophobicity. This also suggests that the hydrophilicity is directly related to the magnitude of the adsorption energy of a single molecule on the surface. Recent Monte Carlo simulations based on model potentials [23] have shown that the contact angle of water droplets on the graphite surface is linearly related to the pre-determined single-molecule adsorption energy. This confirms our conclusions obtained using first-principles calculations with "no artificial parameters".

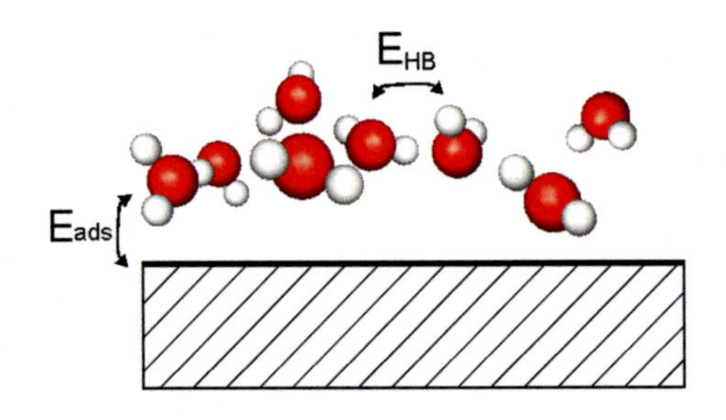

Fig. 8.20 Hydrophilicity and hydrophobicity is determined by competition between water-surface adsorption and hydrogen-bonding interactions

8.6 Vibration, Transformation, Decomposition, and Diffusion of Water Structures on Surfaces

In this section, we will discuss various interesting dynamic processes of water on the surface. Whether in real-life production or in surface science experiments in ultra-high vacuum cavities at low temperatures, water molecules actually "move", vibrate, rotate, diffuse, decompose, and transform between different structures of water. This provides the stage and the challenge for a deeper understanding of the interaction between water and surfaces at the molecular scale.

8.6.1 Vibrations of Water on the Surface

Like on Pt(111) surfaces, water molecules in small clusters and bilayer structures have various modes of intermolecular vibrations, molecular rotations and translations, similar to the situation in Fig. 8.10. We have simulated the vibrational processes of water on Ru, Rh or other surfaces via first-principles molecular dynamics and obtained their vibrational spectra. The vibrational spectra can be directly compared with experimental data and help us to identify the molecular scale structure at the surface and interface.

The vibrational spectra calculated for each adsorption structure are listed in Table 8.5: including unimolecular, bimolecular, and bilayers (H-up and H-down) on Pt(111); bilayers (H-up and H-down) on Pd(111), Rh(111), and Au(111); as well as H-up, H-down, and semi-decomposed bilayers on Ru(0001). Usually, these spectral peaks can be divided into 3 regions: (1) the low-energy region below 120 meV, which mainly corresponds to the rotation and translation of water molecules; (2) near 200 meV, which corresponds to HOH shear motion; and (3) the high-energy region from 350 to 470 meV, which corresponds to OH stretching vibrations. Compared with the available experiments, the calculated vibrational spectra of $\sqrt{3} \times \sqrt{3}R\ 30°$ ($\sqrt{3}$) bilayer ice on Pt(111) and on Au(111) are in good agreement with the HREELS, HAS experiments, indicating that the configuration found by the calculations is the experimental one. On the Ru(0001) surface, the vibrational spectrum of H-up appears to be in the best agreement with the experiment, implying that the experiment may

Table 8.5 Vibration energy of water structure adsorbed on dense metal surface (unit: meV)

Surfaces	Bilayers	Translation and rotation						δ_{HOH}	$\nu_{O\text{-}Hb}$	$\nu_{O\text{-}H}$
Ru(0001)	H-up	34	40	50	67	87	119	200	378,424	462
	H-down	20	48	61	73	89	111,129	196	347,440	440
	Half-disso	20	32	53	77		117,129	186,196	300–380,428	
	Expt.[a]		48		69	87	114	189	364,422,442	457
	Expt.[b]								384,427	457
Rh(111)	H-up	18	44		61	89	111,129	198	349,422	466
	H-down	20	44		75	89	133	200	347,420	440
Pd(111)	H-up	14	40	53	67	89	109,117	198	374,424	466
	H-down	20	42	57	71	89	111,123	202	380,426	444
Pt(111)	Monomer	16		40	61	89	113,121	190		440
	Dimer	20	32	44	65	85	105,133	198	347	432,452
	H-up	18	32	53	69	87	107,119	198	388,432	467
	H-down	16	34	57	69	91	111,119	196,202	384,424	438
	Expt.[c]	17	33	54	65	84	115,129	201	424	455
Au(111)	H-up	17	36				108	201	400,444	466
	H-down	18	36		77		105	202	402,436	468
	Expt.[d]		31				104	205	409	(452)[e]

Notes [a]Literature [19]. [b]Literature [6]. To reflect H_2O, the vibrational energy of D_2O has been multiplied by an isotopic factor of 1.35. [c]literature [20].[d]Literature [21]. [e]Literature [22]. Data on Ag(111)

not be the structure proposed by Feibelman for the semi-decomposition. This is discussed in more detail in Sect. 8.6.4.

From the calculated vibrational spectrum data, it is possible to estimate the effect of the zero-point motion of the atoms on the adsorption energy. For the first bilayer on Pt(111), the estimated zero-point vibrational energy ($E_{zp} = \sum_i \hbar\varpi_i/2$) is about 90 meV. compared to 120 meV in the bulk ice Ice Ih [28], so the zero-point energy correction would make the bilayer more stable by 30 meV (relative to the bulk ice). The zero-point energy contribution can be similarly estimated for different water structures on other surfaces, mostly in the range of 40–100 meV.

8.6.2 Transformation of the Water Structure on the Surface

As mentioned earlier, water has an abundance of structures on its surface. These structures are somewhat stable, and they may transform into other structures when conditions change. For example, at very low temperatures (<20 K) and low coverage, water molecules can adsorb as single molecules on metal surfaces [7, 29–31]. At slightly higher temperatures, these isolated water molecules cluster by diffusion to form small clusters of bimolecules, trimolecules, tetramolecules, pentamolecules, hexamolecules, etc. [32, 33]. Six-molecule clusters already have the characteristics of folded hexagonal rings in bulk ice [34]. Further increases in temperature and coverage result in the formation of $\sqrt{3}$ bilayer Iceland, or even thin ice structures with one to hundreds of bilayers [35, 36].

The double layer of $\sqrt{3}$ has at least two structures, H-up and H-down, if the pointing of the H atom is taken into account. Both of them may exist in practice. On Pt(111), the calculated energy barrier for the transition from H-up to H-down is 76 meV. So the interconversion of these two structures may occur easily due to the influence of the thermodynamic fluctuation, the surface potential, or the STM probe. Also, when growing a multilayer, the H in the lowermost bilayer pointing downward toward the surface may also be converted to point upward in order to efficiently connect the upper and lower bilayers. Very interestingly, if we can effectively control the conversion of both H-up and H-down states and can effectively maintain them, is this really a recording instrument based on individual small molecules? This is because the H pointing of H-up and H-down can correspond to the '1' and '0' states in binary, respectively. Let us venture to imagine that if one day we obtain some technology to facilitate reading, writing and computing of the data programmed from H-up and H-down, it will be a real molecular computer—an ice computer!

There are two other bilayer structures of $\sqrt{39} \times \sqrt{39}R\ 16.1°$ ($\sqrt{39}$) and $\sqrt{37} \times \sqrt{37}R\ 25.3°$ ($\sqrt{39}$) on Pt(111). They are grown at a somewhat higher temperature (135 K) [37, 38]. Our calculations indicate that roughly at 3 BL the bilayer of $\sqrt{39}$ converts to the structure of $\sqrt{3}$. This transformation was also found to occur experimentally with electron beam irradiation or growth to 5 BL [38].

A further increase in temperature (150 K) will cause the multilayers to desorb, leaving only a single bilayer on the Pt(111) surface at around 170 K [39, 40]. A further increase in temperature would also desorbed the water bilayer, leaving the surface possibly with single molecules and small clusters, and they are likely to be mostly present near the location of surface defects. A one-dimensional water chain structure may form on the steps [41].

Some molecules may decompose during the transformation of the water structure. On Ru(0001), the entire lamellar structure may also transform into a semi-decomposed layer structure. Our calculations indicate that the potential barrier for this transformation is not very small, at 0.62 eV. A higher potential barrier in low-temperature experiments may prevent this transformation from occurring.

8.6.3 Proton Transport at the Surface

It was mentioned earlier that a small number of water molecules break down in the $\sqrt{39}$ structure on Pt(111), generating H_3O^+ and OH^--like molecules. We found that in the first $\sqrt{39}$ bilayer protocell, three of the 32 H_2O molecules (9%) decomposed. All of the decomposed water molecules came from the lower water molecules that interacted directly with the surface, forming OH^- 2.1 Å away from the surface, and H_3O^+ in the upper layer. As the membrane thickens, fewer water molecules decompose. In the structure of two $\sqrt{39}$ bilayers, only 2 out of 64 molecules (3%) decomposed. And in three bilayers, only 1 molecule (1%) out of 96 H_2O decomposed. The decomposed H_2O molecules do not appear in the structures of $\sqrt{37}$ and $\sqrt{3}$ 3. The small number of H_2O molecules decomposing in $\sqrt{39}$ originates from the effect of surface interaction and the periodic compression to which the ice structure is subjected in the horizontal direction [42, 43].

The decomposition of a small number of water molecules in the ice phase $\sqrt{39}$ into $H_3O^+$and OH^- ions results in an abundant proton transport process at the surface. Proton transport at the surface is different from that in bulk water because water molecules in the bulk state are connected to four near-neighbor molecules by hydrogen bonds [44, 45], whereas each H_2O molecule at the surface is connected to only three near-neighbor molecules by hydrogen bonds. Through molecular dynamics simulations, we observed four typical proton transport processes in the $\sqrt{39}$ bilayer structure at 130 K [46]: (1) transport of H_3O^+; (2) transport of OH^- ions; (3) binding of H_3O^+ and OH^- ions; and (4) non-stop hopping of H between the two molecules. All these processes originate from the transfer of one H from one H_2O, H_3O^+ or OH^- molecule to another along the direction of hydrogen bond.

Figure 8.21 records the variation of OH bond lengths with time in the simulation. The top side shows the dynamics of the transport of H_3O^+. At the beginning, $H_3O_A{}^+$ and H_2O_B are two near-neighbor molecules in the $\sqrt{39}$ ice bilayer. The proton (H_A) that they share through hydrogen bond jumps from a position close to O_A to a position close to O_B at 0.5 ps. Thus $H_3O_B{}^+$ and H_2O_A are generated, which corresponds to the transport of the hydrated ion H_3O^+ from the A position to the B position. But this newly generated $H_3O_B{}^+$ is in a sub-stable state. It then releases one of its own protons (H_B) to the near-neighbor O_CH^- molecule at the 0.8 ps moment (lower panel of Fig. 8.21). Thus the $H_3O_B{}^+$ and O_CH^- molecules are synthesized into two H_2O molecules. We also observe that H is constantly jumping between the two near-neighbor O atoms, causing the water molecule to break up into H_3O^+ or OH^- molecules at one moment and then recombine again at another (Fig. 8.22). From these processes we find that (1) when the proton transport process occurs, the distance between the two near-neighbor O is typically 2.5 Å, in agreement with other calculations [44, 45] and experiments [47, 48]. (2) The proton transport process is fast, typically completed in 0.2 to 0.5 ps. This can be confirmed by the very small transport potential (~40 meV) found in our molecular dynamics simulations. This potential barrier is even lower if the quantum motion process of H is taken into account.

Figure 8.23 reflects the process of OH^- transport: the transfer of H from an H_2O molecule to an OH^-. We find that the surface is involved in this process. The variation of the distance between Pt–O and O–H in two molecules (initially H_2O_A and O_BH^-) with time is shown in the figure. At 1.1 ps, the proton jumps from O_A to O_B, forming a new O_BH bond. At the same time, O_A (now in O_AH^-) has moved closer to the surface and is tightly bound by the surface at an equilibrium distance of 2.1 Å. So the metal surface plays a unique role in proton transport at the surface by binding and supporting these molecules.

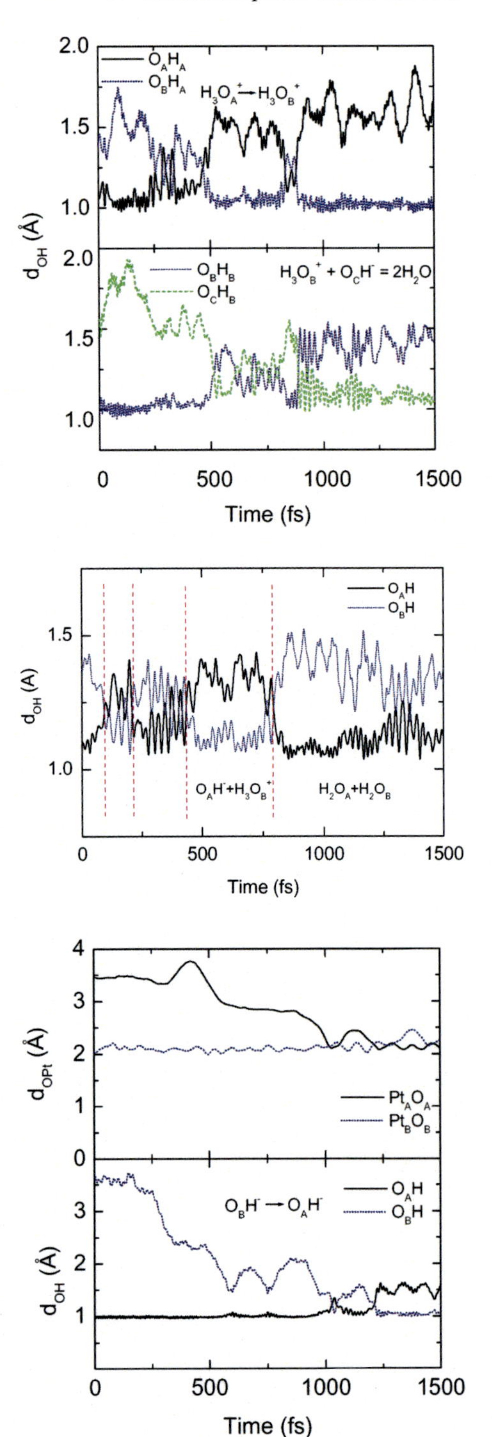

Fig. 8.21 Variation of OH bond length with simulation time, showing the process of H_3O transport and H_2O synthesis

Fig. 8.22 Variation of OH bond length with simulation time, showing the non-stop hopping process of H between two molecules

Fig. 8.23 Variation of OH bond length with simulation time, showing OH transport. The surface is involved in this process

8.6.4 Water Decomposition on Ru(0001)?

Feibelman found that the structure in the ice phase of $\sqrt{3}$ on Ru(0001) with half of the water molecules decomposed has lower energy than the bilayer structure in the molecular state (H-up and H-down) (Fig. 8.24) and the position of the O atoms is in better agreement with the LEED experiment [4]. The distance between the two O atoms normal to the surface, measured in the LEED experiment, is very small (0.1 Å), forming an almost flat D_2O molecular layer. However recent SFG vibrational spectroscopy measurement experiments [6] rejected Feibelman's proposal and the authors concluded that it is a double layer of molecular states on Ru(0001).

We simulated the vibrational spectra of three structures (D-up, D-down, semi-decomposed) on Ru(0001) based on molecular dynamics methods [49], as shown in Fig. 8.25. We find that for the semi-decomposed structure there is no free OH vibrational peak located at 336 meV, while this vibrational peak already appeared when growing a BL layer in the experiment [6]. So the structure in the experiment may not be the semi-decomposed one proposed by Feibelman. A detailed analysis shows that the best agreement with the experimental data is with the D-up bilayer structure over the entire range of vibrational energies.

The reason for the lack of decomposition of water on Ru(0001) may be as follows. The energy potential barrier for the decomposition of the bilayer structure in the molecular state into a semi-decomposed structure that we have calculated is 0.62 eV. However, the adsorption energy of the bilayer structure in the molecular state is 0.53 eV, so although the adsorption energy of the semi-decomposed structure is much lower, they have already desorbed from the surface before the water molecules have reached the decomposed state, resulting in the semi-decomposed water layer structure not being observed experimentally [49]. This process is clearly illustrated in Fig. 8.26.

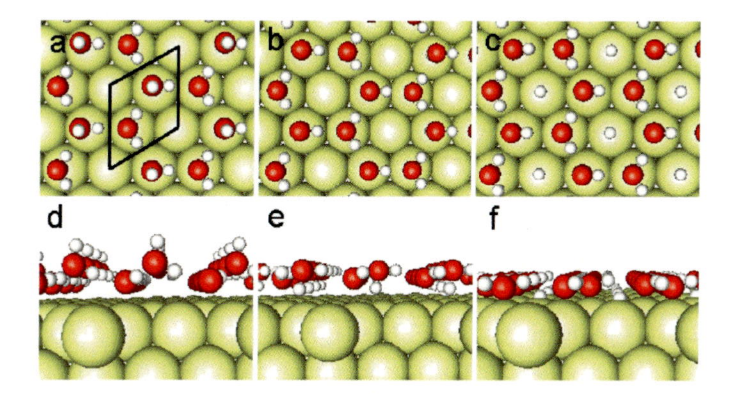

Fig. 8.24 **a** D-up, **b** D-down, and **c** semi-disintegrated D_2O water layer structures (top view) on Ru(0001). **d**, **e**, and **f** are corresponding side views. The proto-cells of $\sqrt{3}$ are also shown in figure

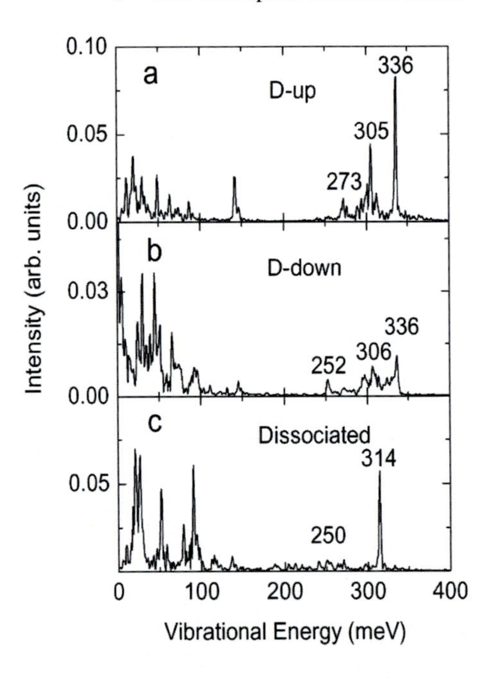

Fig. 8.25 Vibrational spectra of **a** D-up, **b** D-down, and **c** semi-decomposed D_2O water layers on Ru(0001)

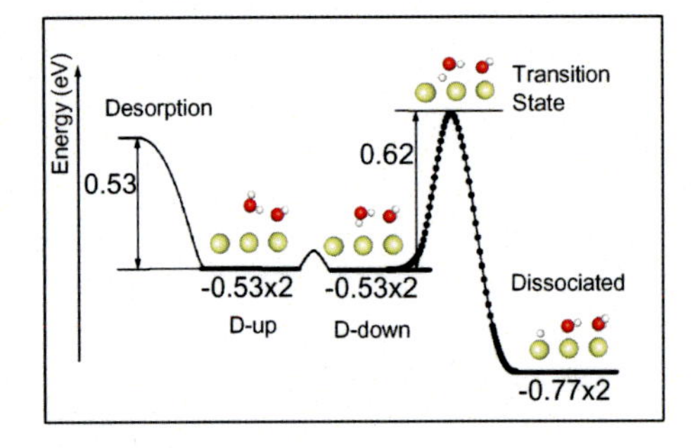

Fig. 8.26 Energy level diagram of the water structure on Ru(0001). The decomposition and desorption potentials have also been labelled in the figure

8.6.5 Diffusion of Water on a Surface

The diffusion of water on a surface is very important and interesting issue. Usually water molecules form each structure by diffusion. It was long thought that the diffusion potential barrier of a single molecule of water on a metal surface is very low (30 meV on Pt(111) [50]), but recent STM experiments have found by measuring the diffusion rate of water on Pd(111) that the diffusion potential barrier of water molecules on Pd(111) is not very low, 126 meV [7]. Our calculation about the diffusion potential class of a single H_2O molecule on Pd(111) is 150 meV, which is slightly larger than the experimental result. The calculations found the path of diffusion (minimum energy pathway, MEP) to be the path through the bridge site, as shown in Fig. 8.27. This indicates that the water molecule diffusion potential can

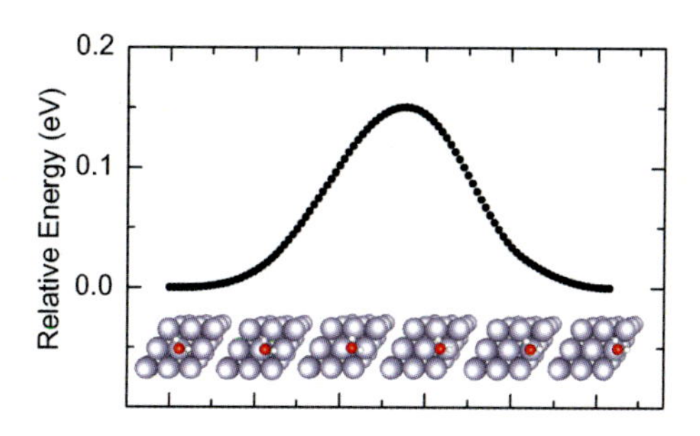

Fig. 8.27 Schematic diagram of water molecule diffusion on Pd(111)

be approximated by the difference in adsorption energy at the top and bridge sites. On Pd(111), this energy difference is: 304 − 146 = 158 meV. Similarly, the single molecule diffusion potential barrier calculated on Ru(0001) is 308 meV, which differs very little from the energy difference of 409 − 92 = 317 meV on the top and bridge positions. In this way, we can also estimate the single-molecule diffusion potential on other surfaces from the energy difference between the top and bridge sites: 282 meV on Rh(111), 168 meV on Pt(111), and 73 meV on Au(111).

Interesting diffusion phenomena were found in STM experiments [7]. The diffusion of water bimolecules in Pd(111) is four orders of magnitude faster than that of single molecules, corresponding to a diffusion potential barrier of about 92 meV. Since the bimolecule $(H_2O)_2$ adsorbs on the surface with one H_2O close to the surface and the other slightly farther away, the diffusion of $(H_2O)_2$ can be divided into two processes, one for the rotation of the water molecules above and the other for the rotation of the water molecules below. For the former process, the calculated potential barrier is very small, about 5 meV. While for the second process, the calculated barrier is not much different from the single-molecule diffusion, 157 meV. That's also the case for the potential barrier for the simultaneous diffusion of both molecules to the next adsorption site. This is not consistent with the experimental estimate. We consider that if the upper and lower molecules can easily exchange positions relative to the surface, then the bimolecule may diffuse through the following path: first the upper molecule rotates, then the upper and lower exchange positions, then the upper molecule (i.e., the original lower molecule) rotates. However, the calculated exchange potential for the upper and lower molecules is 220 meV, which is higher than the potential for the direct rotation of the lower molecule. This inconsistency between experiment and theory may require us to take into account the effects of other physical processes, like the quantum motion of H, when dealing with the diffusion of double water molecules using density functional theory.

8.7 Water Splitting Induced by Plasmon

Plasmonic metal clusters such as Au, Ag, Cu, and Al can concentrate and channel the energy of solar light into the absorbates after plasmon excitation, which have been prevalently utilized in chemical and solar energy conversion, especially in plasmon-driven photocatalysis including O_2 dissociation [51] and H_2O splitting [52].

8.7.1 *Plasmon-Induced Water Splitting on Au/TiO$_2$ Nanoparticles*

Plasmonic metal nanostructures combining with oxide semiconductors as a co-catalyst for water splitting have gained increasing attention because of their enormous light harnessing capability and easy tunability of plasmon excitations [53]. One typical example is the titania-supported gold nanoparticles (Au/TiO_2 NPs) [54]. Upon the excitation wavelength of gold plasmon band, hot electrons with enough energy to overcome the Schottky barrier can be injected from gold NPs to TiO_2, leading to relatively high catalytic performance. On the other hand, the direct injection of hot electrons from plasmonic gold NPs to water molecules, driving water splitting in a Schottky-free junction, was also observed [55]. High catalytic activity of plasmonic metal clusters is mainly attributed to: (i) field enhancement (FE) induced by elevated electric field near the nanostructure [56]; and (ii) electron transfer to foreign molecules by nonradiative plasmon decay [57].

The golden nanocluster Au_{20}, as shown in Fig. 8.28, has a large electronic energy gap of 1.77 eV and a unique tetrahedral structure. High surface area provides more adsorption sites for water molecules, indicating possible high activity of photocatalytic water splitting. The ultrafast dynamics and atomic-scale mechanism of water splitting on Au_{20} cluster stimulated by femtosecond laser pulse were investigated based on ab initio real-time time-dependent density functional theory (rt-TDDFT) molecular dynamics. Under the laser pulse with a field strength of 2.3 V/Å and a photon energy of 2.81 eV (see Fig. 8.28b), corresponding to the dominant peak of the Au_{20}–water adsorption spectrum, the dynamic response of water molecules around Au_{20} cluster is displayed in Fig. 8.28c. During the first 10 fs, all O–H bonds in water molecules oscillate near the equilibrium length. Then in response to laser excitation, two water molecules dissociate with O–H bond lengths increasing from around 1 Å to more than 3 Å. Comparison with the case of liquid water without the presence of Au_{20} cluster, where water molecules remain intact under the same laser irradiation (Fig. 8.28d), confirms that the water splitting is mediated by the Au_{20} nanocluster. As shown in Fig. 8.28e, such a process is found to be assisted by rapid proton transport in a Grotthuss-like mechanism [58], forming hydronium ions. By analyzing the time-evolved change in the occupation of the Kohn–Sham states, electron transfer from the Au_{20} cluster to anti-bonding orbital of water is revealed. Furthermore, position-dependent FE for the Au_{20} cluster shows the same trend as the reaction rate of water splitting, which indicates that the plasmon-induced FE dominantly determines the process, while the commonly assumed electron transfer plays a less important role.

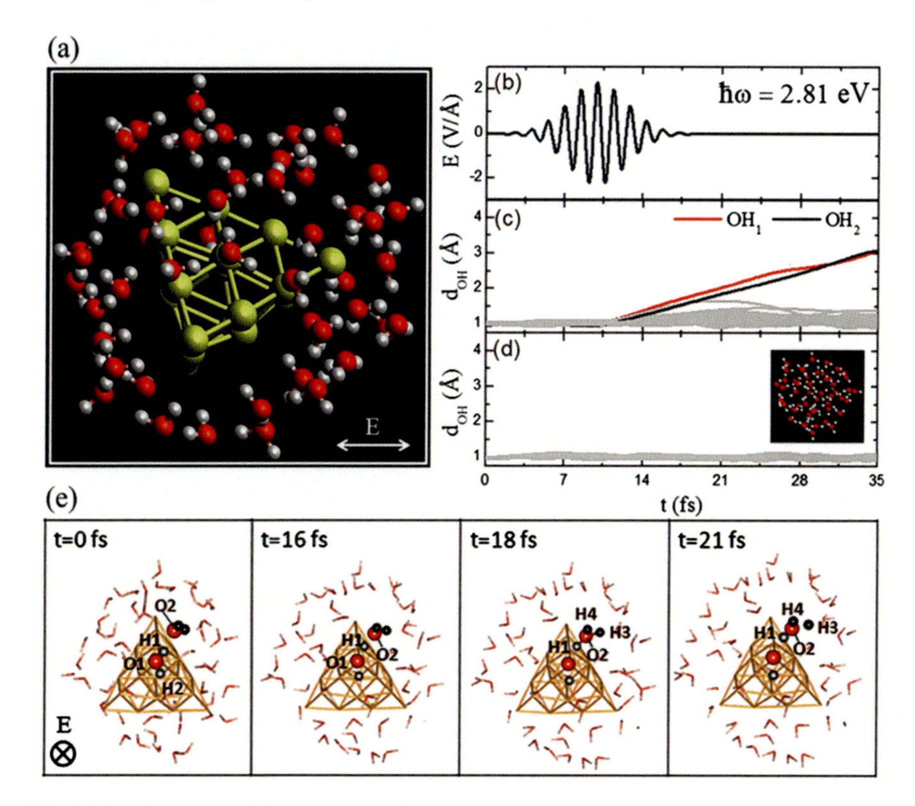

Fig. 8.28 **a** Snapshot of the Au_{20} cluster in water, where yellow, red, and grey spheres represent gold, oxygen, and hydrogen atoms, respectively. The arrow denotes the polarization direction of the laser field. **b** Time evolution of the laser field with field strength $E_{max} = 2.3$ V/Å and frequency $\hbar\omega = 2.81$ eV. Under this laser pulse, time-evolved O–H bond lengths d_{OH} of all water molecules with **c** and without **d** Au_{20} cluster are shown. **e** Atomic configurations at time $t = 0$, 16 fs, 18 fs, and 21 fs

8.7.2 Plasmon-Induced Water Splitting on Ag-Alloyed Pt Single-Atom Catalysts

Given the high atom utilization efficiency, unique electronic structure, precisely identified active site, and excellent catalytic activity and selectivity, single-atom catalysts (SACs) have emerged as a new frontier in heterogeneous catalysis including photocatalytic water splitting in recent years [59].

Meng et al. investigate plasmon-induced water splitting on the Ag cluster doped by a single Pt atom at the single-molecule level using real-time time-dependent density functional theory (rt-TDDFT). Since the tetrahedral Ag_{20} nanoparticle is structurally stable among many other clusters. After the relaxation of the adsorption geometry of the representative high-symmetry configurations, the most stable atomic structures for $Ag_{19}Pt–H_2O$ and $Ag_{20}–H_2O$ are shown in Fig. 8.29a, b. And

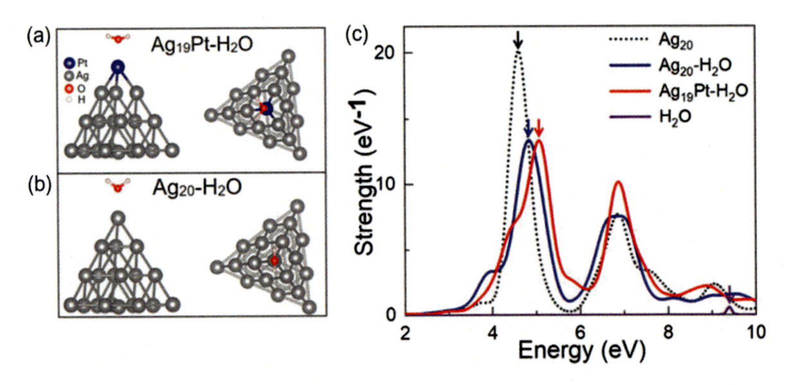

Fig. 8.29 Atomic configuration and absorption spectra. The top and side views of **a** $Ag_{19}Pt–H_2O$ and **b** $Ag_{20}–H_2O$ after geometry optimization. **c** Absorption spectra of Ag_{20}, $Ag_{20}–H_2O$, $Ag_{19}Pt–H_2O$, and H_2O, respectively. The colored arrows denote the position of the corresponding absorption peak

Fig. 8.29c showed their absorption spectra. The absorption peak of freestanding H_2O is located at > 8 eV, corresponding to the transition from the highest occupied molecular orbitals (HOMO) and the lowest unoccupied molecular orbitals (LUMO) of water. Compared to the highest absorption peak located at 4.62 eV for Ag_{20}, $Ag_{19}Pt–H_2O$ and $Ag_{20}–H_2O$ complexes both display a blue shift where the major absorption peak moves to 5.07 and 4.83 eV, respectively, which means that the $Ag_{19}Pt$ cluster has a stronger electronic coupling between the metal cluster and H_2O molecule than the Ag_{20} cluster.

Through the analysis of non-adiabatic molecular dynamic trajectories, it was found that water may split within 40 fs after photoexcitation in the $Ag_{19}Pt–H_2O$ case, while the O–H bond is only slightly elongated and does not break in the simulations for the $Ag_{20}–H_2O$ system (Fig. 8.30). Under the same field strength at $E_{max} = 0.5$ V/Å, a charge transfer $\Delta Q = 0.92e$ from the metal cluster to the H_2O molecule takes place for the $Ag_{19}Pt–H_2O$ cluster, more than the $Ag_{20}–H_2O$ of $0.75e$, implying that the introduction of the Pt atom enhances the amount of charges transferred to water.

The time-evolved transition coefficients from all occupied states to the orbital related are calculated, as shown in Fig. 8.31, to confirm the underlying charge transfer mechanisms for the two systems. It reveals that the main contribution to the HOMO level stems from the HOMO-1 level, which makes a dominant contribution to the excitation to the LUMO and LUMO + 1 state, and means the channels of indirect charge transfer opened via inelastic electron tunneling. Then, charge transfer from the HOMO-8 to LUMO + 11 states, which is mainly contributed by the metal cluster and water molecule respectively, implying a direct charge transfer via localized surface plasmon resonances. In summary, the introduction of the Pt atom opens up more channels of charge transfer including intramolecular, indirect, and direct charge transfer pathways, resulting in efficient charge transfer and subsequent water splitting [60]. These results provide a new microscopic picture of solar water splitting and may facilitate the design of high-efficiency single-atom photocatalysts.

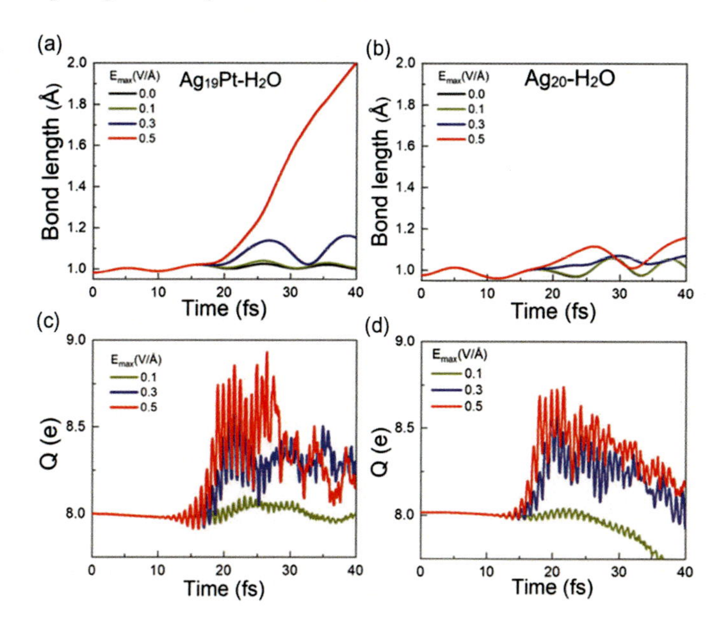

Fig. 8.30 Ultrafast dynamic responses after photoexcitation. Time evolved bond length of O–H for **a** $Ag_{19}Pt–H_2O$ and **b** $Ag_{20}–H_2O$ under different field strengths E_{max}. Time-evolved total charge (Q) located on H_2O for **c** $Ag_{19}Pt–H_2O$ and **d** $Ag_{20}–H_2O$ under different field strengths

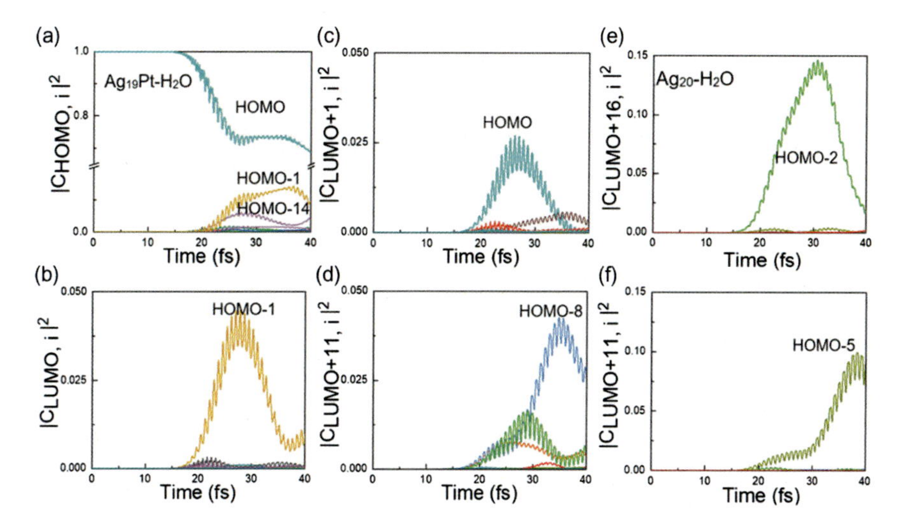

Fig. 8.31 Ultrafast electron dynamics. Time-evolved transition coefficients from all the occupied states i to the Kohn–Sham state of **a** HOMO, **b** LUMO, **c** LUMO + 1, and **d** LUMO + 11 for $Ag_{19}Pt–H_2O$, respectively. The index i (i = 0–18) denotes the numerical order of the occupied states, while LUMO + j (j = 0–19) corresponds to the unoccupied states. Time-evolved transition coefficients from all the occupied states i to the state **e** LUMO + 11 and **f** LUMO + 16 for $Ag_{20}–H_2O$, respectively. The index i (i = 0–13) denotes the numerical order of the occupied states, while LUMO + j (j = 0–19) corresponds to the unoccupied states. Only the important orbitals that significantly contribute to the density change are labeled

In this chapter, we have systematically studied the adsorption of water on densely packed surfaces of transition and noble metals and various related problems using first-principles density functional theory. From the results of the calculations we conclude the following.

(1) The interaction between water and metal surfaces is mainly through the formation of chemical bonds between the lone pair of electrons of the water molecule and the surface state. So the bonding between water and the surface is rather localized, concentrating mainly on the underlying molecule in direct contact with the surface, like the lower molecule in a bimolecule and the lower molecule in a bilayer. It is true that there is also long-range polarization, but the effect is smaller.
(2) Hydrogen bonding between water molecules, whether in small clusters or in layered structures, is usually enhanced by adsorption. This is inconsistent with Pauling's principle. Charge transfer and changes in the bonding environment are responsible for the enhanced hydrogen bonding at the surface. This enhancement may be a unique property of the hydrogen bonding system.
(3) The structure of water in the adsorbed state remains essentially similar to the corresponding states in the gaseous and bulk states. However, the water molecule will adjust its structural features such as elongation of OH bonds and widening of HOH bond angles when adsorbed. This adjustment is more pronounced on surfaces with strong interactions with water, like Ru(0001), Rh(111), etc.; on weakly interacting surfaces like Au(111), the adjustment is minimal. This stems mainly from the competition between the two basic interactions of surface and water, and water and water.
(4) From a microscopic point of view, surface hydrophilic interactions are determined by whether it is the surface-water interaction or the water-water interaction that wins the competition. Further, from the point of view of electronic interactions, they are ultimately determined by localized charge transfer and long-range charge polarization interactions. Applying this image, we obtain a surface hydrophilicity order of Ru > Rh > Pd > Pt > Au, in the same order as their d-band electron filling. This is a direct indication of the effect of localized charge coupling on the hydrophilic properties of the surface.
(5) In general, the vibrational spectra obtained from the calculations confirm the above structure and interaction images. The vibrational spectra, especially the OH vibrational frequencies, provide data for later experimental verification of these structures.
(6) The proton transport process on the surface differs from that in the bulk state in that it does not require H_2O molecules to be hydrogen-bonded to four near-neighbor molecules. Each water molecule on the surface is connected to only three near-neighbor molecules via hydrogen bonds, and the shared protons complete the transport process within 0.2–0.3 ps via hydrogen bonds. The surface is also involved in the proton transport process.
(7) The vibrational spectrum of the molecular bilayer of the D-up structure on Ru(0001) is in the best agreement with the experimental data, indicating that

the semi-decomposed structure may not be present in some experiments. The reason for this is that the decomposition potential is higher than the desorption potential, and the water molecules are already desorbed from the surface between decompositions. Water may decompose due to irradiation damage only when irradiated by a stronger probe beam or electron beam. Similarly, a similar conclusion was made for the representative open surface Cu(110).

(8) The calculated diffusion potential barrier for a single water molecule on Pd(111) is 150 meV, which is slightly larger than the experimental measurement (127 meV). However, the bimolecular diffusion potential barrier is not consistent with experiment. The diffusion potential barrier of water molecules on the surface can be estimated from the difference in adsorption energy at the top and bridge positions.

Density functional theory gives us an understanding of the interaction between water and surfaces on the atomic scale and even on the electronic scale. We believe that this understanding can be extended to other surfaces in general.

References

1. P.A. Thiel, T.E. Madey, The interation of water with solid surfaces: fundamental aspects. Surf. Sci. Rep. **7**, 211 (1987)
2. M.A. Henderson, The interation of water with solid surfaces: fundamental aspects revisited. Surf. Sci. Rep. **46**, 1 (2002)
3. D. Doering, T.E. Madey, The adsorption of water on clean and oxygen-dosed Ru(001). Surf. Sci. **123**, 305 (1982)
4. P.J. Feibelman, Partial dissociation of water on Ru(0001). Science **295**, 99 (2002)
5. P.J. Feibelman, Vibration of a water adlayer on Ru(0001). Phys. Rev. B **67**, 035420 (2003)
6. D.N. Denzler, C. Hess, R. Dudek, S. Wagner, C. Frischkorn, M. Wolf, G. Ertl, Interfacial structure of water on Ru(0001) investigated by vibrational spectroscopy. Chem. Phys. Lett. **376**, 618 (2003)
7. T. Mitsui, M.K. Rose, E. Fomin, D.F. Ogletree, M. Salmeron, Water diffusion and clustering on Pd(111). Science **297**, 1850 (2002)
8. S. Meng, E.G. Wang, S.W. Gao, Water adsorption on metal surfaces: a general picture from density functional theory calculations. Phys. Rev. B **69**, 195404 (2004)
9. A. Michaelides, V.A. Ranea, P.L. de Andres, D.A. King, General model for water monomer adsorption on close-packed transition and noble metal surfaces. Phys. Rev. Lett. **90**, 216102 (2003)
10. H. Ogasawara, B. Brena, D. Nordlund, M. Nyberg, A. Pelmenschikov, L.G.M. Pettersson, A. Nilsson, Structure and bonding of water on Pt(111). Phys. Rev. Lett. **89**, 276102 (2002)
11. J. Ren, S. Meng, Atomic structure and bonding of water overlayer on Cu(110): the borderline for intact and dissociative adsorption. J. Am. Chem. Soc. **128**, 9282 (2006)
12. J. Ren, S. Meng, A first-principles study of water on copper and noble metal (110) surfaces. Phys. Rev. B **77**, 054110 (2008)
13. S. Haq, C. Clay, G.R. Darling, G. Zimbitas, A. Hodgson, Growth of intact water ice on Ru(0001) between 140 and 160K: experiment and density-functional theory calculations. Phys. Rev. B **73**, 115414 (2006)
14. I. Hamada, S. Meng, Water wetting on representative metal surfaces: improved description from van der Waals density functionals. Chem. Phys. Lett. **521**, 161 (2012)

15. M. Forster, R. Raval, A. Hodgson, J. Carrasco, A. Michaelides, c(2x2) water-hydroxyl layer on Cu(110): a wetting layer stabilized by Bjerrum defects. Phys. Rev. Lett. **106**, 046103 (2011)
16. A. Michaelides, A. Alavi, D.A. King, Different surface chemistries of water on Ru{0001}: from monomer adsorption to partially dissociated bilayers. J. Am. Chem. Soc. **125**, 2746 (2003)
17. I. Villegas, M.J. Weaver, Infrared spectroscopy of model electrochemical interfaces in ultrahigh vacuum: evidence for coupled cation-anion hydration in the Pt(111)/K^+ Cl^- System. J. Phys. Chem. **100**, 19502 (1996)
18. J. Li, D.K. Ross, Evidence for two kinds of hydrogen bond in ice. Nature **365**, 327 (1993)
19. P.G. de Gennes, Wetting: statics and dynamics. Rev. Mod. Phys. **57**, 827 (1985)
20. S. Smith, C. Huang, E.K.L. Wong, B.D. Kay, Desorption and crystallization kinetics in nanoscale thin films of amorphous water ice. Surf. Sci. **367**, L13 (1996)
21. P. Löfgren, P. Ahlström, D.V. Chakarov, J. Lausmaa, B. Kasemo, Substrate dependent sublimation kinetics of mesoscopic ice films. Surf. Sci. **367**, L19 (1996)
22. S. Meng, E.G. Wang, S.W. Gao, A molecular picture of hydrophilic and hydrophobic interactions from ab initio density functional theory calculations. J. Chem. Phys. **119**, 7617 (2003)
23. T. Werder, J.H. Walther, R.L. Jaffe, T. Halicioglu, P. Koumoutsakos, On the water-carbon interaction for use in molecular dynamics simulations of graphite. J. Phys. Chem. B **107**, 1345 (2003)
24. P.A. Thiel, R.A. DePaola, F.M. Hoffmann, The vibrational spectra of chemisorbed molecular clusters: H_2O on Ru(001). J. Chem. Phys. **80**, 5326 (1984)
25. K. Jacobi, K. Bedürftig, Y. Wang, G. Ertl, From monomers to ice—new vibrational characteristics of H_2O adsorbed on Pt(111). Surf. Sci. **472**, 9 (2001)
26. G. Pirug, H.P. Bonzel, UHV simulation of the electrochemical double layer: adsorption of $HClO_4/H_2O$ on Au(111). Surf. Sci. **405**, 87 (1998)
27. A.F. Carley, P.R. Davies, M.W. Roberts, K.K. Thomas, Hydroxylation of molecularly adsorbed water at Ag(111) and Cu(100) surfaces by dioxygen: photoelectron and vibrational spectroscopic studies. Surf. Sci. **238**, L467 (1990)
28. E. Whalley, in *The hydrogen bond*, vol. 3, ed. by P. Schuster, G. Zundel, C. Sandorfy. North-Holland, Amsterdam (1976), pp. 1425–1470
29. S. Andersson, C. Nyberg, C.G. Tengstål, Adsorption of water monomers on Cu(100) and Pd(100) at low temperatures. Chem. Phys. Lett. **104**, 305 (1984)
30. H. Ogasawara, J. Yoshinobu, M. Kawai, Water adsorption on Pt(111): from isolated molecule to three-dimensional cluster. Chem. Phys. Lett. **231**, 188 (1994)
31. M. Nakamura, M. Ito, Monomer and tetramer water clusters adsorbed on Ru(0001). Chem. Phys. Lett. **325**, 293 (2000)
32. H. Ogasawara, J. Yoshinobu, M. Kawai, Clustering behavior of water (D_2O) on Pt(111). J. Chem. Phys. **111**, 7003 (1999)
33. M. Nakamura, Y. Shingaya, M. Ito, The vibrational spectra of water cluster molecules on Pt(111) surface at 20 K. Chem. Phys. Lett. **309**, 123 (1999)
34. K. Morgenstern, J. Nieminen, Intermolecular bond length of ice on Ag(111). Phys. Rev. Lett. **88**, 066102 (2002)
35. M. Morgenstern, J. Muller, T. Michely, G. Comsa, The ice bilayer on Pt(111): nucleation, structure and melting. Z. Phys. Chem. **198**, 43 (1997)
36. N. Materer, U. Starke, A. Barbieri, M.A. Van Hove, G.A. Somorjai, G.-J. Kroes, C. Minot, Molecular surface structure of ice(0001): dynamical low energy electron diffration, total-energy calculations and molecular dynamics simulation. Surf. Sci. **381**, 190 (1997)
37. A. Glebov, A.P. Graham, A. Menzel, J.P. Toennies, Orientational ordering of two-dimensional ice on Pt(111). J. Chem. Phys. **106**, 9382 (1997)
38. S. Haq, J. Harnett, A. Hodgson, Growth of thin crystalline ice films on Pt(111). Surf. Sci. **505**, 171 (2002)
39. S.K. Jo, J. Kiss, J.A. Polanco, J.M. White, Identification of second layer adsorbates: water and chloroethane on Pt(111). Surf. Sci. **253**, 233 (1991)

40. G.B. Fisher, J.L. Gland, The interaction of water with the Pt(111) surface. Surf. Sci. **94**, 446 (1980)
41. M. Morgenstern, T. Michely, G. Comsa, Anisotropy in the adsorption of H_2O at low coordination sites on Pt(111). Phys. Rev. Lett. **77**, 703 (1996)
42. S. Meng, L.F. Xu, E.G. Wang, S.W. Gao, Vibrational recognition of hydrogen-bonded water networks on a metal surface—reply. Phys. Rev. Lett. **91**, 059602 (2003)
43. S. Meng, E. Kaxiras, Z.Y. Zhang, Water wettability of close-packed metal surfaces. J. Chem. Phys. **127**, 244710 (2007)
44. D. Marx, M.E. Tuckerman, J. Hutter, M. Parrinello, The nature of the hydrated excess proton in water. Nature **397**, 601 (1999)
45. M.E. Tuckerman, D. Marx, M. Parrinello, The nature and transport mechanism of hydrated hydroxide ions in aqueous solution. Nature **417**, 925 (2002)
46. S. Meng, Dynamical properties and the proton transfer mechanism in the wetting water layer on Pt(111). Surf. Sci. **575**, 300 (2005)
47. F. Bruni, M.A. Ricci, A.K. Soper, Structural characterization of NaOH aqueous solution in the glass and liquid states. J. Chem. Phys. **114**, 8056 (2001)
48. A. Botti, F. Bruni, S. Imberti, M.A. Ricci, A.K. Soper, Solvation of hydroxyl ions in water. J. Chem. Phys. **119**, 5001 (2003)
49. S. Meng, E.G. Wang, C. Frischkorn, M. Wolf, S.W. Gao, Consistent picture for the wetting structure of water/Ru(0001). Chem. Phys. Lett. **402**, 384 (2005)
50. A.B. Anderson, Reactions and structures of water on clean and oxygen covered Pt(111) and Fe(100). Surf. Sci. **105**, 159 (1981)
51. B. Seemala, A.J. Therrien, M. Lou, K. Li, J.P. Finzel, J. Qi et al., Plasmon-mediated catalytic O_2 dissociation on Ag nanostructures: hot electrons or near fields? ACS Energ. Lett. **4**, 1803 (2019)
52. L. Yan, J. Xu, F. Wang, S. Meng, Plasmon-induced ultrafast hydrogen production in liquid water. J. Phys. Chem. Lett. **9**, 63 (2018)
53. S. Linic, P. Christopher, D.B. Ingram, Real-time, local basis-set implementation of time-dependent density functional theory for excited state dynamics simulations. J. Chem. Phys. **129**, 054110 (2008)
54. S.C. Gomes, R. Juarez, T. Marino, R. Molinari, H. García, Influence of excitation wavelength (UV or visible light) on the photocatalytic activity of titania containing gold nanoparticles for the generation of hydrogen or oxygen from water. J. Am. Chem. Soc. **133**, 595 (2011)
55. H. Robatjazi, S.M. Bahauddin, C. Doiron, I. Thomann, Direct plasmon-driven photoelectro-catalysis. Nano Lett. **15**, 6155 (2015)
56. J.H. Kang, D.S. Kim, Q.H. Park, Local capacitor model for plasmonic electric field enhancement. Phys. Rev. Lett. **102**, 093906 (2009)
57. L. Yan, F. Wang, S. Meng, Quantum mode selectivity of plasmon-induced water splitting on gold nanoparticles. ACS Nano **10**, 5452 (2016)
58. M. Rini, B.Z. Magnes, E. Pines, E.T.J. Nibbering, Real-time observation of bimodal proton transfer in acid-base pairs in water. Science **301**, 349 (2003)
59. H.-Y. Zhuo, X. Zhang, J.-X. Liang, Q. Yu, H. Xiao, J. Li, Theoretical understandings of graphene-based metal single-atom catalysts: stability and catalytic performance. Chem. Rev. **120**, 12315 (2020)
60. Y.M. Zhang, D.Q. Chen, W. Meng, S.F. Li, S. Meng, Plasmon-induced water splitting on Ag-alloyed Pt single-atom catalysts. Front. Chem. **9**, 2296 (2021)

Chapter 9
Water Adsorption on Non-metallic Surfaces

Having meticulously studied the structure, dynamics, and trends of water adsorption on various types of metal surfaces, we begin to consider the more general case of water adsorption on some model solid surfaces such as SiO_2, graphite, table salt, etc.

There are a large number of studies focusing on water on the surface of non-metallic solids such as oxides. Representative oxides include TiO_2, MgO, mica, SiO_2, Fe_2O_3, $SrTiO_3$, etc [1–6]. More traditional surface analysis tools such as low-energy electron diffraction (LEED), ultraviolet photoelectron spectroscopy (UPS), and electron energy loss spectroscopy (EELS) have been used extensively to study these systems, and recently, due to the development of real-space imaging techniques such as scanning tunneling microscopy (STM) and thin film sample preparation techniques, a large number of studies have also focused on weakly conducting oxides and oxides on conducting substrates thin layers. Due to the specificity of salt dissolution, which we put into Chap. 11 to narrate, in this chapter we only discuss the adsorption of water on some oxide and graphite layers.

9.1 Water Adsorption on Simple Oxide Surfaces

Oxides are generally bound by strong chemical bonds (covalent and ionic bonds). Symmetry breaking results in the oxide surface exhibiting exposed metal ions or unsaturated other suspended bonds and surface defects, and it is generally believed that water interacts more strongly with the oxide surface than metal. This strong interaction often leads to water adsorption in a dissociated state [7]; however, it is often debated under what conditions and what coverage water molecules would dissociate [8–10]. On more inert oxide surfaces, where water can adsorb without decomposing, strong interactions lead to a more ordered surface water structure that exhibits properties different from those of the bulk state. For example, as early as 1995, based on AFM imaging of surface structure, Hu et al. [1] found that water on mica surfaces exhibits ice-like behaviors at room temperature, which is named "room

S. Meng and E. Wang, *Water*,
https://doi.org/10.1007/978-981-99-1541-5_9

temperature ice". This phenomenon was of great interest, and similar behaviors were later found in different surface systems.

The decomposition of water remains a controversial topic even in the simplest of oxide systems. On a perfect defect-free MgO(001) surface, Giordano et al. [4] found via density function theory that individual water molecules do not decompose, but in the case where water molecules are spread over the entire MgO surface, lengths of OH bonds of 2/3 water molecules are greatly elongated, to a value about 1.05 Å, which already exhibits decomposition properties compared to the OH bond length of 0.963 Å in gaseous water molecules. Recently, the adsorption and decomposition behavior of water on MgO films and its interfacial system with metals has received much attention. Kawai et al. have meticulously studied the different decomposition behavior of individual water molecules under thermal and electronic excitation using STM. It was shown that thermal excitation makes the water molecule partially decompose into OH and H. However, electrons tunneled into the LUMO state of the water molecule can make it decompose completely into adsorbed O and H atoms [5].

TiO_2 surfaces are highly valued surfaces. TiO_2 is widely available with a low cost, while it is also an environmentally friendly material, whose electronic properties including chemical stability and photocatalytic activity can be modulated by regulating its oxygen content. As early as 1972, Fujishima and Honda [2] found that when illuminated under the ultraviolet portion of solar irradiation, TiO_2 can make water molecules decompose and produce oxygen gas on its surface, while the negative electrode (usually Pt) produces hydrogen gas. In 1976, Carey [3] discovered that TiO_2 could non-selectively oxidize (degrade) various organic substances under light and completely mineralize them to produce CO_2 and H_2O. These discoveries pointed to a solution to the current energy problem, while they also pointed to a solution for pollution decomposition and environmental treatment. Therefore, the behavior of water adsorption and chemical reactions on TiO_2 surfaces has been studied in great detail for a half century [7–12].

The first STM study of the TiO_2 surface was performed by Diebold et al. [13, 14]. They found that the $TiO_2(110)$ surface exhibit an alternative bright and dark chain-like structure, where the dark lines correspond to oxygen chains and the bright lines correspond to Ti chains. Another remarkable feature on this surface is the frequent presence of bright spots on the dark lines, which they assigned to oxygen vacancy defects. However, Suzuki et al. [15] subsequently found that these defects could be eliminated using electron beam bombardment of the surface and therefore could not be oxygen vacancies, but instead residual hydrogen atoms in the vacuum adsorbed on the oxygen chains. Subsequently, Schaub et al. [16] in Denmark found two bright spots on the surface and thought that one corresponded to an oxygen vacancy and one to an H adsorbate. Unfortunately, they have made a mistaken assignment. This mistake caused confusion in the field for several years until the correct assignment was given by Bikondoa et al. in 2006 [17]. Subsequently, a clearer understanding of the adsorption and decomposition of water on TiO_2 surfaces was obtained.

Similar to the scenario on MgO surface, some believed that water does not decompose at TiO_2 at low coverages, while it does at high coverages [7, 8]. But there is a great debate as to whether individual water molecules decompose on rutile $TiO_2(110)$ surfaces. This is partly due to the fact that the two states of molecular adsorption and decomposition of water molecules are close in energy and are closely related to the selection of parameters such as the surface layer crystal thickness in DFT calculations, resulting in a variety of different results [10]. The decomposition reaction processes of water on TiO_2 surface, and the diffusion and release of reaction products O and H have been carefully studied. It has been found that chains of hydrogen bonds can be formed between water molecules. Water molecules are easily decomposed in the presence of oxygen vacancies, and oxygen atoms fill the vacancies while hydrogen atoms adsorb on oxygen atoms to form a pair of OH. Recently, the decomposition mechanism and process of small molecules such as water and methanol on TiO_2 surface under light conditions have also been investigated by a combination of STM real space observation and ultrafast spectroscopy experiments [18]. The current problems to be solved include the low photocatalytic conversion efficiency and the contradiction between longer lifetime and higher conversion efficiency.

Similarly, the adsorption and dynamic behavior of water on ZnO surfaces has generated extensive research interests; for example, Meyer et al. found that the water chains on ZnO surfaces exhibit a dynamic behavior with constant decomposition and recombination of water molecules [19]. As an example, we focus below on the adsorption of water on the common SiO_2 surface, a simple model of the glass surface which we use on a daily basis [20–23].

9.2 Water Structures on Silica Surfaces

Silicon dioxide is a common oxide in daily life and is a major component of quartz, sands, and glasses. It exists in both amorphous and crystalline structures. Although amorphous forms are more common, there is a regular ordering on the surface local structure, so a mixture of specifically oriented crystal surfaces is often used to represent amorphous surfaces. There are many different structural phases of silicon dioxide crystals, such as α-quartz, β-quartz, α-cristobalite, β-cristobalite, tridymite, octahedral zeolite, and more than a dozen of other phases. Because of its relatively simple structure and orthorhombic crystal structure, we first study the adsorption of water on the surface of β-cristobalite.

The unit cell of β-cristobalite (100) surface is square in shape where O or Si atoms are exposed to the surface with dangling bonds. These dangling bonds can be saturated with H or OH groups in the presence of water vapor or air, so we only consider surface structures where the surface is completely saturated with OH, as shown in Fig. 9.1. Due to the large number of OH groups present on the surface of β-cristobalite (100), water molecules adsorb on the surface by interacting with these surface groups mainly through hydrogen bonds. Several possible modes of adsorption of water molecules are presented in Fig. 9.1, with the A, B, and C configurations showing water molecules interacting with SiO_2 through three, two and one hydrogen bonds, respectively.

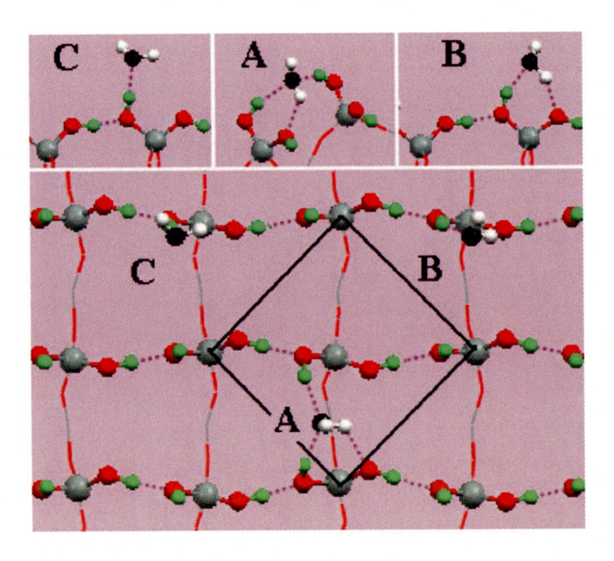

Fig. 9.1 The geometry (A, B, C) for a single water adsorption on β-cristobalite (100). Both top view (lower panel) and side view (upper panels) are shown. The black lines denote the size of unit cell in the calculation

In the A conformation, the water molecule provides an H-bond to the surface OH group (water as the hydrogen bond donor) and accepts a hydrogen bond from the two OH groups attached to each of the two Si atoms on the surface (water as the hydrogen bond acceptor), forming a triple H-bond conformation. The OH groups in the water molecule that are not H-bonded are perpendicular and pointing away from the surface, which provides the adsorption sites in case of high water coverage or multilayer water adsorption. Due to the difference in the number of hydrogen bonds, we found the triple H-bonded configuration to be the most stable position for water adsorption on this surface. DFT calculations yielded adsorption energies up to 622 meV, averaging out to 207 meV/H bond per hydrogen bond. This is much more stable than the double hydrogen bond configuration (adsorption energy of 509 meV) adsorbed at the top bridge position (configuration B, with two OH groups belonging to the same surface Si atom). The double H-bond adsorption at the adjacent bridge position (two OH groups belonging to two different surface Si atoms) is slightly weaker, with an adsorption energy of only 422 meV. This suggests that internal interactions between two OH groups belonging to the same Si atom increase the adsorption of foreign water molecules. All of these are more stable than the single H-bond configuration (configuration C) that adsorbs at the OH top position, which has an adsorption energy of only 339 meV and thus cannot be stabilized. In addition, we note that the OH group on the SiO_2 surface is more stable as an H-bond donor, and the formed H-bond is stronger than that as an H-bond acceptor. The hydrogen bond length (OO distance) formed between the water molecule and the surface O atom is in the range of 2.82–3.04 Å, which is longer than the hydrogen bond length of 2.76 Å in bulk ice Ih. After adsorption, the structure of the water molecule changes little, the bond length remains essentially unchanged, and the bond angle increases slightly (from 104.9° to 105–106°).

Water dimers are very strongly adsorbed on the surface of β-cristobalite (100). The water dimer adsorbs at the bridge sites between the Si atoms on the surface, and in addition to its own H-bonds, each water molecule forms two hydrogen bonds with the surface OH group, where the hydrogen bond acceptor in the dimer provides two H-bonds and the hydrogen bond donor accepts two H-bonds. The adsorption conformation is shown in Fig. 9.2. The hydrogen bond donor interacts more strongly with the surface and its height from the surface is 0.5 Å lower than the hydrogen bond acceptor, similar to the adsorption of a water dimer on a metal surface. Due to the presence of multiple hydrogen bonds and the synergistic interaction between the hydrogen bonds, the H-bonds between the adsorbed water dimers are greatly strengthened and its OO bond length is only 2.53 Å, which is much shorter than the hydrogen bonds in the gaseous free water dimer (2.98 Å). The water dimer is very stable, with an adsorption energy of 748 meV per water molecule.

At half-full layer coverage (i.e., a water molecule to surface Si atom ratio of 0.5), water can adsorb as isolated dimers (Fig. 9.2a) or form a chain of isolated water molecules (Fig. 9.2b). DFT calculations show that isolated water dimers adsorb more stable, while the chain of water dimers has an adsorption energy ~ 100 meV/water smaller than the dimer.

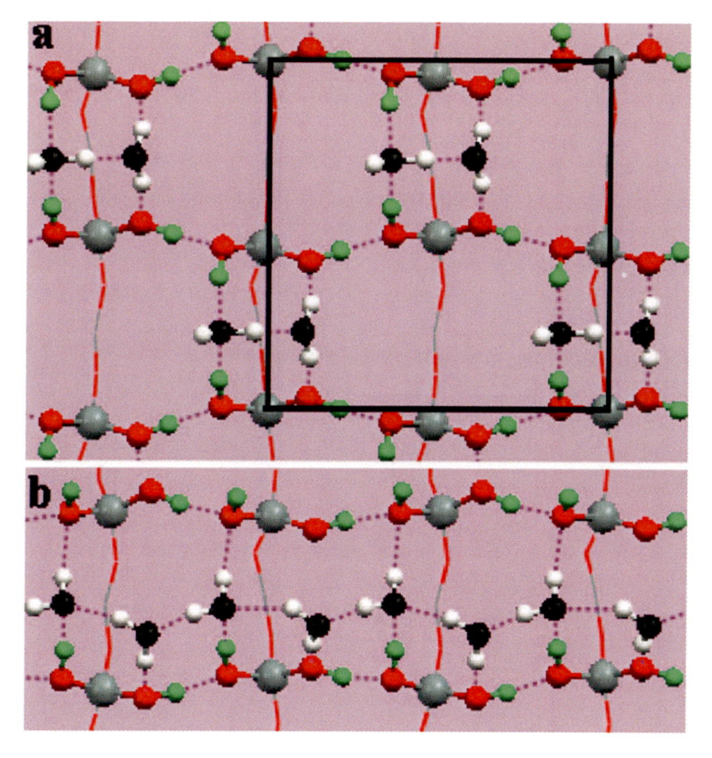

Fig. 9.2 The adsorption structure of water dimer on β-cristobalite (100)

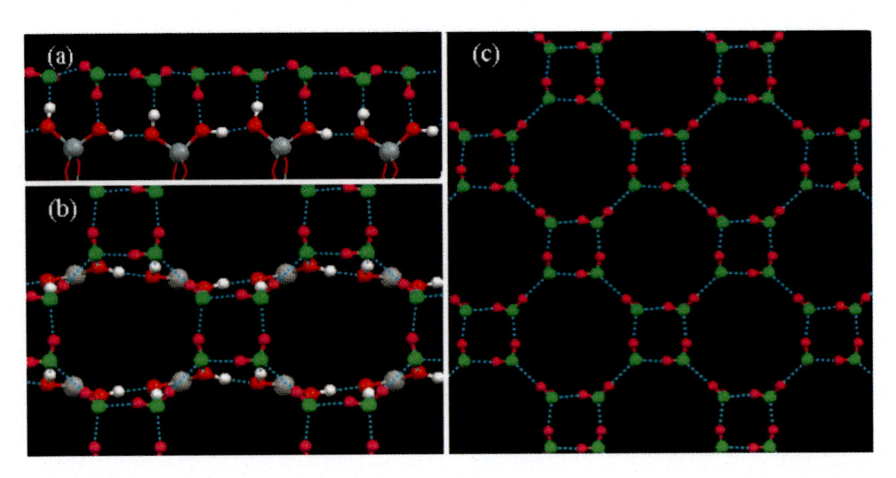

Fig. 9.3 The structure of "tessellation ice" upon water adsorption on β-cristobalite (100) at the full coverage

More interestingly, in the full coverage case, the water molecules self-assemble on the β-cristobalite (100) surface to form a very stable new surface structure, see Fig. 9.3. Here the water molecules form two types of rings, tetragonal and octagonal, which are staggered to form a peculiar pattern of squares like those on the floor surface. We refer to this surface ice structure as "tessellation ice" [20]. Its discovery suggests that the surface confinement can form a new structure different from that of bulk ice, with new properties.

In tessellation ice, all water molecules form a hydrogen bond with the surface OH group and three hydrogen bonds with other water molecules, so all hydrogen bonds are saturated. The adsorption energy is 712 meV per water molecule, which is close to the binding energy of bulk-phase ice (720 meV). In fact, after taking into account the zero-point energy correction, the binding energy of water in tessellation ice is about 30 meV greater than that in the bulk phase.

The strength of hydrogen bonds in tessellation ice is different due to its unique bonding structure. Analysis shows that the hydrogen bonding in the interior of the water-square is stronger, while the hydrogen bonding connecting the square is weaker. This can also be seen by the bond lengths of the two hydrogen bonds: OO length of the former hydrogen bond is 2.82–2.96 Å, while that of the latter hydrogen bond is 3.16–3.30 Å. These two different strengths of hydrogen bonds can be identified by infrared spectroscopy: OH vibrational energies around 400 meV in the infrared spectrum correspond to strong hydrogen bonds inside the square, while OH vibrational peaks around 450 meV correspond to the weak hydrogen bonds connecting the squares.

Due to the high binding energy of the tessellation ice, it exhibits extraordinary stability. Molecular dynamics simulations show that this tessellation ice phase retains its ice structure at room temperature without melting. This may imply that the surface ice has a higher melting point than room temperature, thus aiding biotechnology, infiltration, catalysis, and other applications. Recent experimental work [24, 25] implies

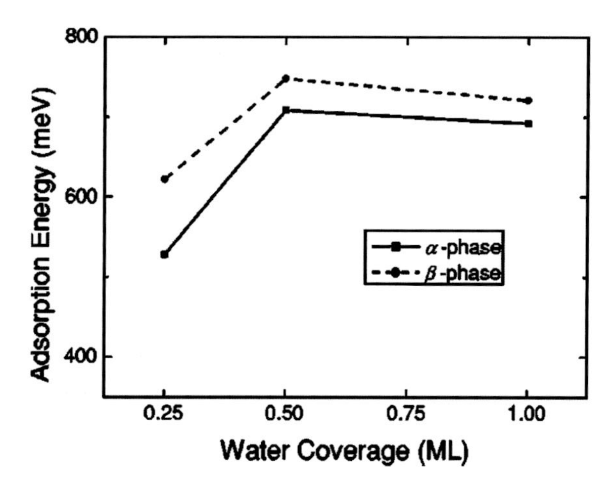

Fig. 9.4 The energy as a function of coverage for water adsorption on α-quartz and β-cristobalite (100)

that water has an ordered structure on the SiO_2 surface, which might correspond to the tessellation ice structure we found.

The adsorption of water on the surface of α-cristobalite (100) is very similar to that of β-cristobalite (100). The difference is that the α-cristobalite lattice constant is smaller and the adsorption energy of water molecules is generally 20–30 meV smaller than on the β-cristobalite (100) surface, while the trend of the adsorption energy as a function of water coverage is consistent with water adsorption on cristobalite [23]. See Fig. 9.4.

On the surface of β-cristobalite (111), the surface lattice has a hexagonal structure and water molecules are adsorbed on the vacancies between the three OH groups, forming a hydrogen bond with each of these three OHs, with the water molecule being the acceptor for two of the hydrogen bonds and the donor for the third one, see Fig. 9.5. Only two of the three hydrogen bonds are stronger due to the distant spacing of the surface OH groups. The water molecule adsorption energy is 701 meV per molecule. The lattice constant of β-cristobalite (111) is large (10.2 Å) compared to the ice surface lattice constant (4.52 Å). So the first layer of water at this surface does not form an ice structure, but exists as discrete water molecules. No hydrogen bonds are formed between any two water molecules. This is also a rather unique surface water layer structure.

9.3 Water Adsorption on Graphite/Graphene Surfaces

The interaction between graphite and water is very important. Many carbon materials such as fullerenes, carbon nanotubes, carbon nanocones, and graphite can be seen as consisting of a single layer of graphite, so-called graphene, curled or overlapped. So monolayer graphite is often used as a prototype for studying carbon materials. One often extracts suitable empirical parameters for calculating the properties of

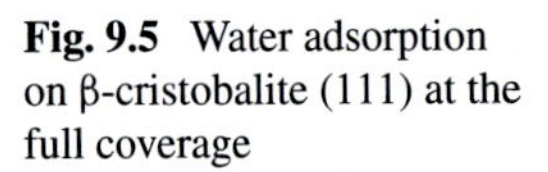

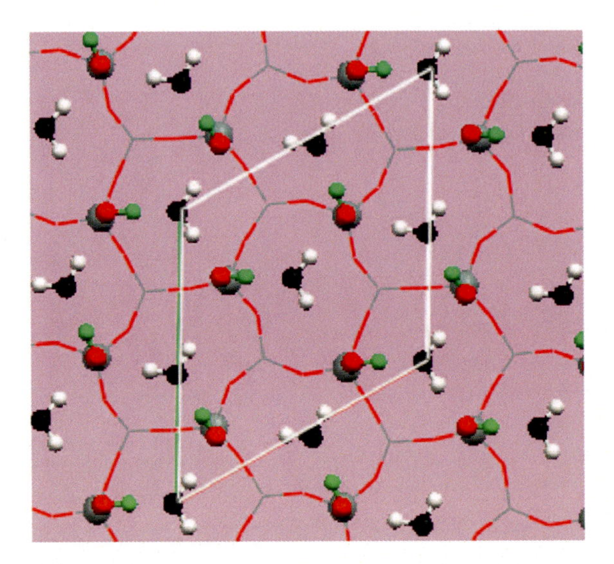

Fig. 9.5 Water adsorption on β-cristobalite (111) at the full coverage

other carbon materials by calculating the interaction of molecules with graphene. Thus the interaction of graphene with water represents a large class of interactions between carbon materials and water. It relates to many phenomena such as wetting, heterogeneous nucleation of ice, water formation on dust and dirt, properties and functions of carbon nanomaterials in biological environments, structural and phase transition behavior of water at the nanoscale, etc. For example, atmospheric rain is generally formed when water molecules accumulate near the condensation point to form droplets, and then the droplets grow as they fall down to form rain. Graphite is a model of a condensation nucleus. Rocket nozzles are also generally composed of carbon materials, while water is a common combustion product, so the interaction of water with these carbon materials is important for the corrosion of rocket nozzles.

The contact angle of water on graphite has been measured by a number of methods [26, 27]. In 1940, Fowkes measured a contact angle of 86° on graphite. A similar conclusion was reached by Morcos, who measured a contact angle of 84°. Schrader, on the other hand, obtained a relatively small contact angle of $42 \pm 7°$. More recently, Luna et al. have measured contact angles of up to 30°. These very different conclusions may stem from impurities on the graphite surface as well as the adsorption of other substance on graphite. From theoretical calculations, Werder et al. [26] used an empirical potential-based molecular dynamics approach to simulate the adsorption behavior of water droplets on graphite to calculate the contact angle of water droplets. Since there are various different model potentials describing water and various parameters for the Lennard–Jones potential describing the interaction of water molecules with carbon, after a series of molecular dynamics simulations of the equilibrium shape of water droplets, Werder et al. [26] found that the contact angle of water droplets on graphite is linearly related to the adsorption energy of individual water molecules on graphite. This work bridges a gap from the macroscopic to the microscopic adsorption of water on graphite surface.

9.3.1 Interaction of Water Molecules and Benzene

One way to figure out how graphite interacts with water is to start with the interaction of the simplest building block of graphite, the benzene ring, with water. The interaction of the model system, a single water molecule and the benzene ring, was first calculated using a DFT method by Ma et al. [28] using the exchange correlation function of PBE, BLYP, and the hybrid ones PBE0, B3LYP. The plots of adsorption energy versus adsorption distance are shown in Fig. 9.6. For comparison, the accurate adsorption energy curves obtained from the ΔCCSD(T) method are also plotted in the figure. It can be seen from the figure that PBE underestimates the adsorption energy by about 50 meV near the equilibrium position and by almost 90 meV at a distance of 2.8 Å. It is interesting to note that although PBE severely underestimates the adsorption energy, it gives the correct adsorption distance. This means that PBE gives relatively accurate results, but the accuracy is usually not sufficient. Its generalized version, PBE0, gives almost exactly the same adsorption energy profile. Compared to PBE, BLYP underestimates the adsorption energy even more severely; and the adsorption structure is not given correctly. In contrast, its generalized version, B3LYP, does slightly better, but still does not correctly characterize the system. In general, these exchange correlations underestimate the interaction energy and do not describe the system well, which is typical in weakly interacting systems. The hybrid exchange correlations give only little, if any, improvement over the original GGA version. This is because essentially none of the existing forms of exchange–correlation functionals in DFT can correctly describe weak interactions such as the van der Waals force. The hybrid functional only improves the exchange energy part, however weak interactions such as the van der Waals forces are entirely correlation effects, so the hybrid functionals fail to improve the performance of the traditional GGA.

As a contrast, the performance of several quantum chemical methods, including the Hartree–Fock, MP2, and CCSD methods, is also given in Fig. 9.6. Both the Hartree–Fock and MP2 methods are selected to discuss the adsorption energy curves in the complete basis vector limit, while the CCSD method is discussed using a ΔCCSD adsorption energy curve similar to that of the ΔCCSD(T) method because it is too computationally intensive. All these adsorption energy curves, as well as the ΔCCSD(T) curve as a reference standard, are depicted in Fig. 9.6. Firstly, it is clear from the figure that similar to BLYP and B3LYP, the Hartree–Fock method also largely underestimates the adsorption energy of water on benzene. This result does not come as a surprise considering that the Hartree–Fock method has no correlation effect. For the MP2 and CCSD methods, the adsorption energy curves of these two methods are very close to the reference curve, which means that they can describe this system better. But meanwhile both methods have their own errors. Similar to the other methods, their errors decrease as the distance between the water and benzene

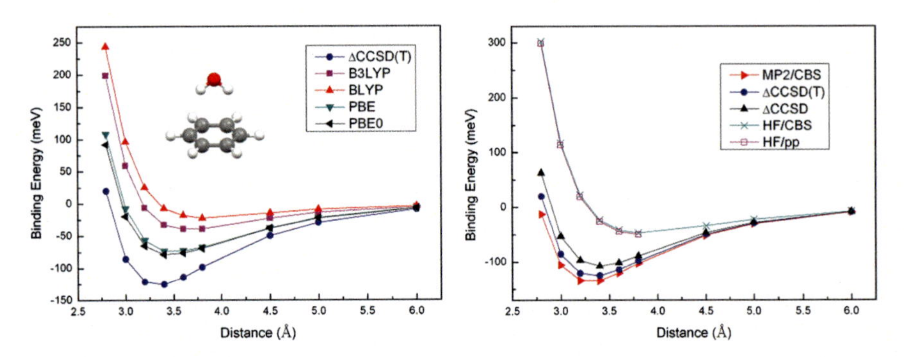

Fig. 9.6 The energy-distance curve for water adsorption on a benzene molecule in different computational schemes. The inset shows the geometry of water molecule on benzene

molecules increases. The MP2 method overestimates the adsorption energy by about 9 meV near the equilibrium position and by about 32 meV at 2.8 Å. In contrast, the CCSD method underestimates the entire adsorption energy curve by about 18 meV near the equilibrium position and by about 42 meV at 2.8 Å. This indicates that triple excitation is still important in the coupled cluster method, especially when the water and benzene molecules are in close proximity to each other, thus cannot be neglected. Overall, the error of the MP2 method is smaller than that of the CCSD method, and the value is more accurate. This observation is similar to other systems with hydrogen bonding interactions. The important reason should be the presence of error canceling in the MP2 method, which makes the final result closer to the exact value. Finally using the diffusion Monte Carlo method, it was found that it not only gives the correct adsorption energy and adsorption distance, but also the shape of the whole curve is correctly described. Thus the quantum Monte Carlo method can accurately describe this system both qualitatively and quantitatively. Since this system is relatively small, the statistical error of the quantum Monte Carlo method is about 4 meV. The obtained adsorption energy curves are in high agreement with the CCSD(T) curves near the equilibrium position as well as at a distance from each other. At closer distances between water and benzene molecules, such as at 2.8 Å, the two methods give an error of about 20 meV. The diffusion Monte Carlo method gives an overly steep repulsive wall. Overall, although many approximations are made, such as the fixed-node approximation, the pseudopotential, and the single Slater-Jastrow trial wave function, the diffusion Monte Carlo method is still very accurate, especially near the equilibrium adsorption position.

9.3.2 Interaction of Water Molecules with Graphene

In 2000, Feller and Jordan [29] used the MP2 method to calculate the adsorption energy of water on clusters of different sizes and extrapolated to the infinity cluster

Fig. 9.7 The finite-sized graphene flakes to model the infinite graphene structure in the quantum chemistry calculations

case. Since the MP2 method is generally only used to calculate isolated molecular cluster systems, it is not yet theoretically capable of dealing with zero-energy-gap metallic or semimetallic systems such as graphene by nature, although periodic versions capable of dealing with solids and introducing k-space sampling have also been developed in recent years. In general, one can only pick a number of fused-benzene clusters to model graphene (as in Fig. 9.7) and extrapolate the adsorption energy in the case of infinitely large clusters, i.e., graphene, by the adsorption energy of clusters of different sizes. They chose aug-cc-pVxZ, x = D, T, Q, and extrapolated to the complete basis limit. Its adsorption structure is obtained as a single-leg (one-leg) structure by structural optimization of the optimized smaller clusters. The final adsorption energy obtained is about 250 meV. It is generally believed that the adsorption energy may be overestimated due to the limitations in basis and cluster size. The one-leg adsorption structure is also only the most stable structure for small clusters, while is metastable for large clusters.

Lin et al. [30] used the MP2 method to optimize the structure of larger clusters under the aug-cc-pVDZ basis and found that the two-legged (two-leg) structure is more stable than the single-leg. Since the MP2 method is too computationally intensive, they also used the DFTB-D method at the same time. Both methods give very close adsorption energies of about 120 meV, much smaller than the results in the previous paper. However, the basis vectors used here are relatively small and are not extrapolated to the complete basis limit, and the selected cluster sizes are limited.

In the above works, the authors used molecular clusters to model graphene. Since at the cluster edges the carbon atoms have dangling bonds, they need to be saturated with hydrogen atoms. Sudiarta and Geldart [31] proposed that hydrogen atoms at the cluster edges would have an effect on the calculation results. They used two elements, hydrogen and fluorine, to saturate the carbon atoms at the cluster edges separately and improved the extrapolation to infinite clusters to exclude as much as possible the effect of saturated atoms at the edges on the adsorption energy. They concluded that the influence at the edges mainly originates from electrostatic interactions and van der Waals interactions, and that more accurate interaction energies can be obtained with extrapolation to infinite clusters after excluding these two contributions. Using this method, the adsorption energies obtained by extrapolation to hydrogen-saturated and fluorine-saturated clusters, respectively, are very close, indicating that the influence at the edges is eliminated. They eventually obtained adsorption energies of about 100 meV. They used a basis set of 6–31G (d = 0.25), which is a relatively small basis set and the MP2 energy is not necessarily reliable.

Using the MP2 method, the interaction energies between water and graphene obtained varies from 100 to 250 meV, a difference of about a factor of two, due to differences in cluster size, basis size, and extrapolation methods. Others have calculated this system using other methods. Sanfelix et al. [32] used the DFT method and selected PW91 form of exchange correlation for the interaction between graphene and water molecules using periodic supercells. However, since DFT does not correctly describe the van der Waals interactions, it cannot correctly describe this weakly interacting system and the adsorption energy obtained is close to 0. Ruuska and Pakkanen [33], on the other hand, used the Hartree–Fock method. The object of their study remains the molecular clusters. Since the Hartree–Fock method has no correlation energy, it also does not describe the system correctly. The final adsorption energy they obtained was only ~ 30 meV and the adsorption height of the water molecules was about 0.8 Å higher than the previous MP2 results, which is typical for Hartree–Fock calculations. S. Xu et al. used a hybrid approach to study larger clusters. They took B3LYP exchange–correlation form for the description of the carbon atoms in the central region of the cluster which are closer to the water molecules, while the surrounding carbon atoms that are farther away from the water molecules were described using the DFTB approach. To compensate for the van der Waals effect, they used an empirical correction. They obtained an adsorption energy of about 90 meV.

As we can see from the above discussion, theoretical calculations give very different conclusions for a variety of reasons, with adsorption energies ranging from a few tens to several hundred meV. The measurement of the adsorption energy is also very difficult experimentally. This is mainly because in experiments water molecules can easily form clusters through hydrogen bonding interactions, and then generally experiments measure the energy of the water molecule clusters to desorb into water molecules rather than the energy of individual water molecules to desorb. In addition, the measurement of such weak interactions is itself very problematic.

Therefore it become necessary to determine the adsorption energy and structure of individual water molecules on graphene by high-precision first-principles theoretical calculations. The following study is carried out using a diffusion quantum Monte Carlo method using the CASINO software package with the trial wave function set to a Slater-Jastrow shape [34]. In performing the evolution of the imaginary-time Schrödinger equation, the time step is chosen to be 0.0125 a.u. The pseudopotential is chosen to be of the Dirac–Fock type, where the carbon and oxygen atoms freeze the internal helium nuclei with nuclear radii of 0.58 and 0.53 Å, respectively, and the hydrogen atom has a nuclear radius of 0.26 Å. The single-electron orbitals are then obtained through the PWSCF package, where the energy truncation of the plane wave is 4082 eV and the exchange association energy is chosen to be of the LDA form. Since the quantum Monte Carlo method is still not able to reliably calculate the forces on the atoms, there is no way to optimize the structure. For this reason the structures obtained from the DFT-PBE calculations are used for the calculation of the total energy, and since the weak interactions of the system distort the structures of the water molecules and graphene very little, the adsorption energies obtained for the different structures do not differ much.

Possible high symmetry geometries were chosen to compare the adsorption energies, i.e. one-leg and two-leg structures. In this case, the one-leg structure is the most stable structure of water molecules on benzene molecules and other small aromatic clusters (at the MP2 level), while the two-leg structure is the most stable structure of water molecules on larger aromatic clusters. Figure 9.8 shows them schematically, where (a) shows the two-leg structure and (b) shows the one-leg structure; the top two panels are side views and the bottom two panels are top views; the gray spheres represent carbon atoms, the red spheres represent oxygen atoms, and the white spheres represent hydrogen atoms. It can be clearly seen that in the two-leg structure, the oxygen atom is directly above the center of the six-carbon ring, and the two hydrogen atoms point symmetrically to the two carbon atoms in the graphene. The symmetry of this structure is relatively high; the oxygen atom has no degrees of freedom in the plane parallel to the graphene (x–y plane) and only one degree of freedom in the direction perpendicular to the graphene plane (z direction). Therefore, only this one degree of freedom in the z-direction needs to be considered in the calculation of this structure, i.e., only the one-dimensional adsorption energy profile needs to be calculated. In the actual calculation, the oxygen atom is fixed directly above the center of the six-carbon ring (the carbon atom is fixed while the hydrogen atom is fully relaxed), and then the adsorption energy of the water molecule is calculated at intervals along the z-direction to find the most stable adsorption height of the water molecule by the adsorption energy curve. On the other hand, in the one-leg structure, the oxygen atom is almost at the top position of the carbon atom, but slightly off, with one hydrogen atom pointing to the one below the oxygen atom and the other hydrogen atom almost parallel (slightly upward) to the graphene plane. In this structure the oxygen atom has a lower symmetry than the two-leg structure, it is only symmetric in the y direction, i.e. the oxygen atom has no degrees of freedom, but in both the x and z directions the oxygen atom has degrees of freedom. This means that for this structure we need to calculate the two-dimensional adsorption energy surface. This is more computationally intensive. In the calculations it is found that the variation of the adsorption energy with the y coordinate is insignificant over a wide range when the height of the oxygen atom is kept constant. Figure 9.9 plots the adsorption energy as a function of the y-coordinate of the oxygen atom for the one-leg configuration of the water molecule for a PBE with a height of about 3.6 Å. Notice that the y-coordinate of the oxygen atom is about 1.42 Å when the water molecule is in the center of the six-carbon ring, while the adsorption energy fluctuates by only 1 meV as the y-coordinate changes from 0 to 0.6 Å, with the adsorption energy maximum occurring at about 0.2 Å, and it is also found that as the height of the water molecule changes, the location where the maximum value appears is almost constant.

Therefore, when calculating the one-leg structure, the y-coordinate of the oxygen atom is fixed at ~ 0.2 Å, and a series of points along the z-direction are selected to

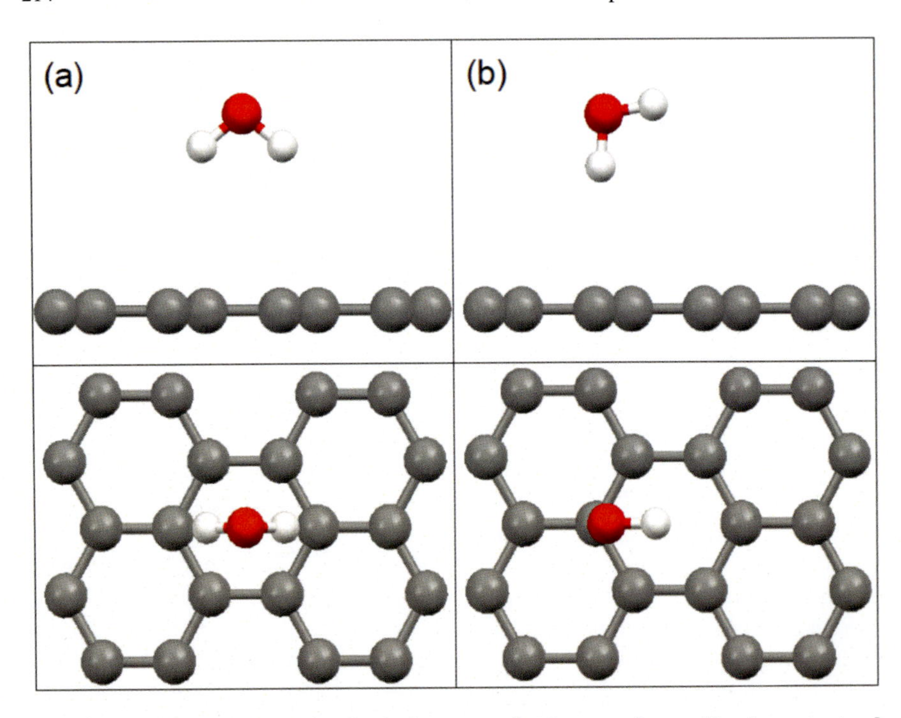

Fig. 9.8 The adsorption geometry of a single water molecule on graphene. **a** Two-leg geometry; **b** one-leg geometry

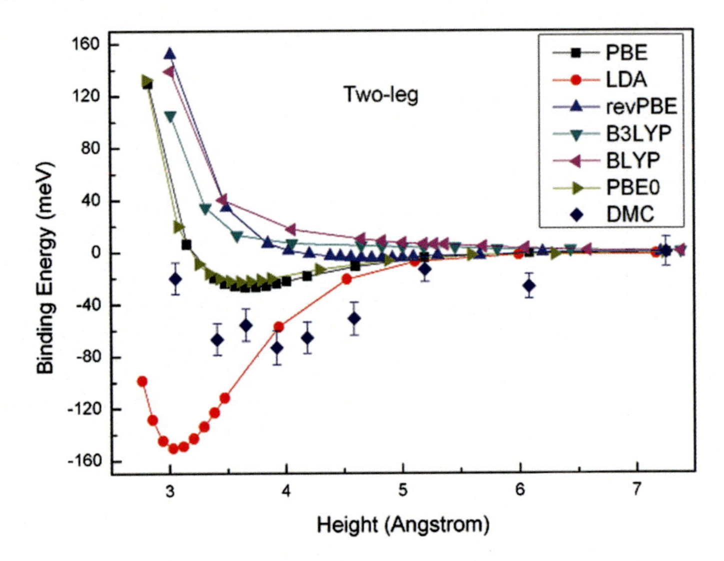

Fig. 9.9 The adsorption energy as a function of adsorption height, for water on graphene in a two-leg geometry

calculate the adsorption energy curve and to find the most stable adsorption height. This reduces a three-dimensional energy surface to two one-dimensional adsorption energy curves, which greatly reduces the workload and also gives us the information we need.

The final adsorption energy curves obtained are shown in Fig. 9.9 (two-leg) and Fig. 9.10 (one-leg). We first look at the adsorption energies obtained for the different exchange-associated forms in the DFT approach. The LDA gives a maximum adsorption energy of about − 150 meV for both configurations, with the one-leg structure being about 10 meV smaller than the two-leg structure, much higher than all the other exchange-associated forms. This is to be expected. Previous experience with calculations tells us that LDA generally overestimates the interaction energy for van der Waals interaction systems, such as rare gas atomic dimers. The adsorption energy for PBE is much smaller, giving about 30 meV for both structures, with the one-leg structure being 4 meV larger than the two-leg structure. The usual experience suggests that PBE generally underestimates the van der Waals interaction energy. Typically, the true value of the interaction energy is between the LDA and the PBE calculation. PBE0 is a hybrid functional version of the PBE that introduces a portion of exact exchange term. In this system, we can see that for both structures PBE0 gives results very close to PBE. This suggests that the introduction of the exact exchange does not improve the performance of DFT method on this system, most likely because the main interaction in this system originates from the correlation term and not from the exchange term. The revPBE functional is a slightly modified version of the parameters of PBE, which gives the energy curves that are closer to those computed by Hartree–Fock. In this system it gives very weak adsorption energies of about 7 meV. BLYP is another very popular form of exchange–correlation function, which in general underestimates interaction energies and overestimates structural parameters such as bond lengths than PBE. Here BLYP gives a completely repulsive interaction energy profile, meaning that there are no stable or metastable adsorption sites at all. B3LYP is a hybrid functional version of BLYP that is very popular in calculations in quantum chemistry and the calculations are often more accurate than other exchange correlations. Similar to the case of PBE0 and PBE, here B3LYP is almost identical to the BLYP curve.

In terms of adsorption energies, the different forms of exchange correlations in DFT give a variety of different adsorption energies from complete repulsion to 150 meV attraction energy, which is also in agreement with previous experience. This suggests that the density generalized method is not a correct description of this system. Essentially this is because the various exchange correlations in DFT are based on local or semi-local approximations, and they have no way of correctly describing the completely non-local van der Waals interactions. Previous literature has shown that the attraction given by the various exchange correlations in density generalized functions usually originates from the exchange term, but the real interaction should be a correlation effect, so even if some exchange correlation in an individual system correctly describes the interaction energy, that is just a coincidence. This is a good illustration of the limitations of the DFT approach.

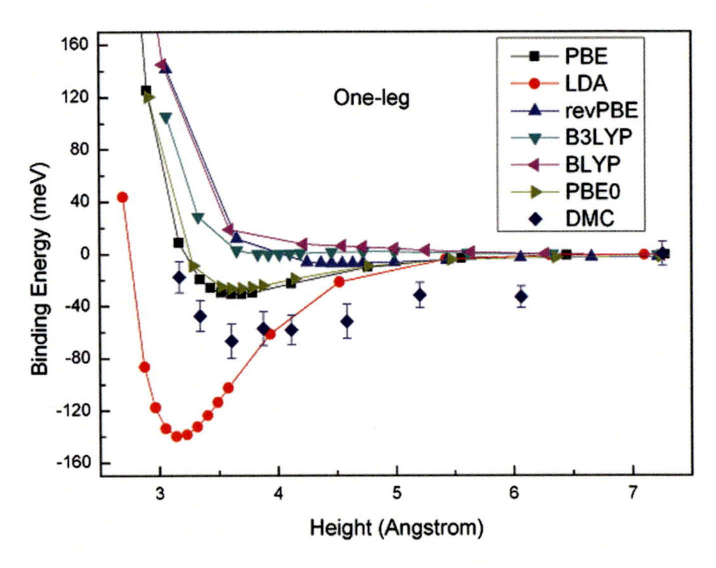

Fig. 9.10 The adsorption energy as a function of adsorption height, for water on graphene in a one-leg geometry

Let's compare again the energy differences between the two different structures. The various adsorption energy curves for both structures are very similar. Of all these exchange–correlation energies, only LDA gives the conclusion that two-leg is more stable than one-leg; several other functionals either give one-leg as more stable than two-leg or do not give any stable adsorption positions. Obviously, there is no reason to believe that a particular form of exchange–correlation will be more accurate than the others, so that the DFT method cannot tell which structure is more stable either. But in general, all exchange–correlation forms give an energy difference of less than 10 meV between the two structures, so we can assume that the two structures are almost degenerate in energy.

Below we discuss the adsorption structure parameters. Let's first look at the two-leg structure. The higher the adsorption energy, the lower the adsorption height of the water molecule, and the more pronounced the changes in OH bond length and HOH bond angle. Since two-leg is a symmetric structure, both hydrogen atoms interact symmetrically with graphene. The one-leg structure, on the other hand, is different. Since its two hydrogen atoms are not symmetrical, we see that the OH bond length of the one parallel to the graphene plane hardly changes, while the OH bond length of the one pointing to the carbon atom changes. This indicates that unlike the two-leg structure, only the hydrogen atom pointing to the carbon atom in the one-leg is involved in the interaction with the graphene, while the other hydrogen atom contributes almost nothing. With the change in the adsorption energy, we can also observe a similar pattern of change in the structural parameters as in the two-leg structure. This shows a link between the adsorption structure and the adsorption energy.

Next we discuss the results of the calculations of the diffusion Monte Carlo method. Since this is a relatively large system (53 atoms, 208 electrons), the Monte Carlo method is very computationally intensive, and the statistical error (error bar) is about 15 meV due to the limitation of computational resources. Due to this error bar, we cannot strictly give the adsorption energy for the diffusion Monte Carlo. Within the error bar, the estimated adsorption energy is around 80 meV. And the adsorption energy curves for the two structures are very close, which means they are almost degenerate. From the comparison of the diffusion Monte Carlo curves and the various DFT curves, none of the curves of the exchange–correlation functionals fall within the error range of the Monte Carlo, which for once shows that none of the various different exchange-correlations in the DFT method can accurately describe the system. And the quantum Monte Carlo curve lies between the LDA and PBE curves, which is also similar to the various calculations we have seen in the past.

Finally we look at charge transfer. In general, charge transfer and redistribution is very useful information for studying system interactions. On the left of Fig. 9.11 is the charge transfer diagram for the two-leg structure, where white and yellow represent increase and decrease of electron density. We can clearly see that there is no charge overlap in the intermediate region between the graphene and water molecules, i.e., there are no chemical bonding interactions. This is a very important feature of the physisorption system. Its interactions arise entirely from non-local interactions due to electron polarization, and not from the overlapping of electron clouds as in the case of chemical bonding. This phenomenon occurs because, after being affected by a water molecule, graphene have a significant charge redistribution only on the carbon atoms nearest to the water molecule, but there is also a charge transfer and redistribution on the carbon atoms in the proximity. This large range of charge redistribution creates this interesting image. It also shows that this system cannot be described by simple nearest-neighbor interactions; it must take into account the interactions of water and the whole graphene. This is directly related to the fact that the energy gap of graphene is zero.

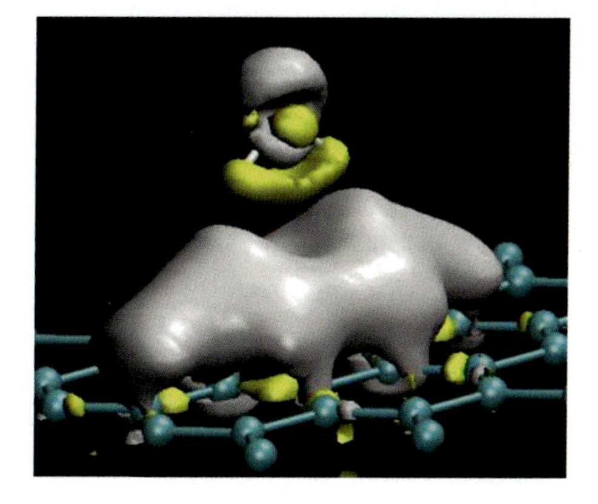

Fig. 9.11 The charge density difference induced by water adsorption on graphene in a two-leg geometry. Gray (yellow) area indicates electron gain (loss)

9.3.3 Interaction of the Water Layers with Graphene

Multiple water molecules adsorbed on graphene will form clusters and layers of water. In general, the hydrogen bonding between water molecules plays a decisive role in the formation of the water structure as the interaction between water molecules and graphene is weak and much smaller than the hydrogen bonding energy. Since the formation of the water structure optimizes the hydrogen bonding between water molecules as much as possible, the interaction between each water molecule and graphene in the water layer is somewhat weaker than in the case of single molecule adsorption.

Li et al. [35] used DFT with van der Waals exchange–correlation functionals to calculate the adsorption scenario of a single water layer on a graphene layer on free and metal substrates. They found that the adsorption energies of both free OH upward and downward water layers on graphene are essentially the same, 542 and 540 meV/H_2O, respectively; the adsorption energy of the water layer essentially comes from the hydrogen bonding of the water molecules $300 \times 3/2 = 450$ meV. They also found that the adsorption of the water layer opens a small energy gap in graphene of about 40 meV. Surprisingly, while aqueous adsorption can hardly shift the Dirac energy level position of free graphene, i.e., it cannot dope it electronically; for graphene on a metal substrate, aqueous adsorption can dramatically change its Dirac energy level position by up to 120 meV, i.e., it can be adjusted by changing the structure of the adsorbed water layer electronic and transport properties of graphene by changing the structure of the adsorbed water layer.

Galli et al. simulated the dynamic distribution of the water layer on graphene by first-principles calculations [36]. In particular, Li and Zeng [37] simulated the infiltration of the water layer on graphene using the van der Waals modified DFT method (DFT + D). Performing molecular dynamics simulations by first-principles, they obtained the wetting angle of the water cluster as 87° (Fig. 9.12), which agrees well with the experimental value of about 86° ~ 93°, fully demonstrating the reliability and validity of first-principles in calculating some macroscopic properties of water.

9.4 Water Splitting on Metal-Free Photocatalysts

One profound matter related to water at surfaces is hydrogen production, arising from photocatalytic water splitting, which provides a promising way to harvest solar energy and gain clean renewable energy. The most essential topic of photocatalytic water splitting is the interactions between water molecules and photocatalytic materials.

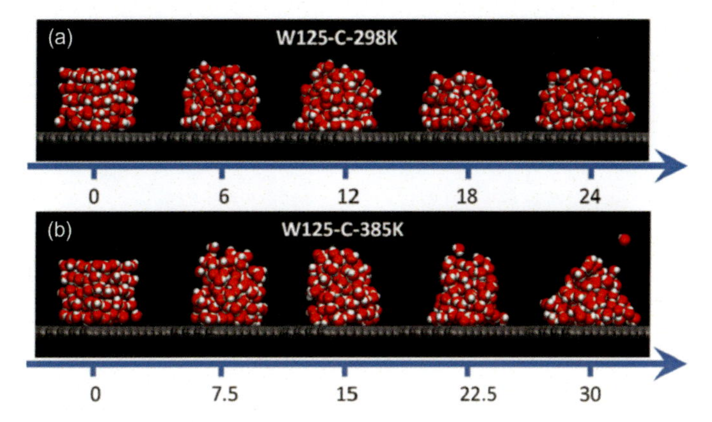

Fig. 9.12 A first-principles molecular dynamics simulations of a cluster comprising 125 water molecules on graphene. The horizontal axis is time (unit: ps). The cluster takes a cubic shape at time zero. **a** The simulation temperature is 298 K, achieving a contact angle of 96°; **b** The simulation temperature is 385 K, achieving a contact angle of 87°

9.4.1 TiO_2 Catalyzes Water Splitting

Rutile (R-TiO_2) is the most stable form of titanium dioxide in nature, has a narrower band gap than anatase, and is considered to be a typical model for metal oxide systems used in water chemistry research. As mentioned before, water chemistry on the surface of rutile (110) has been studied extensively by various techniques such as scanning tunneling microscopy, infrared reflection–absorption spectroscopy, and two-photon photoelectron spectroscopy. So far, many of the basic aspects of adsorption have been established, for example, the adsorption of water molecules to the five-coordination Ti^{4+}(Ti_{5c}) site of R-TiO_2(110) [38]. However, the dissociation of H_2O at the Ti_{5c} site (H_2O_{Ti}) is still poorly understood, no evidence of H_2O_{Ti} dissociation was found on R-TiO_2(110) surface by direct STM imaging at low temperatures.

The photochemistry of H_2O_{Ti} monomers on the R-TiO_2(110) surface was investigated using STM by Tan and co-workers for the first time [39]. As shown in Fig. 9.13, before ultraviolet (UV) light illumination, adsorption of about 0.02 ML H_2O on R-TiO_2(110) at 80 K results in molecular H_2O_{Ti} monomers on the surface, which are marked by black circles. Upon 400 nm light irradiation, a clear observation of HO–H bond cleavage is detected, leading to the formation of surface OH_b and OH_t (OH on Ti_{5c} sites) species. However, in most cases, the authors found that the OH_t groups produced from H_2O dissociation are missing, leaving behind OH_b on the surface. Even OH_t radicals stay on the surface, they appear at the Ti_{5c} sites far away from the original sites of H_2O_{Ti} monomers adsorption. In addition, the authors found that the efficiency of H_2O_{Ti} monomer dissociation is independent of the photon energy or the relative absorptivity of the R-TiO_2(110) surface at different wavelengths but dependent on irradiation time. Moreover, with 266 and 355 nm irradiation, the photoactivity

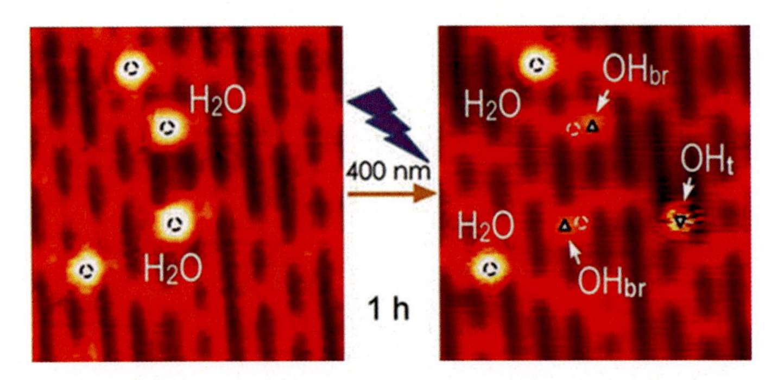

Fig. 9.13 STM images showing the photocatalytic dissociation of H_2O on R-TiO_2(110) after irradiating the H_2O monomer covered R-TiO_2(110) surface for 1 h (black circles, adsorbed H_2O_{Ti} monomers; white circle, sites for dissociated H_2O adsorption; triangles, OH_b; inverted triangles, OH_t)

does not scale with the photon flux as well. Then, these authors proposed that photocatalytic H_2O dissociation on R-TiO_2(110) is a thermalized holes-induced oxidation process, as shown in the formula 9.1, and the net H_2O dissociation occurs only via the exciton interfacial proton-coupled electron transfer mechanism in which the hole of an exciton is located at the occupied VB levels contributed from the $1b_1$ orbital of an H_2O_{Ti} molecule.

$$H_2O_{Ti} + O_b + h^+ \rightarrow \cdot OH(gas) + OH_b \tag{9.1}$$

However, according to the results of ultraviolet photoelectron spectroscopy (UPS), the HOMO of H_2O_{Ti} on R-TiO_2(110) lies about 4 eV below the edge of the TiO_2 valence band. Such a large energy mismatch raises doubt about the direct holes transfer to H_2O_{Ti} or the interfacial exciton generation with the holes directly located at the H_2O_{Ti} molecules.

Earlier, Migani and Blancafort calculated by DFT and proposed that the desorption of OH_t radicals on R-TiO_2(110) could occur only when the irradiation wavelength is $\leq$ 349 nm [40]. Further, Yang's group conducted careful experiments to study H_2O photochemistry on the R-TiO_2(110) surface by varying H_2O coverage and irradiation wavelength [41, 42]. They found the H_2O dissociation process is only detected under 266 nm irradiation by both STM and TPD methods, suggesting that H_2O dissociation on R-TiO_2(110) is wavelength-dependent or photon energy dependent. All these indicate that the photolysis of H_2O on R-TiO_2(110) is energy dependent, and it is still a challenge to identify possible charge/energy transfers in H_2O photolysis on the TiO_2 surface.

9.4.2 Photocatalytic Water Splitting on g-C_3N_4

A new perspective on photocatalytic water splitting is to adopt metal-free carbon photocatalysts, such as modified graphene, carbon nanotubes (CNTs), g-C_3N_4, etc., owing to their nature of non-corrosion, non-toxicity, easy preparation, and handling. Many of them show comparable catalytic performances to metal-containing photocatalysts for water splitting. Pristine graphene, known as a zero band-gap semiconductor, in general, has no photocatalytic activities. However, its properties can be adjusted by the strategies of heteroatom doping, size modulation, and morphology control, to improve the catalytic activities [43]. For example, graphite oxide, incorporating oxygen atoms into graphene framework to form sp^3 hybridization, opens the bandgap to 2.4–4.3 eV and enables catalyzed hydrogen production under solar irradiation [44]. Another recent development on ternary boron carbon nitride alloy demonstrated that by doping carbon within hexagonal boron nitride (h-BN) nanosheet, the bandgap of the alloy can be tuned by the amount of incorporated carbon and the hydrogen or oxygen evolution from water can be catalyzed under visible light illumination [45].

Among these metal-free photocatalysts, g-C_3N_4 has attracted great attention since it was reported as a potential photocatalyst by Wang and co-workers [46], attributed to its high chemical and thermal stability, inexpensiveness, and nontoxic nature. The band gap of g-C_3N_4 is 2.7 eV, which is suitable for visible light absorption, and the hydrogen evolution rate can reach 10 μmol/h with the assistance of Pt cocatalyst. Theoretical calculations based on the DFT have been widely used to study the adsorption configurations and energy profiles of transition states of water on g-C_3N_4 materials. Although a detailed understanding of the structural and electronic properties of g-C_3N_4 is possible to be gained within the DFT framework, this theory is not viable for revealing the real excitation dynamics of photocatalytic water splitting on g-C_3N_4. For time-dependent processes especially involving excited states, rt-TDDFT and advanced nonadiabatic algorithms are essential [47]. Ab initio rt-TDDFT molecular dynamics investigations of water/g-C_3N_4 interface unveil a three-step mechanism for the photocatalytic water splitting process on g-C_3N_4 [48], including photoexcitation, oxidation transfer, and reduction transfer stages. Under photoexcitation, the hole transfer occurs first from carbon to nitrogen atom in g-C_3N_4 sheet, and then a hole current from nitrogen atom to water leads to OH bond weakening in the molecule. Finally, a reverse hole flow from water to nitrogen results in the proton transfer from water to g-C_3N_4, as shown in Fig. 9.14. Different from traditional scheme [49], the hole transfer plays a key role in the nonadiabatic photocatalytic process and dominates the water splitting stages on g-C_3N_4. Followed by photocatalytic water decomposition, where two dissociated hydrogen atoms approach each other and form H_2 gas molecules, while the remaining OH radicals can form intermediates (e.g., H_2O_2). It is important to reveal the nature of photocatalysis and further improve the catalytic efficiency of g-C_3N_4. It also provides new insights for the characterization and further development of efficient water decomposition photocatalysts from a dynamic perspective.

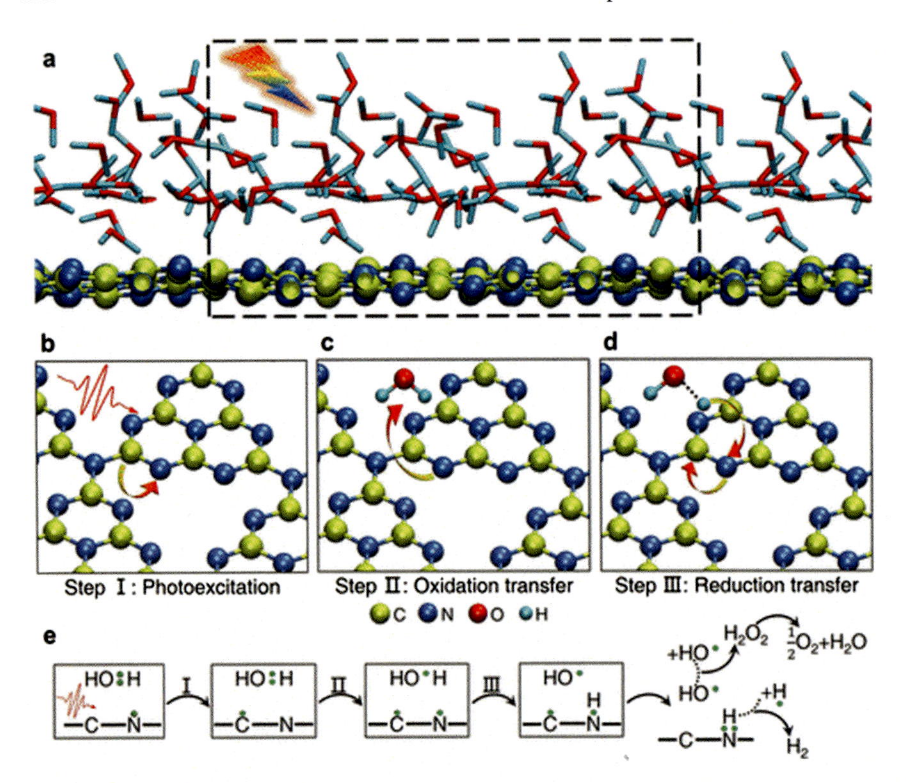

Fig. 9.14 **a** Atomic configuration of the g-C_3N_4/water interface in the simulation supercell and its periodic images. **b–d** Schematic diagrams of three-step photocatalytic water splitting processes on g-C_3N_4. The arrows indicate the direction of hole/H^+ current. **e** General scheme for entangled electron-nuclei mechanism in the three-step model, where green dots denote typical electrons. Possible formations of hydrogen peroxide, oxygen gas, and hydrogen gas are also displayed

9.5 Proton Transport Through Graphdiyne Membrane

Molecular sieving is of great importance to proton exchange in fuel cells, water desalination, and gas separation. Two-dimensional materials, e.g., graphene, graphene oxides, and h-BN, have been shown to have good molecular permeability and selectivity. The molecular conductivity originates from the nanoscale transport channels, including the nanopores on membrane planes and the nanochannel between adjacent layers, which can be adjusted by precisely controlling the size of the transport channel, so as to improve molecular selectivity.

Two-dimensional nanomesh materials have large nanopore density and perfect uniformity, and thus are considered as excellent candidate materials for molecular sieving. Graphdiyne [50], a novel two-dimensional carbon allotrope, as shown in Fig. 9.15a, has attracted great attention because of its exceptional mechanical, physical, and chemical properties. Early classical molecular dynamics (CMD) simulations showed that graphdiyne is impermeability for water molecules [51, 52], while first-principles calculations instead pointed out the possibility of water conductivity when considering the transmembrane interaction [53]. Recently, water transportability of graphdiyne has been studied by extensive FPMD and ab-initio parameterized CMD simulations [54]. It is revealed that water molecules can permeate through graphdiyne membrane with a relatively large flow rate at high external pressure, as shown in Fig. 9.15b. The flow rate can be written as $\Phi = A_0 \exp(-E_a/(k_B T))$. The activation energy E_a can be estimated as $E_a = E_{mem} + E_{water} - PV$, where E_{mem} and E_{water} are the contributions from the membrane and other water molecules through hydrogen bonding, respectively, and PV represents the effect of pressure and temperature. The activation energy can be directly modulated by pressure, leading to a nonlinear dependence of water flow on pressure. Therefore, the activated water flow in graphdiyne can be precisely controlled by adjusting the working pressure. Further analysis of the atomistic dynamics of water molecules shows that a transmembrane hydrogen bond is formed and a unique two-hydrogen-bond structure acts as transient state in the process of water transport [54].

Considering the Grotthuss mechanism of proton hopping, the transmembrane hydrogen bond could be an efficient bridge for transmembrane proton transport. As

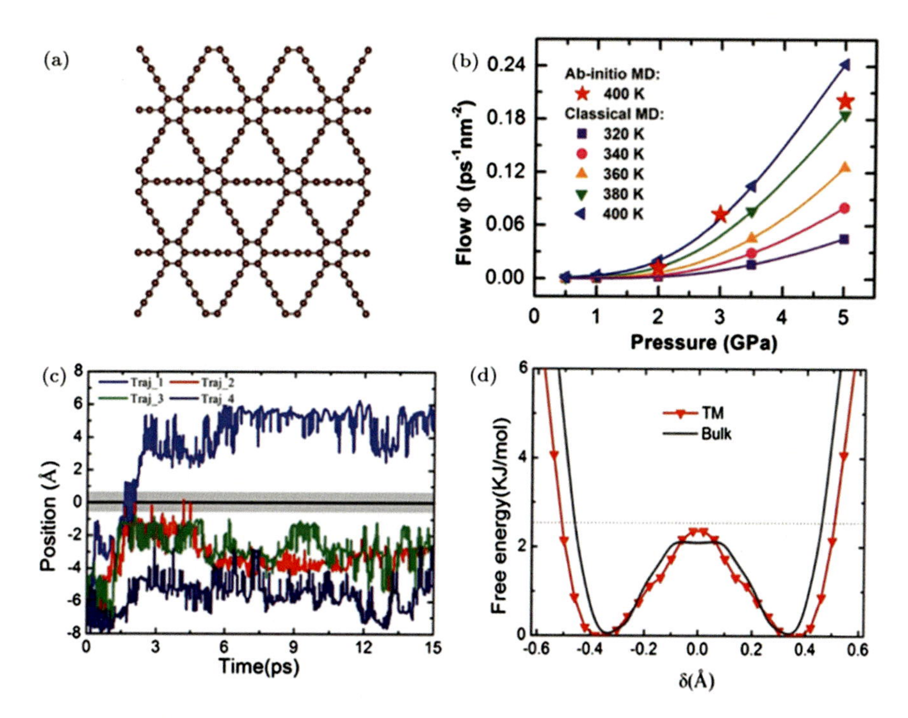

Fig. 9.15 **a** The atomistic structure of graphdiyne. **b** The water flow across graphdiyne versus temperature and pressure. **c** The trajectories of proton diffusion at the water–graphdiyne interface. **d** The free energy barrier for transmembrane (TM) proton transfer and proton transfer in bulk water. The dash line indicates the $k_B T$ at 300 K

shown in Fig. 9.15c, a proton diffuses across graphdiyne membrane in Traj_1 under unbiased conditions, indicating that the thermal fluctuations are the driven force for transmembrane proton transport [55]. Free energy calculations for proton transfer display the energy barrier of transmembrane proton transfer is nearly the same as that of bulk water, and both energy barriers are smaller than $k_B T$. Proton conductivity of graphdiyne, obtained with FPMD simulations under electric field, is 0.6 S cm^{-1}, one order of magnitude larger than that of Nafion [56]. The appropriate nanopore size of graphdiyne can hinder the transport of ions and soluble fuel molecules, and give graphdiyne superior proton selectivity. Therefore, graphdiyne as a proton exchange membrane in fuel cell has a promising application prospect.

References

1. J. Hu, X.D. Xiao, D.F. Ogletree, M. Salmeron, Imaging the condensation and evaporation of molecularly thin films of water with nanometer resolution. Science **268**, 267 (1995)
2. A. Fujishima, K. Honda, Electrochemical photolysis of water at a semiconductor electrode. Nature **238**, 5358 (1972)
3. J.H. Carey, B.G. Oliver, Intensity effects in electrochemical photolysis of water at TiO_2 electrode. Nature **259**, 554 (1976)
4. L. Giordano, J. Goniakowski, J. Suzanne, Partial dissociation of water molecules in the (3 × 2) water monolayer deposited on the MgO(100) surface. Phys. Rev. Lett. **81**, 1271 (1998)
5. H.-J. Shin, J. Jung, K. Motobayashi, S. Yanagisawa, Y. Morikawa, Y. Kim, M. Kawai, State-selective dissociation of a single water molecule on an ultrathin MgO film. Nat. Mater. **9**, 442 (2010)
6. G.S. Parkinson, Z. Novotny, P. Jacobson, M. Schmid, U. Diebold, Room temperature water splitting at the surface of magnetite. J. Am. Chem. Soc. **133**, 12650 (2011)
7. P.J.D. Lindan, N.M. Harrison, M.J. Gillan, Mixed dissociative and molecular adsorption of water on the rutile (110) surface. Phys. Rev. Lett. **80**, 762 (1998)
8. D.A. Duncan, F. Allegretti, D.P. Woodruff, Water does partially dissociate on the perfect TiO_{2}(110) surface: a quantitative structure determination. Phys. Rev. B **86**, 045411 (2012)
9. L.-M. Liu, C. Zhang, G. Thornton, A. Michaelides, Reply to "comment on 'structure and dynamics of liquid water on rutile TiO_2 (110).'" Phys. Rev. B **85**, 167402 (2012)
10. T. He, J.L. Li, G.W. Yang, Physical origin of general oscillation of structure, surface energy, and electronic property in rutile TiO_2 nanoslab. ACS Applied Materials and Interface **4**, 2192 (2012)
11. R. Wang, K. Hashimoto, A. Fujishima, M. Chikuni, E. Kojima, A. Kitamura, M. Shimohigoshi, T. Watanabe, Light-induced amphiphilic surfaces. Nature **388**, 431 (1997)
12. S.U.M. Khan, M. Al-Shahry, W.B. Ingler Jr., Efficient photochemical water splitting by a chemically modified n-TiO_2. Science **297**, 2243 (2002)
13. U. Diebold, J. Lehman, T. Mahmoud, M. Kuhn, G. Leonardelli, W. Hebenstreit, M. Schmid, P. Varga, Intrinsic defects on a TiO_2(110)(1x1) surface and their reaction with oxygen: a scanning tunneling microscopy study. Surf. Sci. **411**, 137 (1998)
14. U. Diebold, The surface science of titanium dioxide. Surf. Sci. Rep. **48**, 53 (2003)
15. S. Suzuki, K. Fukui, H. Onishi, Y. Iwasawa, Hydrogen adatoms on TiO_2(110) − (1 x 1) characterized by scanning tunneling microscopy and electron stimulated desorption. Phys. Rev. Lett. **84**, 2156 (2000)
16. R. Schaub, E. Wahlstroem, A. Ronnau, E. Laegsgaard, I. Stensgaard, F. Besenbacher, Oxygen-mediated diffusion of oxygen vacancies on the TiO_2(110). Surf. Sci. **299**, 377 (2003)

17. O. Bikondoa, C.L. Pang, R. Ithnin, C.A. Muryn, H. Onishi, G. Thornton, Direct visualization of defect-mediated dissociation of water on TiO_2(110). Nat. Mater. **5**, 189 (2006)
18. C.Y. Zhou, Z.F. Ren, S.J. Tan, Z.B. Ma, X.C. Mao, D.X. Dai, H.J. Fan, X.M. Yang, J. LaRue, R. Cooper, A.M. Wodtke, Z. Wang, Z.Y. Li, B. Wang, J.L. Yang, J.G. Hou, Site-specific photocatalytic splitting of methanol on TiO_2(110). Chem. Sci. **1**, 575 (2010)
19. B. Meyer, D. Marx, O. Dulub, U. Diebold, M. Kunat, D. Langenberg, C. Woll, Partial dissociation of water leads to stable superstructures on the surface of zinc oxide. Angew. Chem. Int. Ed. **43**, 6642 (2004)
20. J.J. Yang, S. Meng, L.F. Xu, E.G. Wang, Ice tessellation on a hydroxylated silica surface. Phys. Rev. Lett. **92**, 146102 (2004)
21. J.J. Yang, S. Meng, E.G. Wang, Water adsorption on hydroxylated silica surfaces studied using density functional theory. Phys. Rev. B. **71**, 035413 (2005)
22. J.J. Yang, E.G. Wang, Water adsorption on hydroxylated alpha-quartz (0001) surfaces: from monomer to flat bilayer. Phys. Rev. B. **73**, 035406 (2006)
23. J.J. Yang, E.G. Wang, Reaction of water on silica surfaces. Curr. Opin. Solid State Mater. Sci. **10**, 33 (2006)
24. V. Ostroverkhov, G.A. Waychunas, Y.R. Shen, New information on water interfacial structure revealed by phase-sensitive surface spectroscopy. Phys. Rev. Lett. **94**, 046102 (2005)
25. I.M.P. Aarts, A.C.R. Pipino, J.P.M. Hoefnagels, W.M.M. Kessels, M.C.M. van de Sanden, Quasi-ice monolayer on atomically smooth amorphous SiO_2 at room temperature observed with a high-finesse optical resonator. Phys. Rev. Lett. **95**, 166104 (2005)
26. T. Werder, J.H. Walther, R.L. Jaffe, T. Halicioglu, P. Koumoutsakos, On the water–carbon interaction for use in molecular dynamics simulations of graphite. J. Phys. Chem. B **107**, 1345 (2003)
27. Z. Li, Y. Wang, A. Kozbial, G. Shenoy, F. Zhou, R. McGinley, P. Ireland, B. Morganstein, A. Kunkel, S.P. Surwade, L. Li, H.T. Liu, Effect of airborne contaminants on the wettability of supported graphene and graphite. Nature Materials **12**, 925 (2013)
28. J. Ma, D. Alfe, A. Michaelides, E.G. Wang, The water-benzene interaction: insight from electronic structure theories. J. Chem. Phys. **130**, 154303 (2009)
29. D. Feller, K.D. Jordan, Estimating the strength of the water/single-layer graphite interaction. J. Phys. Chem. A **104**, 9971 (2000)
30. C.S. Lin, R.Q. Zhang, S.T. Lee, M. Elstner, T. Frauenheim, L.J. Wan, Simulation of water cluster assembly on a graphite surface. J. Phys. Chem. B **109**, 14183 (2005)
31. I.W. Sudiarta, D.J.W. Geldart, Interaction energy of a water molecule with a single-layer graphitic surface modeled by hydrogen- and fluorine-terminated clusters. J. Phys. Chem. A **110**, 10501 (2006)
32. P.C. Sanfelix, S. Holloway, K.W. Kolasinski, G.R. Darling, The structure of water on the (0001) surface of graphite. Surf. Sci. **532**, 166 (2003)
33. H. Ruuska, T.A. Pakkanen, Ab initio model study on a water molecule between graphite layers. Carbon **41**, 699 (2003)
34. J. Ma, A. Michaelides, D. Alfe, L. Schimka, G. Kresse, E.G. Wang, Adsorption and diffusion of water on graphene from first principles. Phys. Rev. B **84**, 033402 (2011)
35. X. Li, J. Feng, E.G. Wang, S. Meng, J. Klimes, A. Michaelides, Influence of water on the electronic structure of metal-supported graphene: insights from van der Waals density functional theory. Phys. Rev. B **85**, 085425 (2012)
36. G. Cicero, J.C. Grossman, E. Schwegler, F. Gygi, G. Galli, Water confined in nanotubes and between graphene sheets: a first principle study. J. Am. Chem. Soc. **130**, 1871 (2008)
37. H. Li, X.C. Zeng, Wetting and interfacial properties of water nanodroplets in contact with graphene and monolayer boron-nitride sheets. ACS Nano **6**, 2401 (2012)
38. B. Hammer, S. Wendt, F. Besenbacher, Water adsorption on TiO_2. Top. Catal. **53**, 423 (2010)
39. S. Tan, H. Feng, Y. Ji, Y. Wang, J. Zhao, A. Zhao, B. Wang, Y. Luo, J. Yang, J.G. Hou, Observation of photocatalytic dissociation of water on terminal Ti sites of $TiO_2(110) - 1\times1$ surface. J. Am. Chem. Soc. **134**, 9978 (2012)

40. C.A. Muryn, P.J. Hardman, J.J. Crouch et al., Step and point defect effects on $TiO_2(100)$ reactivity. Surface Science **251** (1991)
41. Q. Guo, C. Xu, Z. Ren, W. Yang, Z. Ma, D. Dai, H. Fan, T.K. Minton, X. Yang, Stepwise photocatalytic dissociation of methanol and water on $TiO_2(110)$. J. Am. Chem. Soc. **134**, 13366 (2012)
42. W. Yang, D. Wei, X. Jin, C. Xu, Z. Geng, Q. Guo, Z. Ma, D. Dai, H. Fan, X. Yan, The effect of hydrogen bond in photo-induced water dissociation: a double edged sword. J. Phys. Chem. Lett. **7**, 603 (2016)
43. Y. Xu, M. Kraft, R. Xu, Metal-free carbonaceous electrocatalysts and photocatalysts for water splitting. Chem. Soc. Rev. **45**, 3039 (2016)
44. T.F. Yeh, J.M. Syu, C. Chen, T.H. Chang, H. Teng, Graphite oxide as a photocatalyst for hydrogen production from water. Adv. Funct. Mater. **20**, 2255 (2010)
45. C. Huang, C. Chen, M. Zhang, L. Lin, X.X. Ye, S. Lin, M. Antonietti, X.C. Wang, Carbon-doped BN nanosheets for metal-free photoredox catalysis. Nat. Commun. **6**, 7698 (2015)
46. X. Wang, K. Maeda, A. Thomas, K. Takanabe, G. Xin, J.M. Carlsson, K. Domen, M. Antonietti, A metal-free polymeric photocatalyst for hydrogen production from water under visible light. Nat. Mater. **8**, 76 (2009)
47. P.W. You, D.Q. Chen, C. Lian, C. Zhang, S. Meng, First-principles dynamics of photoexcited molecules and materials towards a quantum description. WIREs. Comput. Mol. Sci. **11**, 1492 (2020)
48. P.W. You, C. Lian, J.Y. Xu, C. Zhang, S. Meng, Nonadiabatic dynamics of photocatalytic water splitting on a polymeric semiconductor. Nano Lett. **15**, 6449 (2020)
49. H. Ma, J. Feng, F. Jin, M. Wei, C. Liu, Y. Ma, Where do photogenerated holes at the g-C3N4/water interface go for water splitting: H_2O or OH^-? Nanoscale **10**, 15624 (2018)
50. G. Li, Y. Li, H. Liu, Y. Guo, Y. Li, D. Zhu, Precise and ultrafast molecular sieving through graphene oxide membranes. Chem. Commun. **46**, 3256 (2010)
51. M. Xue, H. Qiu, W. Guo, Exceptionally fast water desalination at complete salt rejection by pristine graphyne monolayers. Nanotechnology **24**, 505720 (2013)
52. C. Zhu, H. Li, X.C. Zeng, E.G. Wang, S. Meng, Quantized water transport: ideal desalination through graphyne-4 membrane. Sci. Rep. **3**, 3163 (2013)
53. M. Bartolomei, E. Carmona-Novillo, M.I. Hernandez, J. Campos-Martinez, F. Pirani, G. Giorgi, K. Yamashita, Penetration barrier of water through graphynes' pores: first-principles predictions and force field optimization. J. Phys. Chem. Lett. **5**, 751 (2014)
54. J. Xu, C. Zhu, Y. Wang, H. Li, Y. Huang, Y. Shen, J.S. Francisco, X.C. Zeng, S. Meng, Water transport through subnanopores in the ultimate size limit: mechanism from molecular dynamics. Nano Res. **12**, 587 (2018)
55. J. Xu, H. Jiang, Y. Shen, X.Z. Li, E.G. Wang, S. Meng, Transparent proton transport through a two-dimensional nanomesh material. Nat. Commun. **10**, 3971 (2019)
56. S. Ochi, O. Kamishima, J. Mizusaki, J. Kawamura, Investigation of proton diffusion in Nafion®117 membrane by electrical conductivity and NMR. Solid State Ionics **180**, 580 (2009)

Chapter 10
Macroscopic and Microscopic Pictures of Surface Wetting

10.1 Wetting Phenomena in Nature

From a macroscopic point of view, when a liquid is deposited on a solid surface, due to the existence of forces between the molecules of the liquid and between the molecules of the solid and the liquid, these two forces will eventually tend to balance, making the droplet surface at the edge of the droplet and the plane of the solid at an angle, this phenomenon is called immersion, the angle formed is called the contact angle. Water is the most common liquid. Generally, people refer to surfaces where the contact angle of water droplets is less than 90° as hydrophilic surfaces and surfaces where the contact angle is greater than 90° as hydrophobic surfaces. Wetting plays a vital role in many industrial, biological, physical, chemical, and bionic processes. For example, the Namib Desert in southwestern Africa receives very little rainfall, but the mornings are very foggy and often windy. Whenever it is windy, a desert beetle (Fig. 10.1a) will tilt its body so that its back is facing the wind to collect moisture. Water droplets will flow down the beetle's back to its mouth thanks to a special microstructure on the surface of the beetle (Fig. 10.1b, c), which allows the beetle to collect water in the desert and keep this water from evaporating by producing alternating pro- and anti-hydrophobic configurations on the surface of the beetle [1]. Again, for example, the hydrophobic interaction of amino acid groups at the microscopic scale is the main driver of protein folding in living organisms. It follows that it is important to understand the phenomenon of surface infiltration.

Since wetting is a phenomenon that occurs at an interface, it is strongly related to surface tension (i.e., surface energy). So what is surface tension? In short, surface tension or surface energy is the amount of energy required to form a unit area of the surface. Let's take a liquid as an example, as in Fig. 10.2a, where in the bulk phase each molecule is subjected to forces from surrounding molecules in all directions, resulting in a net force of 0. But molecules that are at the surface of a liquid are not subjected to forces in all directions, and this unevenness of forces, resulting in a net force pointing inward, is what causes surface tension. In terms of energy, because of the forces between molecules, when a molecule is in contact with another molecule,

S. Meng and E. Wang, *Water*,
https://doi.org/10.1007/978-981-99-1541-5_10

Fig. 10.1 **a** A desert beetle in which the protrusions and grooves can be clearly seen on the back. **b** A bulge of the beetle where the peak of the bulge is not stained (still black) while the rest of the beetle is stained. **c** Scanning electron micrograph of the stained area [1]. Reference ruler sizes: **a** 10 mm; **b** 0.2 mm; **c** 10 μm

it will have less energy than it would have in isolation. The number of neighboring molecules on the surface is significantly less than that in the bulk phase. In order to keep its total energy down, it will try to reduce the surface area. It is the surface tension that allows liquids to exhibit different infiltration phenomena on solid surfaces.

As early as 1804, British scientist Thomas Young discovered that droplets have a constant contact angle (θ) when they come into contact with a solid. Infiltration can also be divided into partial infiltration ($\theta > 0°$) and complete infiltration ($\theta = 0°$). For the partial infiltration case, the part of the solid surface that is infiltrated by the liquid is bounded by a three-phase line £ (for Fig. 10.2b, it is a circle). Near the three-phase line £, the structure is very complex and depending on the specific system, and this region is called the core region. However, we can avoid this region to discuss the relationship between the infiltration angle and the surface energy. Thomas Young proposed that when a liquid is in a steady state at the surface, moving the boundary of the droplet (dx) does not make a change in its energy. According to Fig. 10.2b, we know that when £ moves dx, (1) the energy of its bulk phase does not change; (2) the energy of the core region does not change either, given that the core region is simply translated; (3) the solid/gas, solid/liquid, and gas/liquid areas are elevated dx

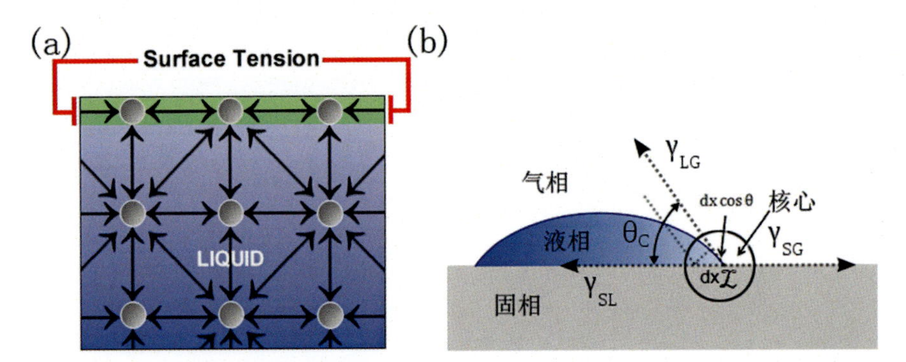

Fig. 10.2 **a** The forces acting on surface liquid molecules and bulk phase liquid molecules. **b** Schematic diagram of a droplet on a solid surface, £ giving the three-phase line perpendicular to the paper

(for the solid/gas interface), $-dx$ (for the solid/liquid interface), and $-dx \cos \theta_c$ (for the gas/liquid interface). At this point it is obtained that:

$$\gamma_{SG} dx - \gamma_{SL} dx - \gamma_{LG} \cos \theta_c dx = 0$$

Cancelling out dx,

$$\gamma_{SG} - \gamma_{SL} - \gamma_{LG} \cos \theta_c = 0 \quad (10.1)$$

This is the very famous Young's equation. Here, γ_{SG}, γ_{SL} and γ_{LG} are the interfacial energies of solid/gas, solid/liquid and gas/liquid, respectively. θ_c is the angle between the tangent line of the droplet at the triple-phase point and the bottom surface, or the contact angle, and its magnitude characterizes the wettability of a solid surface to a liquid.

From Eq. (10.1), it follows that when

$$\gamma_{LG} = \gamma_{SG} - \gamma_{SL}$$

$\cos \theta_c = 1$ ($\theta_c = 0$), which is also known as complete infiltration. Then if

$$\gamma_{SG} > \gamma_{LG} + \gamma_{SL} \quad (10.2)$$

what will happen? In fact, this situation is impossible in thermodynamic equilibrium, because assuming that Eq. (10.2) holds, from an energy point of view, in order to reduce the total energy, the surface of the solid at equilibrium must contain a liquid film, so that the true γ_{SG} is equal to $\gamma_{LG} + \gamma_{SL}$, i.e., making it completely infiltrated.

If we need to deal with the non-equilibrium case, we may get solid/gas interfacial energy γ_{SO} greater than $\gamma_{LV} + \gamma_{SL}$. The difference between them

$$S = \gamma_{SO} - \gamma_{LG} - \gamma_{SL} \quad (10.3)$$

is called the diffusion coefficient. Physically, we can think γ_{SO} is the surface energy of a "dry" surface (ideally without a liquid layer on the surface), while γ_{SG} is the surface energy of a "wet" surface (in practice, there is often a layer of liquid on the surface). For different systems, they vary considerably. For example, for water on a metal oxide surface, $\gamma_{SO} - \gamma_{SG} \sim 300$ ergs/cm^2; while for organic liquids on metal oxide surfaces, $\gamma_{SO} - \gamma_{SG} \sim 60$ ergs/cm^2. How to determine S? We can define S by the thickness of the liquid film after diffusion equilibrium on the surface. The smaller S is, the thicker the film is.

10.2 Classical Wetting Models

Further, it is natural to ask, how does surface wettability relate to the composition of the liquid or solid? Let us first look at solids, which can be simply divided into two categories based on the interatomic forces: (a) Hard solids (strong bonding interactions between atoms through covalent, ionic and metallic bonds). (b) Weak molecular solids (through weaker interactions like van der Waals forces or hydrogen bonds) [2]. Hard solids have a high surface energy because they carry many unsaturated suspended bonds on their surface ($\gamma_{SO} \sim 500 - 5000$ ergs/cm^2), while molecular solids have relatively low surface energy ($\gamma_{SO} \sim 50$ ergs/cm^2). At high-energy surfaces, most liquids can achieve complete infiltration. Why is this? By the Dupré equation.

$$\gamma_{SL} = \gamma_{SO} + \gamma_L - W_{SL}\ (W_{SL} > 0) \tag{10.4}$$

In this context, γ_L is the surface energy of the free surface of the liquid, and W_{SL} is the interaction potential between the liquid and the solid. Equation (10.4) can be understood as follows: first we have a solid surface and a liquid surface, respectively, which are far enough apart so that it has an energy of $\gamma_{SO} + \gamma_L$, when they come into contact with each other, they get a solid–liquid interface, requiring work W_{SL}.

Furthermore, by the same reasoning we get

$$2\gamma_{LG} = W_{LL}\ (W_{LL} > 0) \tag{10.5}$$

In this context, W_{LL} is the interaction potential between the liquids. From Eqs. (10.4) and (10.5), we can obtain the diffusion coefficient defined in (10.3)

$$S = \gamma_{SO} - \gamma_{LG} - \gamma_{SL} = W_{SL} - W_{LL} \tag{10.6}$$

For complete infiltration, it requires $S > 0$, which also means that

$$W_{SL} > W_{LL} \tag{10.7}$$

Consider that there are mainly van der Waals forces between solid and liquid molecules, and between liquid molecules, and that the van der Waals potential is proportional to the polarization rate α_i of substance i, i.e.,

$$W_{ij} = k\alpha_i\alpha_j \tag{10.8}$$

Here, k is independent of substance. Substituting (10.8) into (10.7) yields

$$\alpha_S > \alpha_L \tag{10.9}$$

It can be seen that the high-energy surface can achieve complete infiltration of the liquid, not because the γ_{SO} is relatively large, but because the hard solid has a greater polarizability relative to most liquids.

For low-energy surfaces, on the other hand, often only very weakly polar liquids can be infiltrated. Using the example of the infiltration of n-alkane droplets on a solid surface, for polyethylene solids, this series of alkanes can achieve complete infiltration. However, when placed on the surface of a polytetrafluoroethylene solid, Zisman et al. found that it develops a limited infiltration angle θ_e, and it varies with the chain length of the alkane [3]. Its cosine value varies with the surface tension of the liquid as shown in Fig. 10.3. Although in many cases we cannot directly derive the surface energy corresponding to the liquid at complete infiltration, the critical surface energy γ_c can be obtained by extending a curve similar to the one in Fig. 10.3, so that the angle of infiltration is exactly 0. Normally we would think of γ_c dependent only on the solid itself; in fact, it is also related to the nature of the liquid. However, if one is dealing only with simple liquids (mainly through van der Waals forces), Zisman et al. found γ_c nothing to do with the liquid, but only with the solid itself [3].

What physical parameters of the solid are related to γ_c? Many scientists, including Girifalco [4], Good [5], and Fowkes [6] have done a lot of work trying to answer this question. Here, we will only give a brief discussion. From (10.1), (10.4), (10.5), and (10.8) we have the following relations

$$\cos\theta_c = \frac{\gamma_{SG} - \gamma_{SL}}{\gamma_{LG}} \cong \frac{\gamma_{SO} - \gamma_{SL}}{\gamma_{LG}} \cong \frac{W_{SL} - \gamma_{LG}}{\gamma_{LG}} = \frac{2\alpha_S}{\alpha_L} - 1 \tag{10.10}$$

In the example given above, the shorter alkanes have a greater polarizability relative to the longer ones, which is the reason why θ_c will become larger as n becomes smaller. Analyzing Eq. (10.10), we can obtain that when

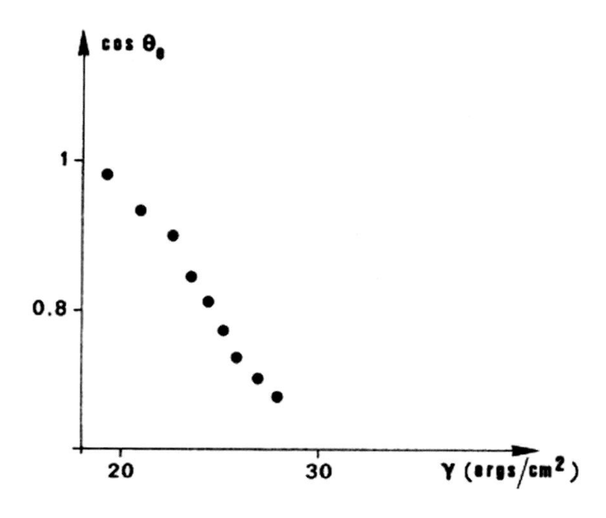

Fig. 10.3 The variation of the cosine of the contact angle θ_e of n-alkanes at equilibrium on the solid surface of polytetrafluoroethylene with liquid surface tension. For this system, γ_c is approximately 18 $ergs/cm^2$

$$\alpha_{Lc} = \alpha_S \tag{10.11}$$

$\theta_c = 0°$. Considering Eqs. (10.8), (10.10), we can obtain the critical surface energy:

$$\gamma_c = \frac{1}{2} k \alpha_S^2 \tag{10.12}$$

Equation (10.12) shows that γ_c is only related to the nature of the solid, increasing as its polarization rate increases. In fact, Eq. (10.12) is only a very rough model and there is a lot of work trying to refine it in various ways, such as:

(a) Since there are actually many other forces besides van der Waals forces that contribute in the solid–liquid interaction, such as hydrogen bonding and Coulomb forces, a further correction to Eq. (10.4) is needed [6]. At this point γ_c would be more or less related to the nature of the liquid.
(b) Even if only van der Waals forces are present in the solid–liquid interaction, Eq. (10.8) is very crude in using the average polarization rate to describe the action potential. Owens et al. pointed out that a more precise approach is to consider the frequency dependence of the polarization rate [7].
(c) Considering that the structure of the interfacial liquid is very closely related to the interfacial interaction between the solid and the liquid, we actually have to consider the interfacial structure of the liquid when studying the infiltration problem.

The theory of infiltration seems to tell us that the contact angle of a liquid on a solid can be accurately measured with experimental. In fact this is not the case. In many cases it has been found that the three-phase line even in $\theta \neq \theta_c$, i.e., in a certain range in the vicinity of θ_c, can exist stably and does not move, which is known as the pinning phenomenon. This means that the angle we measure experimentally may not be θ_c, but within a certain range around it, viz:

$$\theta_r < \theta < \theta_a \tag{10.13}$$

θ_a and θ_r are referred to as the forward and backward contact angles, respectively. They can be obtained by measuring the two different contact angles of a sliding droplet before and after the sliding direction (as in Fig. 10.4). This is the contact angle hysteresis effect. The difference between the forward and backward contact angles can be as much as 10°, or even greater. So what causes this contact angle hysteresis? Three main causes have been identified:

(i) Surface roughness. As early as 1932, Trillat and Fritz found that when the three-phase line £ is parallel to the groove in the system, it is easy to pin [8]. This is the main reason for the contact angle hysteresis.
(ii) chemical inhomogeneity. In 1964, Dettre and Johnson immersed glass liquid beads into solid paraffin and found that the different wetting properties of glass and paraffin can have an effect on the contact angle hysteresis properties [9].

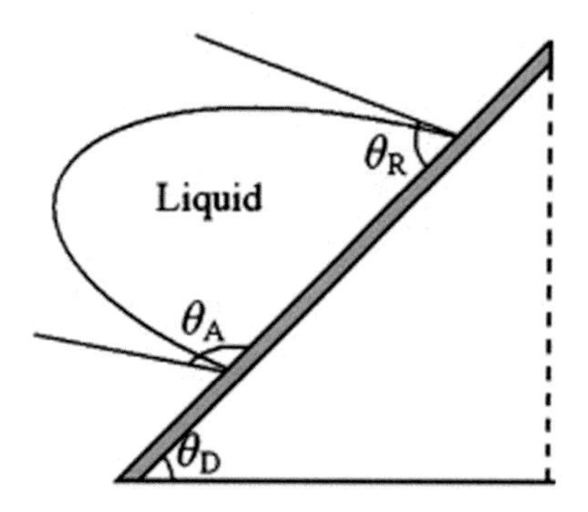

Fig. 10.4 Dynamic contact angle of a droplet sliding downwards on a surface where θ_A and θ_R refer to the forward and backward angles, respectively

(iii) Solutes in liquids (e.g., surfactants and polymers) may deposit on solid surfaces, which can have an impact on contact angle hysteresis [10].

As stated earlier, real solid surfaces are not absolutely smooth, hard, or chemically homogeneous. Therefore we introduce here some models that we hope will help understand the infiltration phenomenon on real solid surfaces. Here we focus on the surface model for striped surfaces. We know that there are microscopic cracks that can easily develop on a surface, and that chemical reactions are more likely to occur in these cracks relative to other places. So it is very important to understand the effect of these cracks on surface infiltration. We have already mentioned that it can have a significant effect on contact angle hysteresis. Here we focus on two striped surface models (Wenzel model [11] and Cassie-Baxter model [12]).

Wenzel model

Wenzel was one of the first few scientists to begin studying the effect of surface roughness on infiltration. Figure 10.5a gives the Wenzel model with droplets filling the grooves of the bar surface model, where we assume that the roughness scale of the surface is much smaller than that of our liquid beads. Imitating the derivation of Young's equation, when the three-phase line is shifted by dx, the surface energy changes by.

$$dE = r(\gamma_{SL} - \gamma_{SG})dx + \gamma_{LG}dx \cos \theta^* \tag{10.14}$$

In the equation, r is the roughness. For a smooth surface, $r = 1$, and for a rough surface, $r > 1$. When in equilibrium, $dE/dx = 0$, Young's equation yields.

$$\cos \theta^* = r \cos \theta_c \tag{10.15}$$

In the above equation, θ_c is the Young's contact angle. From Eq. (10.12), it is obtained that

(1) 1, If $\theta_c < 90°$ (hydrophilic surface), then $\theta^* < \theta_c$.
  2, If $\theta_c > 90°$ (hydrophobic surface), then $\theta^* > \theta_c$.

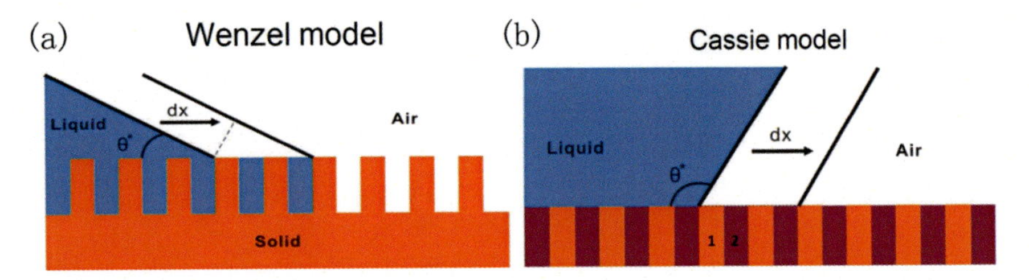

Fig. 10.5 **a** Wenzel model for a striped rough surface; **b** Cassie-Baxter model for a striped rough surface. dx gives the translation distance of the liquid/gas partition line during the derivation

That is to say, the roughness of the surface can change the wettability of the substrate: for the Wenzel model, it makes hydrophilic surfaces more hydrophilic and hydrophobic surfaces more hydrophobic.

Cassie-Baxter model

For chemically heterogeneous surfaces, we can introduce a similar model to describe them. As in Fig. 10.5b, a surface consists of two substances, both of which have their own contact angles, respectively θ_1 and θ_2. We use f_1 and f_2 to denote the proportion of surface area occupied by these two substances ($f_1 + f_2 = 1$). Again we assume that the stripe width is very small relative to the size of the droplet. Similarly to Eq. (10.14), we get that

$$dE = f_1(\gamma_{SL} - \gamma_{SG})_1 dx + f_2(\gamma_{SL} - \gamma_{SG})_2 dx + \gamma_{LG} dx \cos\theta^* \quad (10.16)$$

where the subscripts 1 and 2 refer to substances 1 and 2. At equilibrium, taking into account the Young's equation yields the Cassie-Baxter relationship:

$$\cos\theta^* = f_1 \cos\theta_1 + f_2 \cos\theta_2 \quad (10.17)$$

From Eq. (10.17), we know that the angle at this time is between θ_1 and θ_2.

10.3 Wetting Mechanism at the Atomic Scale

While all the above presentations are the understanding of infiltration on the macroscopic scale, in this subsection we will focus on some recent advances in the study of water infiltration on the microscopic atomic scale. When the scale of inhomogeneity of the surface structure is comparable to the size of a single water molecule, surface infiltration presents very rich and peculiar phenomena, quite different from the classical macroscopic understanding. For example, let us first look at an interesting question: is the water surface hydrophilic or hydrophobic? If we analyze from the macroscopic theory we introduced earlier, when a water drop is dropped onto a flat water layer, in order to ensure the lowest energy, it needs a minimum surface area, and then the water drop will be completely spread out. Recently, however, Fang Haiping [13] from the Shanghai Institute of Applied Physics, Chinese Academy of Sciences, found that "liquid water at room temperature can be nonhydrophilic when it is on a specific surface" through molecular dynamics simulations, meaning that water can be hydrophobic at room temperature, which is contrary to our previous analysis, but was later proved experimentally [14]. Why is there a hydrophobic water layer at room temperature? Further analysis revealed that the surface water layer forms an ordered hexagonal arrangement due to the surface charge, and it is this specific structure that leads to the emergence of the peculiar phenomenon of hydrophobic water layer at room temperature. This phenomenon was also found in some scientific experiments on surfaces at low temperatures as early as 2005 [15].

On the other hand, is water necessarily wettable on surfaces with polar groups? Considering that water is a polar molecule, the conventional wisdom is that surfaces with polarities are easily wetted by water and are hydrophilic surfaces. In reality, modification of polar molecular groups on a surface is an important means of making a surface hydrophilic. Is this really the case? H. Fang et al. used molecular dynamics simulations to show that the wettability of a solid surface is clearly dependent on the dipole length of the polar molecules on the surface [16]. There is a critical value of the dipole length, and when the dipole length of polar molecules on the surface is less than this critical value, the water molecules cannot "sense" the presence of the dipole, regardless of how strong the polarity is. The surface with the polar group is still hydrophobic at this point. But as the dipole moment length increases, the solid surface becomes increasingly hydrophilic.

It can be seen that the surface microstructure can have a significant impact on the surface infiltration properties. Recently we have systematically simulated the infiltration properties of water on the surface of face-centered cubic solids (111) with different lattice constants (Fig. 10.6a). It was found that for hydrophobic surfaces stretched horizontally $\pm 3\%$ the change in the surface wettability to water is small. While for hydrophilic surfaces stretched $\pm 3\%$ there is a very large change in the water infiltration angle at the interface [17, 18]. Moreover, the contact angle does not vary monotonically with the surface lattice constant, but when the surface lattice constant and the bulk phase water O–O distance projection at the interface is equal, the contact angle of the water droplet exhibits a minimum (Fig. 10.6b). Further analysis of the structure of the interfacial water layer reveals that the microstructure of the interfacial dense water layer is disrupted, thus more closely resembling the characteristics of the bulk phase water (Fig. 10.6c). In contrast, on hydrophobic surfaces, the surface lattice essentially does not change the structure of the interfacial water layer. It can be seen that the microstructure of the interfacial water layer has a strong influence on surface infiltration.

All three of the above works use simulations to probe the effect of microstructure on surface infiltration. Then, are there similar experiments? The answer is yes. Using two components with scales comparable to the scale of the molecules in the solution to compose different surfaces, Jeffrey et al. found that the nature of water infiltration on the surface is related to the specific nanostructure composed of the components,

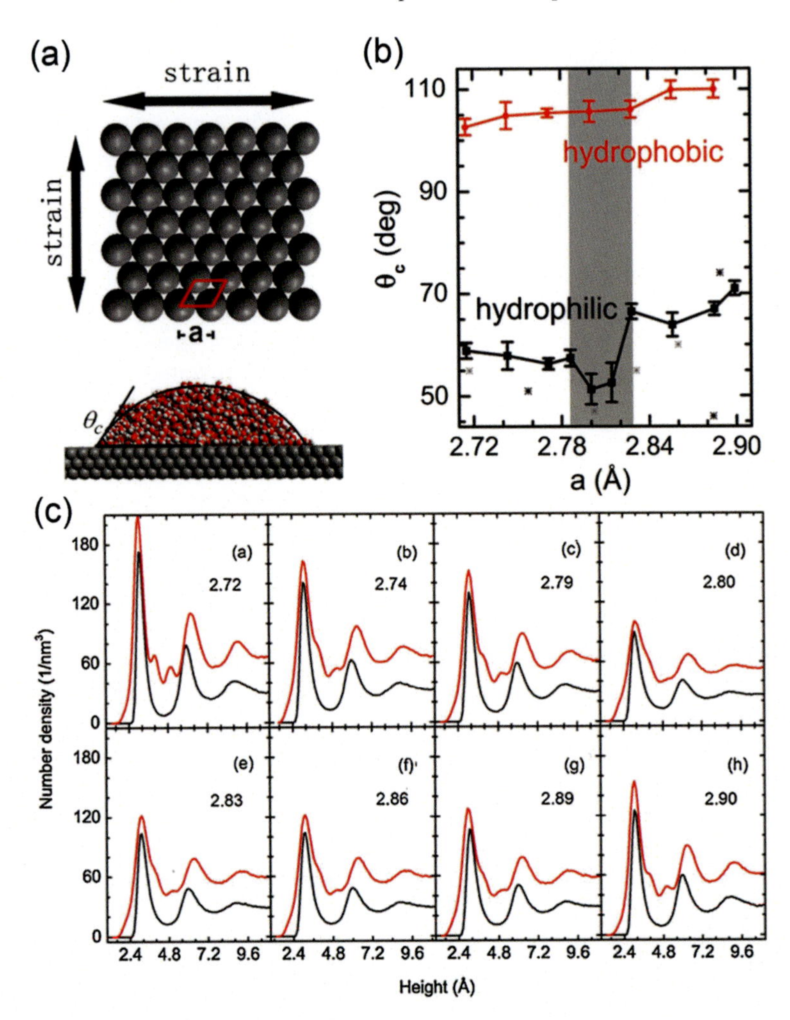

Fig. 10.6 **a** Top: Geometric configuration of the (111) face of a face-centered cubic crystal. Bottom: side view of a water droplet on a model substrate. **b** Variation of the contact angle of water droplets on hydrophilic and hydrophobic surfaces with the lattice constant. The asterisks give the experimental data of the wettability angle of water on different metal surfaces [18]. **c** Variation of the density of hydrogen and oxygen atoms in water molecules with the height for hydrophilic surfaces with different lattice constants. The red line corresponds to hydrogen atoms and the black line to oxygen atoms. The numbers in the figure are the surface lattice constants

in addition to the ratio of the two components [19]. Their findings further illustrates the importance of understanding surface infiltration at the nanoscale; one could even argue that the infiltration properties of a surface cannot be defined until the nanoscale arrangement of surface components and small structures is determined.

All of the previous talk has used macroscopic contact angles to characterize the strength of a substance's wettability to a particular liquid. However, for nanoscale substances such as carbon nanotubes, nanoparticles and protein molecules, it is obviously very difficult to measure the contact angle of a liquid droplet on them, so we need other microscale characterization methods. Here we focus on some of these methods.

The molecular dynamics simulations of water infiltration on self-assembled films are given in Fig. 10.7 [20]. In the system, the self-assembled film molecule is composed of an end-group and a decane linkage, and by changing the end-group, surfaces with different hydrophobicity can be obtained. The density of water molecules on different surfaces at different heights from the surface is given under Fig. 10.7. It can be seen from the figure that water produces a layered structure on the solid surface regardless of the hydrophobicity of the surface. But the magnitude of the density of the first layer of water is not necessarily related to the hydrophobicity of the surface!

Further, Rahul Godawat et al. studied the behavior of water at different hydrophobic surfaces with density fluctuations using molecular dynamics simulations [20]. The number of water molecules in the fixed volume space near hydrophobic (–CH_3) and hydrophilic (–OH) surfaces are given as a function of time in Fig. 10.8a and b, respectively. It can be seen that the average number of water molecules in the fixed volume space of these two different surfaces is about the same, which means that we cannot distinguish them by the average density size. However, we can clearly see that the fluctuationsin the number of water molecules is much

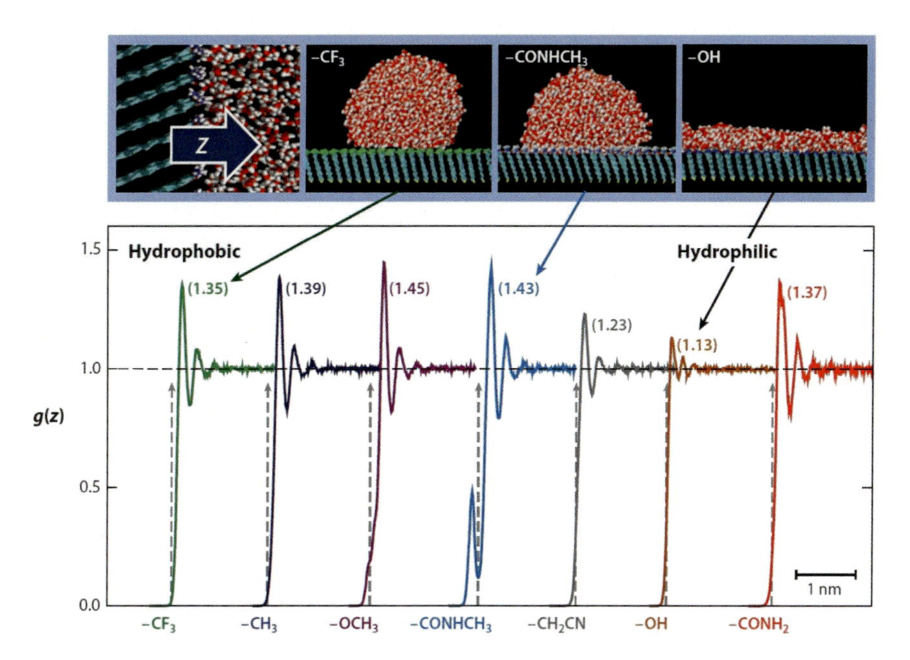

Fig. 10.7 Variation of the density of water on different self-assembled films as a function of their height from the surface. From left to right the self-assembled film becomes increasingly hydrophilic

larger on the hydrophobic surface (–CH_3) than on the hydrophilic surface (–OH). The variation of the logarithm (y-axis) of the probability of the number of water molecules in the circle shown in the right of Fig. 10.8d with the normalized number of water molecules (x-axis) is given on the left of Fig. 10.8d. As can be seen from the plot, the more hydrophilic the surface, the narrower the plot of the function, which means that it is less likely to be compressed. So we can determine the hydrophilicity of interfacial water by its compressibility. Using this method, recently, Acharya et al. successfully calculated the hydrophobicity of various parts of a protein [21].

In addition to the approach using density fluctuations that allows for the characterization of wettability at the microscopic scale, many other approaches have been proposed. For example, in 2003 we found by first-principles calculations [22] that the ratio ω between the hydrogen bonding of the surface water layer and the adsorption energy of water molecules is also a good scale of wettability: the larger ω the more hydrophobic the surface, the smaller ω the more hydrophilic the surface. For modified carbon-based surfaces, the macroscopic contact angle is approximately related to the microscopic ω by $\theta = 108° - 108°/\omega$ [23].

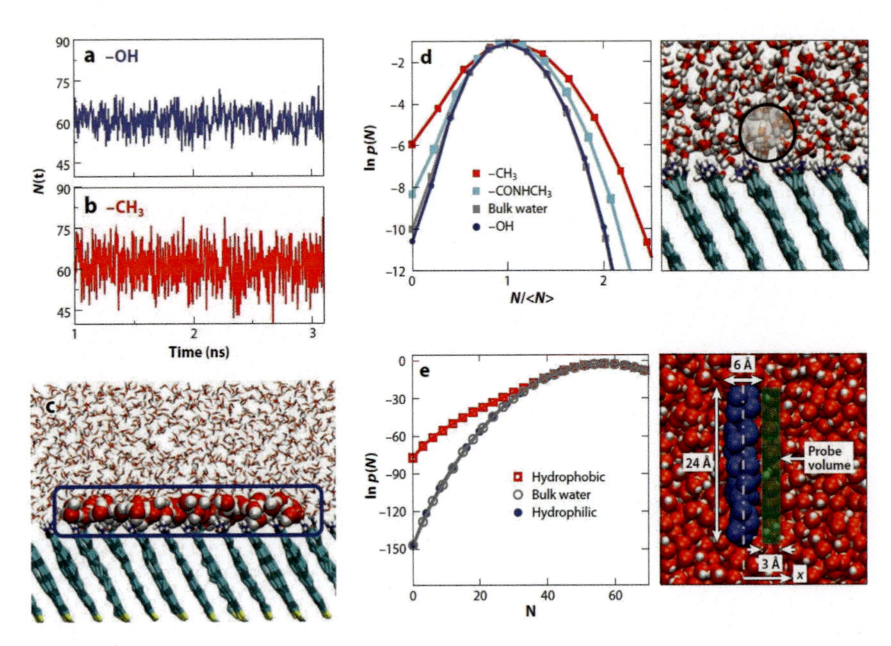

Fig. 10.8 Fluctuations of the number of the water molecules in a 2.5 × 2.5 × 0.3 nm^3 square **c** near hydrophilic (–OH) and hydrophobic (–CH_3) surfaces with time (**a**, **b**). **d** The logarithm of the probability p(N) of observing N water molecules within the circle (r = 0.33 nm) (right) near the self-assembled film. **e** The logarithm of the probability p(N) of observing N water molecules in a larger space (2.5 × 2.5 × 0.3 nm^3) near the self-assembled film [20]

10.4 Practical Applications of Surface Wetting

The phenomenon of infiltration has been studied for hundreds of years, so what is the importance of these studies for real life? We have already mentioned some applications earlier. Apparently nature recognized the importance of the propertiesof infiltration before mankind did, and made use of it. The lotus leaf, for example, must be quite familiar with its property of "coming out of the mud and washing the water". From the point of view of infiltration, the lotus leaf is a superhydrophobic material (Fig. 10.9a). What, then, causes its superhydrophobic properties? Electron microscopic observation (Fig. 10.9b, c) reveals that the upper surface of the lotus leaf is covered with a very large number of tiny papillae, with an average diameter of about 6–8 μm, a height of about 11–13 μm, and the spacing of about 19–21 μm. Among these tiny papillae there were also some larger ones, with an average size of about 53–57 μm. They are also composed of tiny papillae of 6–13 μm. These small and large protrusions on the surface of the lotus leaf are like raised "buns" one after another, and the depressions between the "buns" are filled with air, thus forming an extremely thin layer of air, only nanometer thick, close to the leaf surface. Given that the minimum diameter of water droplets is 1–2 mm, which is much larger than the papillae on the lotus leaf, the droplets, when on the surface of the leaf, is separated by a very thin layer of air and only makes contact with the tops of the "hills" on the surface at a few points. Using the Cassie-Baxter model, we can get from Eq. (10.17) that the water cannot infiltrate the surface of the lotus leaf. Water droplets can form spherical droplets under their own surface tension, and the water spheres adsorb dust while rolling and roll out of the liquid surface, thus cleaning the leaf surface. Similar microscopic structures can be found on the legs of water spiders and on the wings of cicadas.

Nature is the best teacher, and it is based on the understanding of the "lotus leaf effect" that Lei Jiang's group at the Institute of Chemistry, Chinese Academy of Sciences, have successfully made a microstructure-like film from microtubules of alkyl pyridine (Fig. 10.9d) and microspheres/nanofibers of polyethylenes (Fig. 10.9e) [24]. Such films do have superhydrophobic properties. Later this technology was successfully applied in the production of nano self-cleaning ties, tiles, glass, etc. Jiang also produced a "light" regulated superhydrophobic/superhydrophilic "switch" material. He also developed a superhydrophobic/superhydrophilic "switch" material with "temperature" regulation combined with surface chemical modification and surface roughening. These two research results may have future applications in gene delivery, nondestructive liquid transport, microfluidics, and drug delivery. For example, if the second material is used to make clothing, the clothing will be hydrophilic in the summer when the temperature is high, so it will not feel too hot if it is hydrophilic and absorbs sweat; and the clothing will become hydrophobic in the winter when the temperature is low, so it will defend cold and keep warm.

Frost protection is another meaningful application of surface infiltration research [25]. A very important cause of snow damage is freezing. Once frozen it is very difficult to remove, causing damage to electricity, and the accumulated ice can even

Fig. 10.9 Microstructure (**b**, **c**) of the lotus leaf (**a**); scanning electron microscopy images of polyethylene microtubule film (**d**) and polystyrene microsphere/nanofiber film (**e**) prepared by Lei Jiang et al. Both **d** and **e** are superhydrophobic films [24]

crush houses and disrupt rail, road and civil aviation traffic. In daily life, freezing in a refrigerator can make it much less efficient and fritter away energy. Many studies have now found that surface hydrophilic properties have a great relationship with icing, which may become a breakthrough in anti-icing materials research.

Above we have covered many of the basics of infiltration and some of the current research advances. However there is still a lack of clarity in the current understanding of infiltration. This focuses on two main aspects: the role of additives in liquids, and the lack of a complete picture of the microscopic understanding of surface infiltration. These two aspects are themselves related.

In real life production, it is more common for us to see liquids with impurities such as surfactants and polymers in them than pure liquids. How these impurities affect the infiltration of the liquid on the surface is not well understood. For example, as early as 1964, Bascom et al. found that some peculiar liquid structures were observed when the liquid contained volatile impurities [26]. However, there is no conclusive answer to the cause of this phenomenon until now. Liquids containing surfactants and macromolecules can easily appear at various interfaces (solid/liquid, solid/gas, liquid/gas). In practical applications, their influence is very significant. For example, Lelah and Marmur found that very little surfactant changes the wettability of the surface [27].

Another issue we need to address is the microscopic understanding of surface infiltration. We have mentioned some work above, all of which found that classical macroscopic models are already difficult to understand these physical phenomena when microstructure (compared to the molecular scale of liquids) is taken into account. At this point we need to re-model the understanding of surface infiltration from a microscopic perspective, e.g. how much does the structure of interfacial water really affect surface infiltration? These questions are still under further research. It is expected that in the near future, the microscopic scale understanding of surface infiltration will be improved and more exotic and rich infiltration phenomena will be discovered, and simple, effective and low-energy regulation of surface infiltration will be performed to meet the growing needs of people's life and economic development.

References

1. A.R. Parker, C.R. Lawrence, Water capture by a desert beetle. Nature **414**, 33 (2001)
2. H.W. Fox, W.A. Zisman, The spreading of liquids on low energy surfaces. i. polytetrafluoroethylene. J Colloid Sci. **5**, 514 (1950)
3. W.A. Zisman, Relation of the equilibrium contact angle to liquid and solid constitution. Adv. Chem. Ser **43** (1964)
4. L.A. Girifalco, R.J. Good, A theory for the estimation of surface and interfacial energies. i. Derivation and application to interfacial tension. J. Phys. Chem. **61**, 904 (1957)
5. F.M. Fowkes, Determination of interfacial tensions, contact angles, and dispersion forces in surfaces by assuming additivity of intermolecular. J. Phys. Chem **66**, 382 (1962)
6. R.J. Good. *Contact Angle, Wettability and Adhesion.* Advances in Chemistry Series (American Chemical Society, Washington, D. C., 1964), pp. 74

7. N.F. Owens, P. Richmond, D. Gregory et al., *Contact Angles of Pure Liquids and Surfactants on Low-energy Surfaces* (Academic Press, London, 1978), pp. 127
8. J. Trillat, R. Fritz, J. Chim. Phys. **35**, 45 (1937)
9. R.E. Johnson Jr., R.H. Dettre, Contact angle hysteresis. iii. study of an idealized heterogeneous surface. J. Phys. Chem. **68**, 1744 (1964)
10. J. Chappuis, in *Multiphase Science and Technology*, ed. by G.F. Hewitt, J. Delhaye, N. Zuber (Hemisphere New York, 1984), pp. 387
11. R.N. Wenzel, Resistance of solid surfaces to wetting by water. Ind. Eng. Chem. **28**, 988 (1936)
12. A.B.D. Cassie, S. Baxter, Wettability of porous surfaces. Trans. Faraday Soc. **40**, 546 (1944)
13. C.L. Wang, H.J. Lu, Z.G. Wang, P. Xiu, B. Zhou, G.H. Zuo, R.Z. Wan, J. Hu, H.P. Fang, Stable liquid water droplet on a water monolayer formed at room temperature on ionic model substrates. Phys. Rev. Lett. **103**, 137801 (2009)
14. M. James, T.A. Darwish, S. Ciampi, S.O. Sylvester, Z.M. Zhang, A. Ng, J.J. Gooding, T.L. Hanley, Nanoscale condensation of water on self-assembled monolayers. Soft Matter **7**, 5309 (2011)
15. G.A. Kimme, N.G. Petrik, Z. Dohnalek, B.D. Kay, Crystalline ice growth on Pt(111): observation of a hydrophobic water monolayer. Phys. Rev. Lett. **95**, 166102 (2005)
16. C.L. Wang, B. Zhou, Y.S. Tu, M.Y. Duan, P. Xiu, J.Y. Li, H.P. Fang, Critical dipole length for the wetting transition due to collective water-dipoles. Sci. Rep. **2**, 358 (2012)
17. C.Q. Zhu, H. Li, Y.F. Huang, X.C. Zeng, S. Meng, Microscopic insight into surface wetting: relations between interfacial water structure and the underlying lattice constant. Phys. Rev. Lett. **110**, 126101 (2013)
18. R.A. Erb, Wettability of metals under continuous condensing conditions. J. Phys. Chem. **69**, 1306 (1965)
19. J.J. Kuna, K. Voitchovsky, C. Singh, H. Jiang, S. Mwenifumbo, P.K. Ghorai, M.M. Stevens, S.C. Glotzer, F. Stellacci, The effect of nanometre-scale. Nat. Mater. **8**, 837 (2009)
20. R. Godawat, S.N. Jamadagni, S. Garde, Characterizing hydrophobicity of interfaces by using cavity formation, solute binding, and water correlations. Proc. Natl. Acad. Sci. **106**, 15119 (2009)
21. H. Acharya, S. Vembanur, S.N. Jamadagni, S. Garde, Mapping hydrophobicity at the nanoscale: applications to heterogeneous surfaces and proteins. Faraday Discuss. **146**, 353 (2010)
22. S. Meng, E.G. Wang, S.W. Gao, A molecular picture of hydrophilic and hydrophobic interactions from ab initio density functional theory calculations. J. Chem. Phys. **119**, 7617 (2003)
23. S. Meng, Z.Y. Zhang, E. Kaxiras. Tuning solid surfaces from hydrophobic to superhydrophilic by submonolayer surface modification. Phys. Rev. Lett. **97**, 036107 (2006)
24. X.J. Feng, L. Jiang, Design and creation of superwetting/antiwetting surfaces. Adv. Mater. **18**, 3063 (2006)
25. J.Y. Lv, Y.L. Song, L. Jiang, J.J. Wang, Bio-inspired strategies for anti-icing. ACS Nano (2014). https://doi.org/10.1021/nn406522n
26. W. Bascom, R. Cottington, C. Singleterry, in *Contact Angle, Wettability and Adhesion*, ed. by F.M. Fowkes. Advances in Chemistry Series 43 (1964), pp. 355
27. A. Marmur, M.D. Lelah, The spreading of aqueous surfactant solutions on glass. Chem. Eng. Commun. **13**, 133 (1981)

Chapter 11
Hydrated Ions on Surfaces

The interaction of water and ions at the surface is common in many physical [1, 2], chemical and biological phenomena and plays an important role in processes such as ion solubilization and ion transport in electrochemistry, fuel cells, ion channels in cell membranes [3, 4], activation of enzymes and neural signaling in biology. In solution environments and free clusters, the water surrounding the ions forms a three-dimensional (3D) shell structure called the hydration shell of ions [5–10]. It usually consists of two parts: a first shell formed by chemical bonding between ions and water molecules and an outer shell bound to the ionic potential field by hydrogen bonds. The structure and stability of the ion hydration shell determines the chemical reactivity, motility properties and physiological activity of the ions.

So far, most studies on ion hydration shell have focused on the situation in bulk solutions and free cluster environments. However, in electrochemical and biological processes, the role of ion hydration shell at electrode surfaces or biofilm surfaces is at least as important, if not more important. The ion hydration shell at the surface has been little studied and is poorly understood [11, 12]. Part of the reason for this is that the structure and dynamics of ion hydration shell at the surface become more complex due to symmetry breaking and charge transfer effects.

In this chapter we study the formation of hydration shell at hydrophobic solid surfaces by K and Na ions [13, 14]. We find that unlike the three-dimensional hydration shell structures in clusters and liquids, the initial hydration shell on the surface are two-dimensional [13]. Moreover, the two ions form different hydration structures on the surface with different dynamic processes [14]. For convenience, we start with the two-dimensional hydration shell structure of the K ion.

S. Meng and E. Wang, *Water*,
https://doi.org/10.1007/978-981-99-1541-5_11

11.1 Two-Dimensional Hydration Shell of K Ion on the Surface

We have studied a typical system using first-principles molecular dynamics simulations (ab initio MD) [13]: graphite surface C(0001) on K(Na) + nH_2O, $n = 1–10$. Using a C(0001) 4 × 4 primitive cell, taking a graphite layer (i.e., graphene) with a vacuum layer of 17 Å. This has neglected the very weak Van der Waals interaction. Only one K or Na atom is placed in the primitive cell, with a number of water molecules. The corresponding ion coverage is $1.2 \times 10^{14}/cm^2$, which is similar to the coverage in the experiment [15]. The plane wave energy cutoff in the calculations is taken as 300 eV and the k-point sampling is 3 × 3 × 1. The time step for the molecular dynamics simulations is 0.5 fs, and the Nose method is used for constant temperature simulations at 85 K. The hydration shell structure was obtained from the MD simulations at 4–8 ps, and further structural optimization gave the corresponding binding energies.

The structures and adsorption energies of K and Na atoms adsorbed on graphite surfaces are listed in Table 11.1. K and Na are adsorbed more stably on vacancies with adsorption energies of 1.01 eV and 0.75 eV, respectively, at distances of 2.7 Å and 2.4 Å from the surface. At small coverages, alkali metal atoms adsorbed on graphite surfaces are ionized by the surface [16], forming an electric dipole layer (Helmholtz layer). This is clearly shown in Fig. 11.1 by drawing the distribution of electrons in the direction normal to the graphite layer before and after K adsorption. The polarized electrons are mainly distributed in the p_z orbitals of the six nearest neighboring C atoms (Fig. 11.2b). This is because both the K 4 s and Na 3 s orbitals lie essentially above the Fermi energy level, where K(Na) transfer electrons to the graphite surface and turn themselves into ions. A detailed analysis shows that K transfers 0.68 electrons to the C surface, while Na also loses 0.61e.

When water molecules are added to the K(Na)/C(0001) surface, they tend to adsorb around the ions due to the strong ion-water interactions. We look at the case of K ions first and then compare the similarities and differences between K and Na ions. We find that the hydration shell of K ions can be divided into three types in the range of the number of water molecules $n = 1–10$. They have very different binding and vibrational characteristics.

Type I, $n \leq 3$, forms the first shell. Single water molecule is adsorbed on the surface and bound to the K ion through K–O bond with a binding energy of 0.59 eV. The water molecule tends to stand upright on top of the next nearest neighbor C

Table 11.1 Adsorption heights and adsorption energies of K and Na on graphite surfaces

	Vacancy		Top site	
	d/Å	E_{ads}/eV	d/Å	E_{ads}/eV
K/C(0001)	2.70	1.01	2.83	0.95
Na/C(0001)	2.38	0.75	2.62	0.70

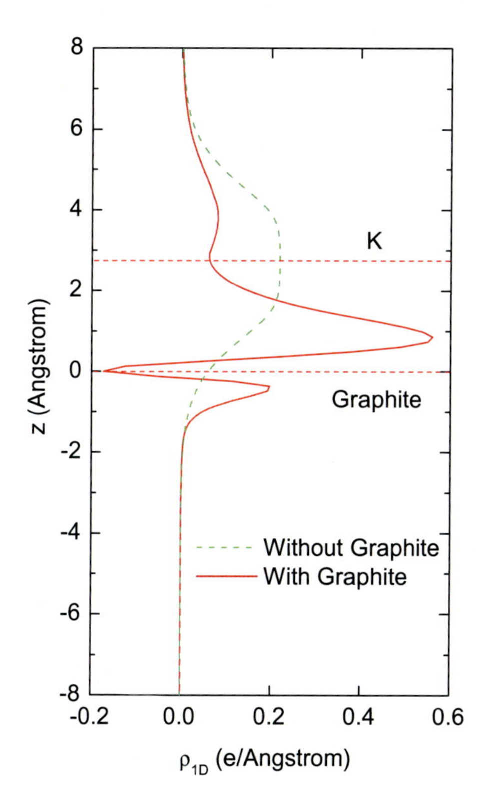

Fig. 11.1 1D charge density distribution along the normal before (dashed line) and after (solid line) the adsorption of K on the graphite surface

atom (Fig. 11.2c) and can rotate almost freely around the K ion with a rotation potential of 10 meV. $n = 2, 3$, the water molecules all surround K. The repulsive forces between O keep them away from each other, forming straight lines (Fig. 11.2d) and triangular (Fig. 11.2e) hydration structures. Thus, as in the free cluster [17, 18], the K ion attracts three water molecules and keeps them in one plane. The slight difference is that here one OH per water molecule is pointing towards the surface, forming an OH…C hydrogen bond. This pointing is due to the surface electric dipole layer caused by charge transfer. OH…C bonds are weak in both the gaseous and water-graphite surfaces [19], but are greatly enhanced by charge transfer. The same hydrogen bonding enhancement effect is found in the $(H_2O)_2$/Pt(111) [20], and $(O_2 + C_2H_4)$/Ag(110) systems [21].

Type II, $4 \leq n \leq 6$, forms the second shell. In the free cluster, a 3D hydration shell is formed at this point. On the surface, however, we find that the hydration circle is still two-dimensional. $n = 4$, the fourth water molecule is still on the surface, and it is not bonded to K but to other water molecules via hydrogen bonding, forming the second shell (Fig. 11.2f). The total energy is 60–270 meV higher for the configuration in which the water molecule is directly bonded to K on top of K or in the plane. This is due to the fact that the formation of two hydrogen bonds ($2 \times 0.35 = 0.70$ eV) is energetically more favorable than the formation of one K–O bond (maximum

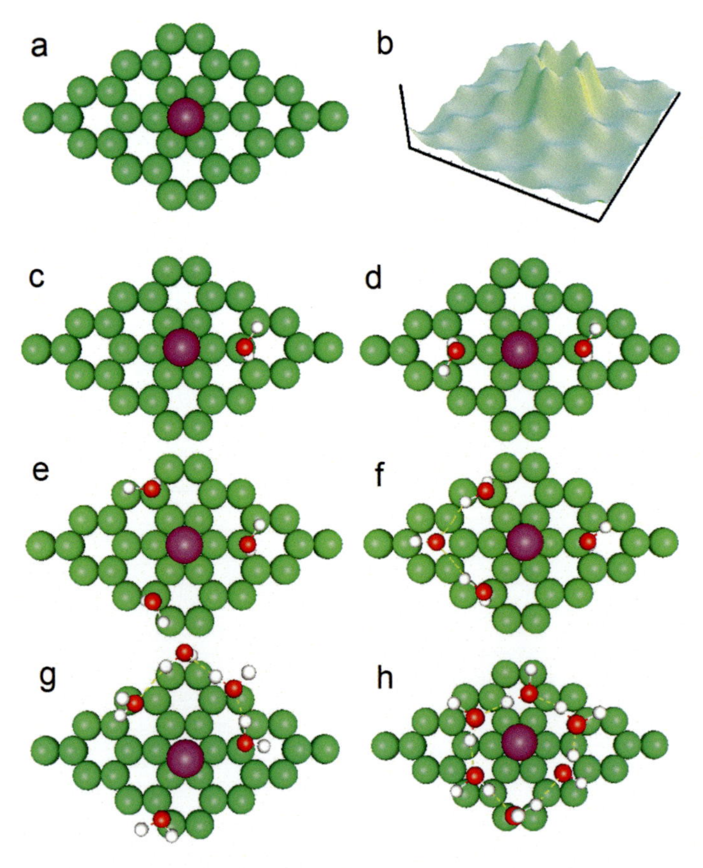

Fig. 11.2 K adsorption sites on the graphite surface (**a**) and charge distribution in the z = 0.85 Å plane (**b**), and the 2D hydration shell structure (**c–h**) for K + nH_2O, $n = 1$–6

0.59 eV). As shown in Fig. 11.3. The formation of a strengthened hydrogen bond with the surface also makes the 3 + 1 structure more stable. Here $n_1 + n_2$ is the number of coordination sites for the ion in the first (n_1) and second (n_2) hydration shells. In addition to the difference in static energy, the contribution of entropy [22] also makes the 3 + 1 structure more stable than the 4 + 0 structure. Our molecular dynamics simulations also show that the 4 + 0 structure quickly converting to the 3 + 1 structure in the simulations. At $n = 5$ and $n = 6$, the 3 + 2 and 3 + 3 2D hydration shell structures are formed, respectively. They remain in the plane as in Figs. 11.2g and 8.2 h. The hydration structure of K + $6H_2O$ is a folded hexagonal ring, and the six molecules $(H_2O)_6$ on the surface [23] are similar to the basic unit of a bilayer ice [24].

Type II, $n \geq 7$, begins to form 3D shells. When $n \to 10$, hemispherical shells start to form. It can be considered as half of the hydration shell in liquid water [18]. This arises from the constraints of the 2D scale of the surface. Table 11.2 lists the hydration structures and the corresponding binding energies obtained in the calculations. They

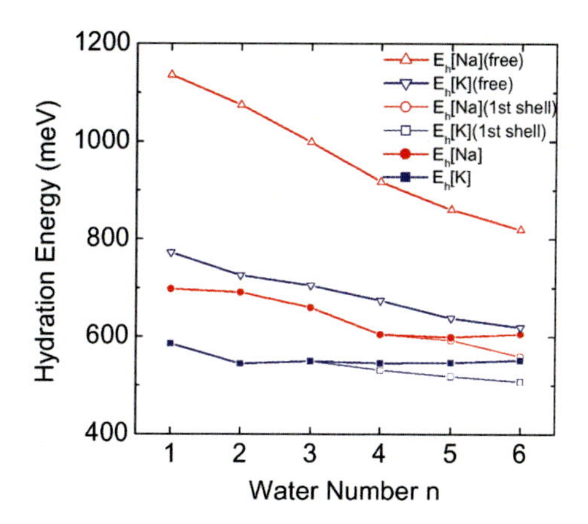

Fig. 11.3 Variation of the hydration energy of K, Na with the number of water molecules on the graphite surface. The thin line corresponds to the structure that allows water molecules to be forced to remain in the first shell. The curves for the free hydration clusters [18, 22] are also plotted in the figure

Table 11.2 Calculated structure ($n_1 + n_2$) and hydration energy (eV/H_2O) of the K, Na/C(0001) hydration shell

n	1	2	3	4	5	6	7	8	9	10
E_h[K]	0.59	0.54	0.55	0.55	0.55	0.55	0.57	0.58	0.59	0.59
$n_1 + n_2$	1 + 0	2 + 0	3 + 0	3 + 1	3 + 1	3 + 3	3 + 4	3 + 5	3 + 6	3 + 7
E_h[Na]	0.70	0.69	0.66	0.60	0.60	0.61	0.60	0.62	0.58	0.61
$n_1 + n_2$	1 + 0	2 + 0	3 + 0	4 + 0	4 + 1	4 + 2,3 + 3	4 + 3	4 + 4	4 + 5	4 + 6

are compared with the energies of hydration structures in free clusters in Fig. 11.3. In general, the hydration energy of free clusters is larger due to the fact that the K ions are only partially ionized (70%) on the surface. From Table 11.2 we see that, unlike in solution where a K ion usually has six nearest neighbor water molecules, the number of water molecules in the first shell on the surface is 3.

11.2 Vibration Identification of Two-Dimensional Hydration Shell

To be able to identify experimentally these two-dimensional hydration shell structures on the surface, we calculated the vibrational spectrum of K + nH_2O (n = 1–6) on the graphite surface, as shown in Fig. 11.4a. The left side of the vibrational spectrum shows the translation and rotation modes of the H_2O molecule, the middle shows the HOH shear mode (200 meV), and the right side corresponds to the OH vibrations inside the molecule. We see that the vibrational spectra for n = 1–3 are very similar, due to the weak interactions between water molecules in the first shell. The symmetric and antisymmetric OH stretching vibrations have frequencies of 440 and 452 meV, respectively, which are lower than in free water molecules (454 and 464 meV [25]),

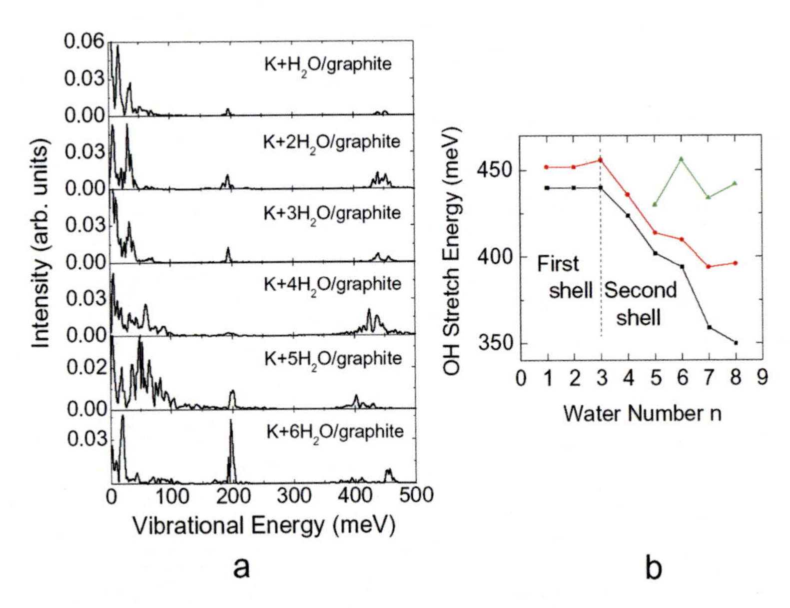

Fig. 11.4 Vibrational spectrum of the 2D hydration shell of K on the graphite surface (**a**) and variation of the OH vibration with the number of water molecules (**b**)

due to the attraction of K ion. $n \geq 4$, more vibrational modes appear in the 80–105 meV range. The redshift of the OH vibrational frequencies is more pronounced and the peaks become wider. The characteristic energies of OH vibrations are 424 and 436 meV for $K + 4H_2O$; 402, 414 and 430 meV for $K + 5H_2O$; and 394, 410 and 456 meV for $K + 6H_2O$.

The evolution of the OH vibrational energy with the increase of the number of H_2O molecules is plotted in Fig. 11.4b. From $n = 1$–8, it is evident that the OH vibrational energy is divided by $n = 4$. The left side remains essentially constant at 440 and 452, and the lowest frequency on the right side gradually decreases. The high frequency peak (450 meV) originates from the free OH vibrations of the edge molecules. The redshift of the OH vibrational peak and the appearance of the high frequency rotation mode (80–110 meV) are both characteristic manifestations of the formation of hydrogen bonds between water molecules. The vibrational spectrum reflects the changes in the structure of the hydration shell in Fig. 11.2.

It is interesting to compare these vibrational spectra with high-resolution electron energy loss spectroscopy (HREELS) experiments [15] (Table 11.3). The structure of the $K + nH_2O/C(0001)$ formed at 85 K and 120 K was measured experimentally. Our translational and rotational modes 32, 51, 71 meV can be compared with the 37, 51, 79 meV in experiments of the structure at 85 K. This implies that at this point the water molecules form the first loop with the K ion only. On the contrary, the peaks of 40–100 meV of the second shell can be compared with the vibrational peaks (40, 51, 85, 105 meV) of the structure at 120 K. Combined with the experiments, these results suggest that heating activates the diffusive movement of water molecules on

Table 11.3 Vibrational spectra calculated for the K hydration shell and compared with the experiment (literature [15])

	Translation and rotation						δ_{HOH}	$\nu_{O\text{-}H}$ (S)	$\nu_{O\text{-}H}$ (AS)
First Shell	16	32		51	71		194	440	452
Expt. (85 K)		37		51	79		203	438	457
Second Shell	18	34	42	51	85	101	198	394, 402, 412, 424, 436	456
Expt. (120 K)			41	51	85	105	203	439	

the surface, which causes the transition from the first to the second hydration shell around the K ion.

11.3 Different Structures and Dynamics of K, Na Ion Hydration Shell

Na ions on graphite surfaces differ from K ions in two ways: (1) Na dissociates less because it requires a higher dissociation energy, and (2) the ion radius of Na (1.13 Å) is smaller than that of K (1.51 Å) [26]. The second point makes Na and water have stronger interactions. These differences result in their different hydration structures and dissolution dynamics on the surface.

Similar to K, the hydration shell Na + nH_2O/C(0001) of Na on the surface can be divided into three types: (1) $n \leq 4$, forming the first shell. (2) $n = 5, 6$, forming the 2D second shell. (3) For larger n, forming a 3D shell close to the bulk state. The hydration energy on Na/C(0001) is also plotted in Fig. 11.3. n = 1, the binding energy is 0.70 eV, which is larger than that of K and water (0.59 eV). n increases as the hydration energy decreases [27]. Also the difference in hydration energy between Na and K becomes smaller. Importantly, unlike the K ion which has three H_2O in its first hydration shell, Na, on the surface, has four H_2O molecules in its first hydration shell. This is because Na and H_2O have stronger interactions. The biggest difference between K and Na ions on the surface is their different dissolution dynamics. Figure 11.5 shows the distance of ions from the surface in different hydration structures. Among all n, K remains on the surface ($d_{KC} \approx 3.2$ Å) and does not dissolve in the water cluster. In contrast, for Na, the Na-surface distance gradually increases as n increases. At $n =$ 10, Na is already almost completely dissolved in water ($d_{NaC} \approx 5$ Å). This difference arises mainly from the competition between hydration and ion-surface interactions. In the case of Na, hydration wins the competition. This 2D hydration structure on the surface, which we report for the first time, and the interesting competition between hydration and surface interactions have general implications for ionic processes at the surface interface and in biological systems.

In conclusion, a study of the structure of hydration shells formed by K, Na ions on graphite surfaces using first-principles molecular dynamics simulations reveals that,

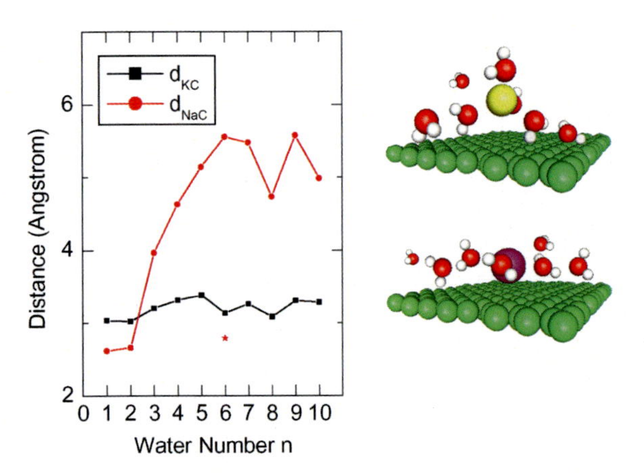

Fig. 11.5 Variation of the distance of K, Na from the graphite surface with the number of water molecules, showing their different dissolution dynamics. The asterisk indicates the 3 + 3 structure of Na, which is almost as stable as the 4 + 2 structure

unlike the situation in free clusters and bulk solutions, the hydration shells of ions on the surface are two-dimensional. This arises from the 2D confinement of the surface and the charge transfer between the ions and the surface. For K and Na, the first hydration shell has three and four water molecules, respectively. As more water molecules are added, the water gradually transforms from a 2D structure to a 3D hydration layer. The K ions remain on the surface and form a hemispherical 3D hydration shell around them, while the Na ions gradually move away from the surface until they are completely dissolved in the water. Their difference is determined by the competition between hydration and ion-surface interactions. The interaction between K and the surface is stronger, while the interaction between Na and the water molecules is stronger. For the purpose of identification, the vibrational spectrum of each hydration structure was also calculated. A comparison with existing experiments shows that there is a transition from the first to the second hydration shell caused by thermal excitation in the experiments.

References

1. P.A. Thiel, T.E. Madey, The interation of water with solid surfaces: fundamental aspects. Surf. Sci. Rep. **7**, 211 (1987)
2. M.A. Henderson, The interation of water with solid surfaces: fundamental aspects revisited. Surf. Sci. Rep. **46**, 1 (2002)
3. D.A. Doyle et al., The structure of the potassium channel: Molecular basis of K^+ conduction and selectivity. Science **280**, 69 (1998)
4. L. Guidoni, V. Torre, P. Carloni, Potassium and sodium binding to the outer mouth of K^+ channel. Biochemistry **38**, 8599 (1999)
5. M.Y. Kiriukhin, K.D. Collins, Dynamic hydration numbers for biologically important ions. Biophys. Chem. **99**, 155 (2002)
6. M.E. Tuckerman, D. Marx, M. Parrinello, The nature and transport mechanism of hydrated hydroxide ions in aqueous solution. Nature **417**, 925 (2002)
7. D. Marx, M.E. Tuckerman, J. Hutter, M. Parrinello, The nature of hydrated excess proton in water. Nature **397**, 601 (1999)

8. W.H. Robertson, E.G. Diken, E.A. Price, J.-W. Shin, M.A. Johnson, Spectroscopic determination of the OH^- solvation shell in the $OH^-(H_2O)_n$ clusters. Science **299**, 1367 (2003)
9. M.F. Kropman, H.J. Bakker, Dynamics of water molecules in aqueous solvation shells. Science **291**, 2118 (2001)
10. A.W. Omta, M.F. Kropman, S. Woutersen, H.J. Bakker, Negligible effect of ions on the hydrogen-bond structure in liquid water. Science **299**, 1367 (2003)
11. Y.K. Cheng, P.J. Rossky, Surface topography dependence of biomolecular hydrophobic hydration. Nature **392**, 696 (1998)
12. S. Baldelli, G. Mailhot, P.N. Ross, G.A. Somorjai, Potential-dependent vibrational spectroscopy of solvent molecules at the Pt(111) electrode in a water/acetonitrile mixture studied by sum frequency generation. J. Am. Chem. Soc. **123**, 7697 (2001)
13. S. Meng, D.V. Chakarov, B. Kasemo, S.W. Gao, Two dimensional hydration shells of alkali metal ions at a hydrophobic surface. J. Chem. Phys. **121**, 12572 (2004)
14. S. Meng, S. Gao, Formation and interaction of hydrated alkali metal ions at the graphite-water interface. J. Chem. Phys. **125**, 014708 (2006)
15. D.V. Chakarov, L. Österlund, B. Kasemo, Water adsorption and coadsorption with potassium on graphite (0001). Langmuir **11**, 1201 (1995)
16. L. Österlund, D.V. Chakarov, B. Kasemo, Potassium adsorption on graphite (0001). Surf. Sci. **420**, 174 (1999)
17. D. Feller, E.D. Glendening, D.E. Woon, M.W. Feyereisen, An extended basis set ab initio study of alkali metal cation–water clusters. J. Chem. Phys. **103**, 3526 (1995)
18. O. Borodin, R.L. Bell, Y. Li, D. Bedrov, G.D. Smith, Polarizable and nonpolarizable potentials for K^+ cation in water. Chem. Phys. Lett. **336**, 292 (2001)
19. D. Feller, K.D. Jordan, Estimating the strength of the water/single-layer graphite interaction. J. Phys. Chem. A **104**, 9971 (2000)
20. S. Meng, L.F. Xu, E.G. Wang, S.W. Gao, Vibrational recognition of hydrogen-bonded water networks on a metal surface. Phys. Rev. Lett. **89**, 176104 (2002)
21. S.W. Gao, J.R. Hahn, W. Ho, Adsorption-induced hydrogen bonding in the CH group. J. Chem. Phys. **119**, 6232 (2003)
22. J. Kim, S. Lee, S.J. Cho, B.J. Mhin, K.S. Kim, Structure, energetics, and spectra of the aquasodium (I): thermodynamic effects and nonadditive interactions. J. Chem. Phys. **102**, 839 (1995)
23. K. Morgenstern, J. Nieminen, Intermolecular bond length of ice on Ag(111). Phys. Rev. Lett. **88**, 066102 (2002)
24. D.L. Doering, T.E. Madey, The adsorption of water on clean and oxygen-dosed Ru(001). Surf. Sci. **123**, 305 (1982)
25. F. Sim, A.S. Amant, I. Papai, D.R. Salahub, Gaussian density functional calculations on hydrogen-bonded systems. J. Am. Chem. Soc. **114**, 4391 (1992)
26. Y. Marcus, Ionic radii in aqueous solutions. Chem. Rev. **88**, 1475 (1988)
27. I. Bako, J. Hutter, G. Palinkas, Car-Parrinello molecular dynamics simulation of the hydrated calcium ion. J. Chem. Phys. **117**, 9838 (2002)

Chapter 12
Microscopic Processes of Salt Dissolution and Nucleation

Table salt is one of the most common compounds in our daily life. It is formed by the combination of Na atoms (ions) and Cl atoms (ions) at the ratio of 1:1. Its crystal structure is shown in Fig. 12.1. It is a face-centered cubic complex lattice, with one primitive cell containing one Na^+ and one Cl^-. This lattice is of great importance in the history of crystallography. In 1915, the Braggs were awarded the Nobel Prize for Physics for the determination of the structure of NaCl and several other compounds by means of X-ray diffraction. As a typical representative of ionic compounds, NaCl crystal was one of the first models to be studied in the band theory of solid. In 1936, W. Shockley calculated the band structure of NaCl using the primitive-cell method, which was a representative work in early electronic structure calculations [1].

The interaction between water and salts has great significance not only in life process [2] but also in the fields of surface science, aqueous solution chemistry, and environmental science [3]. Although a great deal of work has been done by previous authors, many problems remain unsolved due to the complexity of this system; advances in theoretical computation methods have allowed for new perspectives on this system. This chapter attempts to theoretically clarify some of the processes involved in the dissolution and crystallization of salt at the atomic, molecular, and electronic levels, using a combination of classical and first-principle calculations.

12.1 Adsorption of Water on Salt Surface

Adsorption of water molecules is the starting point to study the interaction between water and the surface of salt. In the case of water adsorption on the surface of NaCl (001), Bruch et al. found a (1×1) diffraction structure using helium atom scattering (HAS) [4], while Fölsch et al. obtained a $c(2 \times 4)$ structure consisting of a ring of hexagonal water molecules using low-energy electron diffraction (LEED) analysis [5]. Toennies et al. found that under the irradiation of an electron beam, the (1×1) structure of water molecules can gradually change to a $c(2 \times 4)$ structure [6]. Using

S. Meng and E. Wang, *Water*,
https://doi.org/10.1007/978-981-99-1541-5_12

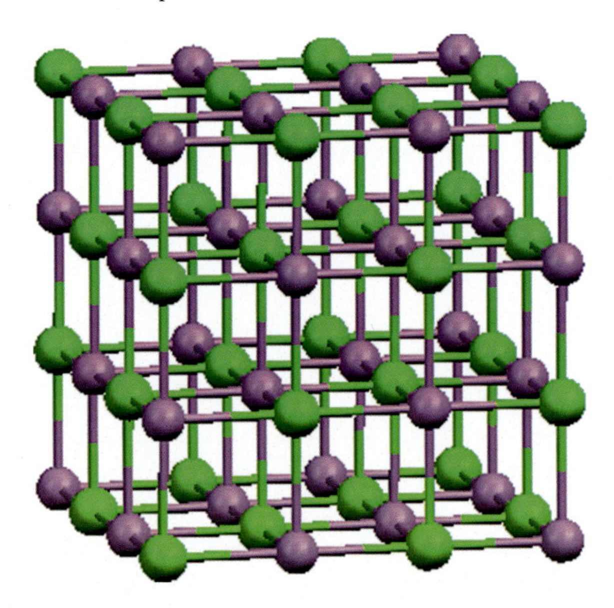

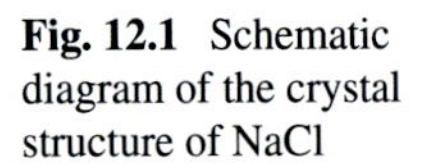

Fig. 12.1 Schematic diagram of the crystal structure of NaCl

infrared spectroscopic analysis, the Ewing group found that water forms a thin liquid-like layer on the NaCl (001) surface at room temperature [7, 8]. In addition, scanning polarization force microscopy (SPFM), a type of atomic force microscopy (AFM), has also been used in recent years to study the adsorption of water molecules on the dissociated NaCl (001) surface at different relative humidity [9]. To date, it is not experimentally possible to make direct observations of the adsorption configuration of individual water molecules on the salt surface of bulk materials. The adsorption of water on ultrathin salt layers on conductive substrates can be observed using scanning tunneling microscopy and some progress has been made [10]. See Chap. 6.

Various methods have been used to study the adsorption of water on the NaCl (001) plane, such as classical molecular dynamics methods [11], semi-empirical methods [12], first-principle calculations based on the cluster model [13], and density functional calculations under the LDA framework [14]. Some stable adsorption conformations of individual water molecules appear to be standing conformations whose polar planes (HOH planes) are perpendicular to NaCl (001) [11, 13], while others are flat conformations whose polar planes are almost parallel to the NaCl (001) plane [14]. However, DFT calculations within the LDA framework typically give excessive adsorption energies that are particularly inappropriate for the depiction of the unique hydrogen bonding interactions between water molecules. Park et al. studied the adsorption of water monomolecules, trimers, 1 monolayer (1 ML), and 1.5 monolayers (1.5 ML) on the NaCl (001) plane using DFT calculations within the GGA framework [15]. Their calculations show that the adsorption energy of the 1.5 ML configuration is much larger than that of 1 ML. Therefore, they predict that the (1×1) configuration of 1 ML may transform into a $c(4 \times 2)$ configuration similar to that of 1.5 ML under appropriate conditions while they do not propose what this transformed configuration would look like. Although there has been quite a bit of theoretical work [11–18], no one has ever studied the nature of bonding of water on the NaCl (001) surface via first principles. At the same time, the atomic mechanism for the transition of the adsorption configuration of water molecules from a (1 × 1) pattern to a $c(4 \times 2)$ pattern observed experimentally under electron irradiation [6] has not been clarified. The effect of the competition between hydrogen bonding interactions of water molecules and interactions between water molecules and the substrate on the adsorption configuration of water molecules is also worthy of further investigation.

The interaction between salt and water was calculated by Yang Yong et al. using the program package VASP based on density functional theory [19]. The electron wave function was expanded using plane waves and the electron-nucleus (ion core) interaction was depicted by Vanderbilt's ultra-soft pseudopotential. The exchange–correlation energy is approximated by the GGA in the Perdew–Wang form (PW91). The NaCl (001) surface is modeled by a "super primitive cell" that repeats periodically in the X, Y, Z directions. This super primitive cell contains five NaCl atomic layers with vacuum layers of height 12.66 Å. The water molecules are placed on one side of the crystal plane, as shown in Fig. 12.2.

The calculated lattice constant of NaCl is 5.67 Å, which agrees well with the experimental value (5.64 Å) [20]. For single water molecules (monomer), dimer and tetramer of water molecules, and the 1 ML and 2 ML adsorption configurations, the

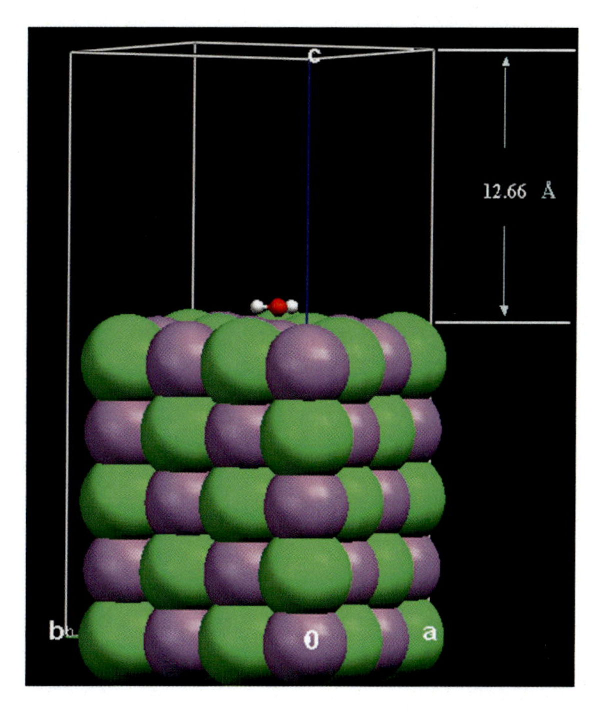

Fig. 12.2 Schematic diagram of the H_2O–NaCl (001) super primitive cell. a, b, and c denote orientations of the X, Y, and Z axes, respectively

planar unit cell of NaCl substrate employed is $p(2\sqrt{2} \times 2\sqrt{2})$; for the adsorption configurations of 1.5 ML and 1.75 ML, the employed planar unit cell of NaCl (001) is $p(2 \times 4)$. Here an adsorption monolayer (ML) is defined in such a way that one water molecule corresponds to one NaCl unit. The summation of the total energy of the system in the Brillouin zone is done using the Monkhorst–Pack scheme with the origin at Γ point. $2 \times 2 \times 1$ and $2 \times 1 \times 1$ k-point partitioning method are used for the planar unit cells $p(2 \times 4)$ and $p(2\sqrt{2} \times 2\sqrt{2})$ respectively. The plane wave energy cutoff $E_{cut} = 300$ eV. The effect of using a higher energy cutoff $E_{cut} = 400$ eV on the total energy of the system is only a few meVs, which is negligible. The Fermi energy level is expanded in a Gaussian shape and the expansion width is 0.2 eV. This set of parameters ensures that the total energy of the system converges to the level of at least 0.5 meV/atom. When performing the structural optimization, the adsorbed water molecules and the top three layers of atoms are relaxed together, while the bottom two layers of atoms are fixed (similar to the bulk atoms). The operation ends when the energy converges to the level of less than 1 meV. In fact, given the errors in the exchange–correlation energy in the calculations, and the inadequacy of the depiction of the van der Waals interaction in density functional theory, the errors in the energy are estimated to be around 10 meV. Thus, the accuracy of the interaction energy given below is preserved to the order of 10 meV.

12.1.1 Adsorption of Single Water Molecule

To gain a deeper understanding of the interaction between water molecules and NaCl (001), Yang Yong et al. first studied the adsorption configuration and energies of individual water molecules on the NaCl (001) surface. More than 40 different adsorption conformations were calculated in order to obtain the most stable conformation in terms of energy. These include the top site of Na^+, Cl^-, the bridge site between Na^+, Cl^-, the hollow site among the four atoms, and the point on the line connecting the two nearest neighbors of Na^+, as shown in Fig. 12.3.

The adsorption energy of a water molecule is given by the following equation.

$$E_{ads} = E\left[(\mathrm{NaCl\ (001)})_{\mathrm{relaxed}}\right] + E\left[(\mathrm{H_2O})_{\mathrm{gas}}\right] - E\left[(\mathrm{NaCl\ (001)} + \mathrm{H_2O})_{\mathrm{relaxed}}\right] \quad (12.1)$$

Here, $E[\mathrm{NaCl(001)} + \mathrm{H_2O}]$, $E[\mathrm{NaCl(001)}]$, and $E[\mathrm{H_2O}]$ are the energies of the adsorption system, the NaCl substrate, and the individual water molecule in the gaseous state after relaxation, respectively. The geometric parameters used to describe the structure of the water molecules are shown in Fig. 12.4.

The geometrical parameters and adsorption energies calculated for some typical adsorption configurations of individual water molecules are listed in Table 12.1. The lowest-energy single molecule adsorption conformation is shown in Fig. 12.5a, b. The water molecule lies obliquely along the Na^+–Na^+ linkage with the polar plane

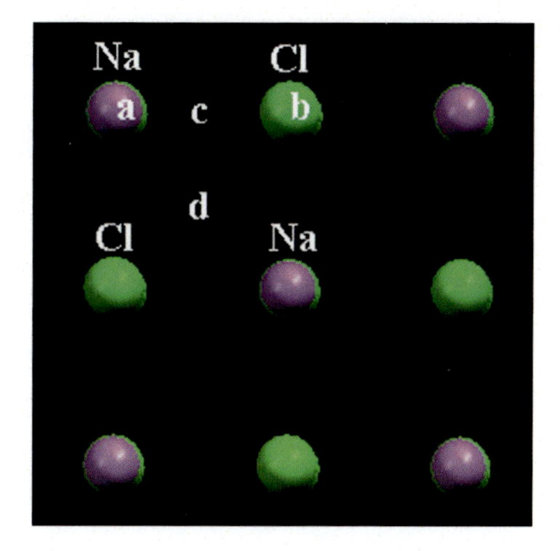

Fig. 12.3 Typical adsorption sites of single water molecules: a, b: top sites of Na^{+}, Cl^{-}; c: bridge sites; d: vacant sites

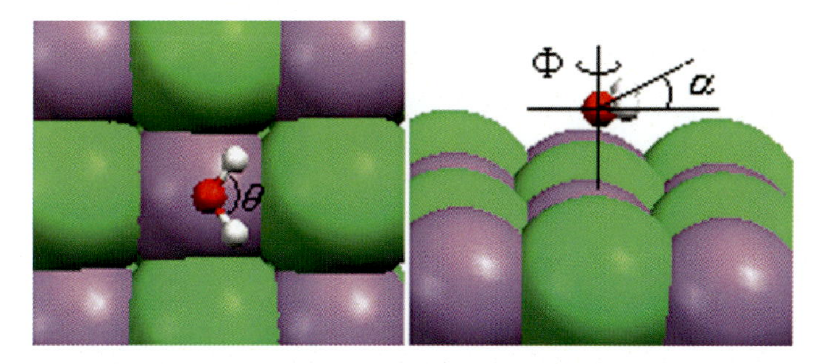

Fig. 12.4 Top (**a**) and side (**b**) views of the configuration of a single water molecule adsorbed on the NaCl (001) surface. The Na^{+}, Cl^{-} ions are represented by purple (second largest) and green (largest) spheres, respectively; the water molecules are represented by a curved ball-and-stick connection model, with the red ones being oxygen atoms and the white (smallest) ones being hydrogen atoms. This representation will be followed in the following views in this chapter

tilting slightly downwards. This conformation generally looks similar to the results given in the two previous works [14, 15], although the oxygen atom shifts more away from the top position of Na^{+} (here $\Delta Oxy = 1.1$ Å, compared to $\Delta Oxy = 0.7$ Å [14] and 0.83 Å [15]). The angle at which the polar plane tilts downward is also much larger (tilt angle $\alpha \approx -27°$, compared to $\alpha \approx -10°$ [14] and $\alpha \approx -18°$ [15]). It is possible that this difference comes from the different forms of exchange–correlation energy: the literature [14] uses LDA, while the literature [15] uses GGA in the form of PBE.

The adsorption of stable configurations contains both O–Na, H–Cl attraction, as can be seen from the structure and energy data in Table 12.1. Calculations show that strong O–Na (H–Cl) attraction exists when the O–Na distance is around 2.4 Å (H–Cl distance is around 2.2 Å). Interestingly, as the water molecules slip along the Na–Na linkage shown in Fig. 12.5a ([110] direction), a series of sub-stable states with

Table 12.1 Adsorption energies and structural parameters of single water molecules on the NaCl (001) surface

Configuration	E_{ads}	O–Na	H^1–Cl	ΔO xy	α	Φ	θ	O–H^1	O–H^2
11.5(a)	0.40	2.385	2.347	1.074	−27	45	104.9	0.980	0.980
11.6(a)	0.33	2.392	3.074	0.182	14	49	105.8	0.975	0.974
11.6(b)	0.37	2.403	2.192	1.048	27	48	104.6	0.987	0.973
11.6(c)	0.30	2.319	2.640	0.452	90	0	108.7	0.977	0.971
11.6(d)	0.34	2.302	2.231	0.831	90	0	109.5	0.983	0.967
11.6(e)	0.28	2.384	3.670	0.000	90	0	106.1	0.970	0.970
11.6(f)	0.17	4.336	2.232	0.013	90	0	104.7	0.982	0.971

Here, H^1 denotes the hydrogen atom in the water molecule that forms hydrogen bond with Cl^-, and H^2 denotes the other hydrogen atom. ΔOxy denotes the distance that the oxygen atom deviates from the exact top position of Na^+ or Cl^-. α is the angle between the polar plane of the water molecule and NaCl (001), Φ is the angle between the dipole moment linkage of the water molecule and the Na-Cl linkage ([001]), and θ is the HOH angle, see Fig. 12.4. The units of energy, length, and angle are eV, Å, and degree (°), respectively

slightly smaller adsorption energies are obtained, as shown in Fig. 12.5c. During such a shift, either the O–Na attraction is reduced or the H–Cl attraction is weakened. The existence of this series of sub-stable configurations originates from the sp^3 hybrid molecular orbitals of the water molecule, see Fig. 4.4d. Such hybridization allows the two lone pairs of electrons to distribute in a plane perpendicular to the HOH polar plane, and the plane of the water molecule takes a near-flat form in order to allow sufficient contact interaction between O and Na to occur. Moreover, considering the H–Cl attraction, the plane of the water molecule slips forward somewhat, making the position of O deviate from the exact top position of Na, which gives the most stable conformation shown in Fig. 12.5a, b: the plane lies obliquely, containing both strong O–Na and H–Cl attraction at the same time.

Other typical adsorption configurations are shown in Fig. 12.6. Some adsorption configurations are achieved mainly through O–Na interactions, as in Fig. 12.6a, c; while some adsorption configurations contain both strong O–Na and H–Cl interactions, as in Fig. 12.6b, d. As seen in Table 12.1, the strength of the O–Na interaction depends not only on the distance between the two atoms, but also on the relative orientation of the water molecule planes. That is related to the orientation of the lone pair of electrons of the water molecule. The configuration with the polar plane lying slightly oblique is always energetically more stable than the adsorption configuration with the polar plane perpendicular to NaCl (001), which can be understood from the perspective of the characteristics of the sp^3 hybridization of the water molecule. This is further confirmed by the electron density differential chart in Fig. 12.5e, f. It can be seen from the chart that the characteristics of covalent bonding exist. Therefore, the conclusion can be made that the interaction between the water molecule and the substrate contains both ionic and covalent bonding features.

The comparison of the two interactions, O–Na and H–Cl, is presented below. Focusing on two typical configurations: in Fig. 12.6e, the water molecule is adsorbed

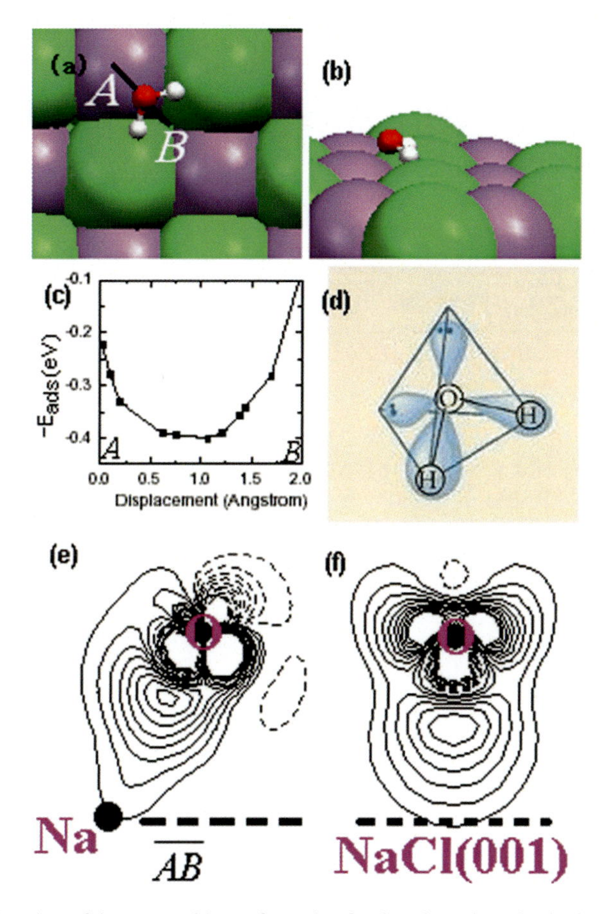

Fig. 12.5 **a** Top view of the most stable configuration for the adsorption of a single water molecule; **b** side view of the most stable configuration for the adsorption of a single water molecule; **c** relationship between the single molecule adsorption energy and the displacement from the top site of Na^+ (along the AB linkage); **d** schematic diagram of sp^3 hybrid molecular orbital of the water molecule; **e** contour plot of the charge density difference $\Delta\rho$. $\Delta\rho = \rho[H_2O/NaCl(001)] - \rho[NaCl(001)] - \rho[H_2O]$, where $\rho[H_2O/NaCl(001)]$, $\rho[NaCl(001)]$, $\rho[H_2O]$ are the total charge density of the system, the charge density of the NaCl (001) substrate, and the charge density of a single water molecule in the gaseous state, respectively. The tangent plane is along the AB linkage and perpendicular to the NaCl (001) surface; **f** similar to (**e**), except that the tangent plane is along the direction perpendicular to the AB linkage and the NaCl (001) surface. The contour lines take the values $\Delta\rho = \pm 0.005 \times ne/\text{Å}^3$, where n = 1, 2, …, 10. The solid line represents $\Delta\rho > 0$, while the dashed line represents $\Delta\rho < 0$.

perpendicular to NaCl (001) at the top site of Na^+, in which case adsorption can be considered as achieved mainly by O–Na attraction; in Fig. 12.6f, the water molecule is adsorbed perpendicular to NaCl (001) through hydrogen at the top site of Cl^-, in which case adsorption can be considered as achieved mainly by H–Cl attraction. As seen in Table 12.1, their adsorption energies are 280 meV and 170 meV, respectively,

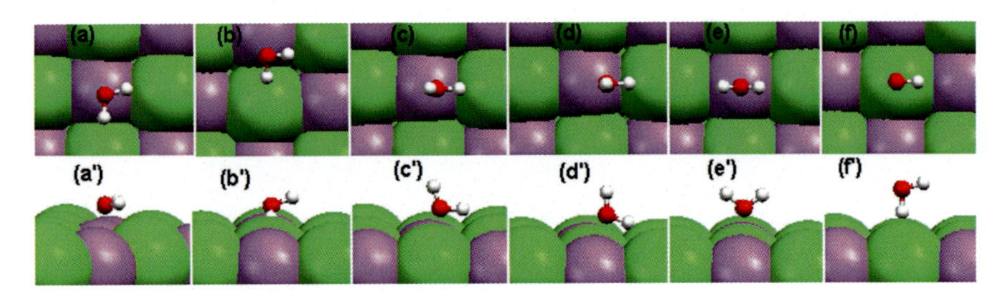

Fig. 12.6 Top (top panel) and side (bottom panel) views of some typical adsorption configurations for the adsorption of single water molecules on the NaCl (001) surface. The adsorption energy of each configuration is indicated below

i.e., the strength of the O–Na attraction is almost twice that of H–Cl. This conclusion is different from an earlier DFT-LDA work [14], where the contribution of H–Cl attraction to the adsorption energy was considered to be dominant.

12.1.2 Adsorption of Water Clusters

Next, Yang Yong et al. studied the adsorption of clusters of water molecules covering less than one monolayer (ML)—dimer and tetramer of water molecules. At this point, hydrogen bonding interactions between the water molecules come into play. Generally, hydrogen bonds are formed when a positively charged H atom and a negatively charged, strongly electronegative atom X (X=N, O, F, Cl, …) with a lone pair of electrons are in close proximity. The hydrogen bonding interaction between water molecules is one of the most important interactions, which brings many unique properties to water. For two water molecules that form a hydrogen bond, the hydrogen donor is called the proton donor, while the hydrogen acceptor is called the proton acceptor. For a free water dimer, the typical distance between O–O is 2.95 Å and the energy of hydrogen bond is 0.24 eV. How does the situation differ when a free dimer is adsorbed onto NaCl (001)? A typical series of adsorption conformations of the water dimer calculated by Yang et al. are shown in Fig. 12.7. The corresponding structural and energy parameters are given in Table 12.2.

The adsorption and interaction energies of water molecules are calculated by the following equations:

$$\begin{aligned} E_{ads} =& \frac{1}{2}\{E[(\mathrm{NaCl}(001))_{relaxed}] + 2 \times [(\mathrm{H_2O})_{gas}] \\ &- E[(\mathrm{NaCl}(001) + dimer)_{relaxed}]\} \end{aligned} \tag{12.2}$$

$$E_{ww} = \frac{1}{2}\{2 \times [(\mathrm{H_2O})_{gas}] - E[dimer]\} \tag{12.3}$$

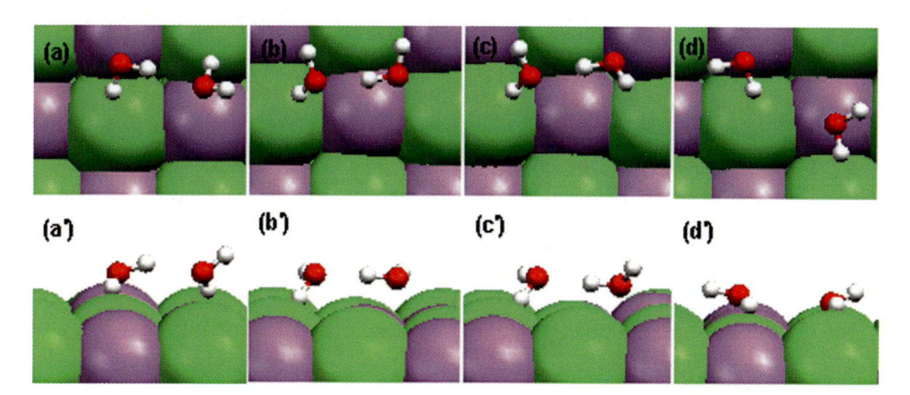

Fig. 12.7 Some typical conformations of a dimer of water molecules adsorbed on NaCl (001). The top panel shows the top view and the bottom panel shows the side view

Table 12.2 Energy and structural parameters for the adsorption of two water molecules on the NaCl (001) surface

Configuration	E_{ads}	E_{ww}	E_{sw}	O^D–Na	O^A–Na	O^D–O^A	H^D–O^A	H^A–Cl
(a)	0.41	0.10	0.31	2.430	2.917	2.873	1.973	2.218
(b)	0.20	0.09	0.11	3.056	–	2.758	1.774	2.201
(c)	0.29	0.11	0.18	2.545	–	2.838	1.848	2.241
(d)	0.35	0.03	0.32	2.440	2.361	3.666	2.962	–

The adsorption energy E_{ads}(eV/HO_2), the interaction energy between the water molecules E_{ww}(eV/H_2O), and the interaction between the water molecules and the substrate E_{sw}(eV/H_2O) are given separately. O^D denotes the oxygen atom in the proton donor, and H^D denotes the hydrogen atom that forms a hydrogen bond with the other water molecule; O^A denotes the oxygen atom in the proton acceptor, and H^A denotes the hydrogen atom in the proton acceptor that $^{-}$forms a hydrogen bond with Cl^-. The unit of distance is Å

$$E_{sw} = E_{ads} - E_{ww} \quad (12.4)$$

Here, E[dimer] refers to the free water molecule dimer, which has the same structure as adsorbed on the NaCl (001) surface. Approximately, $2 \times E_{ww}$ can be thought of as the energy of hydrogen bond in the water molecule $E_{H\text{-}bond}$. E_{sw} in turn, can be thought of as the interaction energy of the water molecule and the substrate. For the tetramer of water molecules, the formula for the calculation is similar.

The most energetically stable water dimer is shown in Fig. 12.7a, a′. The oxygen atoms of both water molecules are close to Na^+, and at the same time a strong hydrogen bond ($E_{H\text{-}bond}$ ~ 0.20 eV) exists between the water molecules. If the two water molecules are allowed to relax freely to their respective ground states without the constraint of hydrogen bonding, the corresponding water-substrate interaction energy is 0.40 eV/H_2O, which is ~ 0.09 eV/H_2O higher than the present energy. This suggests that hydrogen bonding affects the adsorption of water molecules.

In turn, the adsorption configuration of the water molecule affects the strength of the hydrogen bond. The two water molecule dimers shown in Fig. 12.7b, c have almost the same proton acceptor configuration, the main difference being the position of the proton donor. The water molecule proton donor shown in Fig. 12.7b is almost parallel to the NaCl (001) surface, and its projection on the NaCl (001) surface is at the hollow position. In contrast, the water molecule proton donor shown in Fig. 12.7c is tilted (angle ~ 45°) and adsorbed near the closest Na^+. From the O–Na^+ distance and water-substrate interaction energy E_{sw} given in Table 12.2, it can be seen that the O–Na^+ interaction is stronger in Fig. 12.7c than that in Fig. 12.7b. This stronger interaction with the substrate implies that electrons of the proton donor in the dimer transfer from the oxygen atom to the NaCl (001) substrate. This is directly confirmed by the electron density difference plot in Fig. 12.8. As the lone pair of electrons of the proton donor transfer to the substrate, the proton donor becomes less likely to bind the hydrogen atoms forming hydrogen bonds. Therefore, the greater the electron transfer from the proton donor to the substrate, the weaker its binding to the hydrogen atoms that form hydrogen bonds, and in turn the stronger the hydrogen bonds. This is clear when comparing Fig. 12.8a, b.

The substrate also affects the formation of the water molecule dimer. Figure 12.7d shows the adsorption conformation of two water molecules after relaxation. Their initial O–O distance is 3 Å, which becomes 3.7 Å after relaxation on the NaCl (001) surface. The hydrogen bonding energy in this case is only about 0.06 eV, and if the two water molecules are allowed to relax freely without the restriction of the substrate, a strong hydrogen bond will form between them: $E_{H\text{-}bond} = 0.20$ eV. Calculations show that when the initial O–O distance is greater than 4 Å, only very weak hydrogen bonds (with energies of about a few tens of meV) will form between the two adsorbed water molecules, regardless of their relative orientation, in which case the water molecules behave much as they would in monomeric adsorption.

Another typical adsorption scenario is adsorption of water molecule tetramer. This is when four water molecules interact and adsorb near the top site of the nearest neighboring Na^+, Cl^- and are connected to each other with hydrogen bonds, forming a tetrameric ring structure as shown in Fig. 12.9. Each water molecule is connected by two hydrogen bonds, acting as both proton donor and proton acceptor. Furthermore, each water molecule forms an O–Na^+ or H–Cl^- bond with the substrate. Since this structure matches the lattice of the substrate and the hydrogen bonds between the water molecules are close to saturation, this structure is a stable adsorption structure on the NaCl (001) surface. The calculated adsorption energy is 0.47 eV/H_2O, which is much larger than that of the monomer or dimer of water molecules. The water molecule-water molecule interaction energy E_{ww} and the water molecule–substrate interaction energy E_{sw} are 0.25 eV/H_2O and 0.22 eV/H_2O, respectively. E_{ww} is much larger compared to the most stable conformation of the dimer of the water molecule, while E_{sw} is relatively smaller. This is a consequence of the strengthening of the hydrogen bonding structure of the adsorption system. It is worth mentioning that this structure shares a similar structure to that of the water molecule trimer given by the earlier DFT-GGA [15], which corresponds to the structure left after taking out the water molecule at one of the corners (top position of Cl^-) of the water molecule

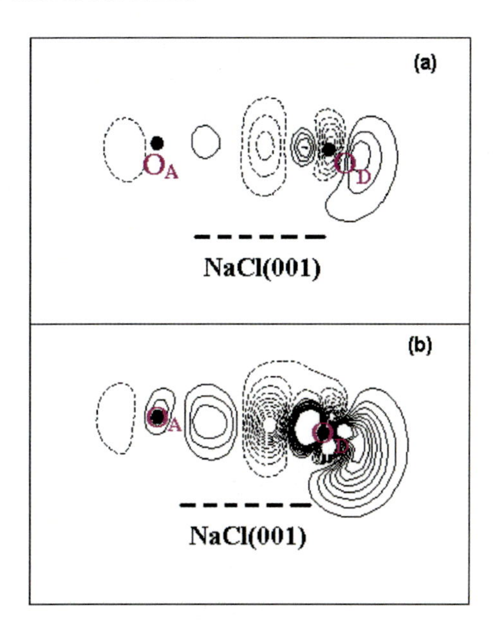

Fig. 12.8 Schematic of the contour of the charge density difference ($\Delta\rho$). Here, the definition of $\Delta\rho$ is $\Delta\rho = \rho[(H_2O)_2/NaCl(001)] - \rho[(H_2O)_A/NaCl(001)] - \rho[(H_2O)_D]$. $\rho[(H_2O)_2/NaCl(001)]$, $\rho[(H_2O)_A/NaCl(001)]$ and $\rho[(H_2O)_D]$ $[(H_2O)_D]$ refer to the total charge density of the system, the charge density of the H-acceptor adsorbed on the NaCl (001) plane, and the charge density of the water molecule as H-donor in the gaseous state, respectively. The cut-in plane is along the [010] direction, through the oxygen of the H-donor, perpendicular to the NaCl (001) surface. **a** corresponds to the dimer structure in Fig. 12.7b; **b** corresponds to the dimer in Fig. 12.7c. O_D, O_A denote the oxygen atoms of the H-donor and H-acceptor, respectively. The contours take the values $\Delta\rho = \pm 0.005 \times ne/\text{Å}^3$, $n = 1, 2, \ldots, 10$. The solid line represents $\Delta\rho > 0$ and the dashed line represents $\Delta\rho < 0$

tetramer here. This stable geometry of the water molecule tetramer makes it the most stable adsorption cluster on the NaCl (001) surface. As will be seen below, this structure (which may be somewhat distorted for different coverages) is the basic building block of the multilayer adsorption structure of water molecules.

12.1.3 Adsorption of Water Monolayers and Multilayers

The adsorption of monolayer and multilayer water molecules onto NaCl (001) will be discussed next. Focusing on the competition between water molecule-water molecule interactions and water molecule–substrate interactions during adsorption, the aim is to further understand the role played by hydrogen bonding in the adsorption process. The adsorption energy E_{ads}, the water molecule-water molecule interaction energy E_{ww}, and the water molecule–substrate interaction energy E_{sw} are calculated as follows:

Fig. 12.9 Schematic representation of the adsorption configuration of a tetramer of water molecules on the NaCl (001) surface. The top panel shows the top view while the bottom panel shows the side view. Hydrogen bonds are represented by dotted lines

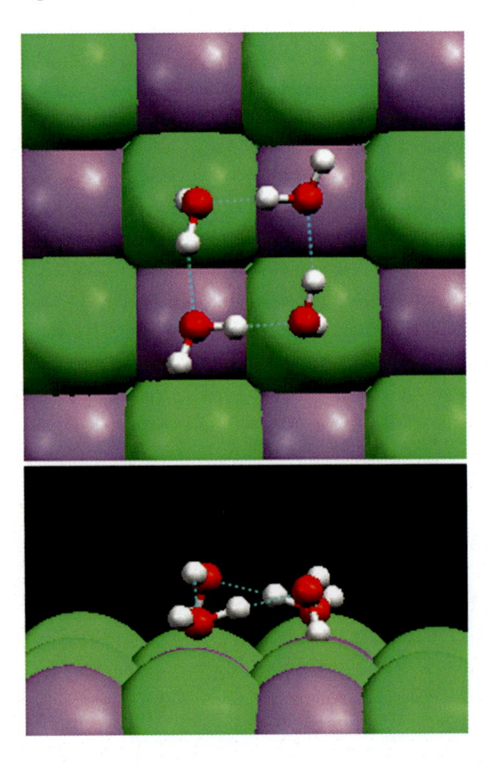

$$E_{ads} = \frac{1}{n}\{E[(\mathrm{NaCl}(001))_{relaxed}] + n \times E[(\mathrm{H_2O})_{gas}] - E[(\mathrm{NaCl}(001) + \mathrm{H_2O})_{relaxed}]\} \quad (12.5)$$

$$E_{ww} = \frac{1}{n}\{n \times E[(\mathrm{H_2O})_{gas}] - E[(\mathrm{H_2O})_{network}]\} \quad (12.6)$$

$$E_{sw} = E_{ads} - E_{ww} \quad (12.7)$$

Here, n indicates the number of water molecules contained in the "super primitive cell" used for the calculation.

For the adsorption of a monolayer, three typical configurations were calculated, including two with the HOH planes lying flat (Fig. 12.10a, b), and one with the HOH plane standing up (Fig. 12.10c). In all three diagrams, all water molecules are adsorbed near the top site of Na^+. These three conformations are also the ones most commonly used in the previous calculations and used to study the 1ML water molecule adsorption. Their adsorption energies are 0.39 eV/H_2O, 0.39 eV/H_2O, and 0.35 eV/H_2O. This is similar to the case of the adsorption of a single water molecule, where the conformation is more stable with the water molecule planes lying flat than standing. This result is contrary to that given in the literature [14], where the order of the adsorption energies is given as (a) > (c) > (b). A detailed comparison of the DFT-LDA [14] and DFT-GGA results is given in Table 12.3. The water molecule-water molecule interaction energy in Fig. 12.10b is larger than those in both Fig. 12.10a, c due to the orientation of the water molecules in Fig. 12.10b and the relative ease with which the nearest neighboring water molecules form hydrogen bonds. On average, the water molecule-water molecule interaction energy for these three configurations is about 0.07 eV/H_2O, which is much lower than that of the water molecule–substrate interaction (~ 0.34 eV/H_2O).

To further understand the role of hydrogen bonding, Yang Yong et al. constructed a chain-like structure of water molecules with a coverage of 1 ML (Fig. 12.10d). Neighboring water molecules are linked head to tail by hydrogen bonds. Surprisingly, the adsorption energy of such a structure (0.48 eV/H_2O) is much larger than that of the three previous 1ML configurations (Fig. 12.10a–c). This suggests that energetically more stable structures exist for the 1ML coverage case, except that they will be significantly different in structure from the typical configurations commonly used in the literature (Fig. 4.9a–c).

To confirm this, the thermal stability of these conformations was studied at 80 K (canonical ensemble) using first-principle molecular dynamics. Interestingly, the chain-like structure shown in Fig. 12.10d spontaneously transforms during a thermal equilibrium process of 2 ps into a tetragonal-hexagonal water molecules, shown in Fig. 12.10e, linked by hydrogen bonds. Based on this, a 2 ps MD simulation was made at 80 K. The final structure obtained is shown in Fig. 12.10f, which essentially maintains the structure of the 4–6-membered ring of the water molecules. The structure and energy parameters of these configurations are listed in Table 12.4. The adsorption energy of the ring structure shown in Fig. 12.10e (0.52 eV/H_2O) is much larger than that of the (1×1) structure shown in Fig. 12.10a–c with the 1ML configuration (maximum is 0.39 eV/H_2O). This energy reduction comes from hydrogen-bonding interactions between water molecules, obtained in optimal complementarity with water molecule–substrate interactions. This structure is more stable than any of the previously reported adsorption structures of 1 ML of H_2O/NaCl (001) in the literature. Similarly, the first-principle molecular dynamics simulation of the structures shown in Fig. 12.10a, b at 80 K shows that both structures slowly evolve into a network structure similar to that in Fig. 12.10e, linked by hydrogen bonds.

A longer molecular dynamics simulation of the structure shown in Fig. 12.10e also shows a 4–5-membered ring structure of the water molecules. This indicates that the hydrogen bonds are not saturated, so this structure is somewhat flexible. The surface primitive cell of this structure is $p(2\sqrt{2} \times 2\sqrt{2})$, which is $c(4 \times 4)$ on a large scale. Hope that this new structure will be experimentally observed in the future.

The adsorption of water on the NaCl (001) surface was also studied at higher coverages, e.g. 1.5 ML, 1.75 ML, 2 ML. The calculated structure and energy parameters are shown in Table 12.4.

Considering the experimental results of Fölsch et al. [5], Yang Yong et al. calculated the adsorption structures of two types of 1.5 ML water molecules in the $p(2 \times 4)$ planar crystals, whose planar primitive cells are $p(2 \times 4)$ and $c(2 \times 4)$, respectively, as shown in Fig. 12.11. After structural optimization (relaxation), Fig. 12.11a

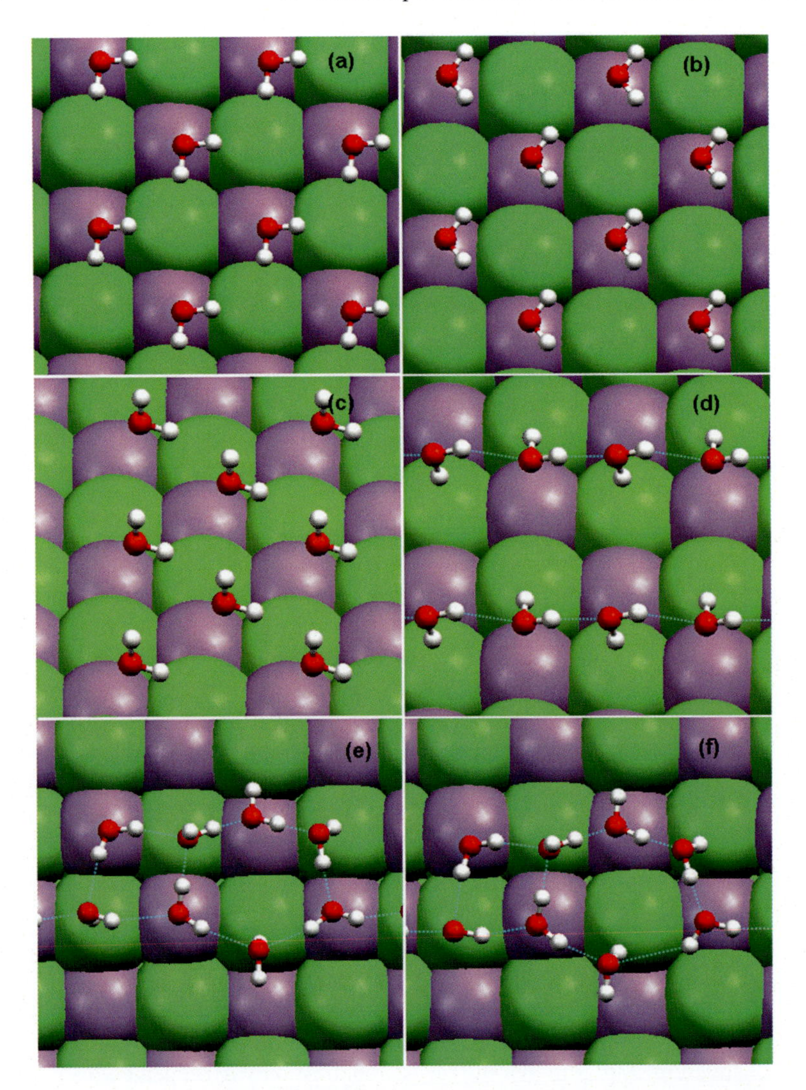

Fig. 12.10 Schematic representation of the adsorption conformation of 1 ML of water molecules on the NaCl (001) surface. Configurations (**e**) and (**f**) are obtained from configuration (**d**) after equilibration for 2 ps at 80 K, and 2 ps of molecular dynamics (MD) simulation after the equilibration, respectively. Similar results were obtained by MD of the configurations (**a**) and (**b**). The hydrogen bonds are indicated by dotted lines

actually contains six hexagonal rings of water molecules connected by hydrogen bonds, while Fig. 12.11b contains three hexagonal rings of water molecules. The water molecules of the $p(2 \times 4)$ primitive cell form a three-layer structure in the surface normal direction, with distances from the oxygen atom to the NaCl (001) surface being $Z_{O\text{–}Nacl(001)} \approx 2.5$ Å, 2.9 Å, and 3.8 Å, respectively, whereas the water molecules of the $c(2 \times 4)$ primitive cell form a two-layer structure in the surface

Table 12.3 Comparison of the adsorption results of DFT-LDA [14] and DFT-GGA (calculated by Yang Yong et al.) for 1 ML of water molecules

Configuration	E_{ads}	E_{ww}	E_{sw}	O–Na	ΔOxy	α	Φ	θ
(a) GGA	0.39	0.05	0.34	2.434	0.713	−10	45	103.2
(b) GGA	0.39	0.14	0.25	2.433	0.142	7	0	105.1
(c) GGA	0.35	0.06	0.29	2.360	0.793	90	0	106.6
(b)[a] LDA	0.63	0.06	0.57	2.404	0.7	−10	45	105
(a')[a] LDA	0.56	0.18	0.38	2.300	0	0	0	105
(c)[a] LDA	0.58	0.07	0.51	2.280	0.6	90	0	105

The top three rows are the results for GGA and the bottom three rows are the results for LDA [14]. The adsorption energy (E_{ads}), water molecule-water molecule interaction energy (E_{ww}), and water molecule–substrate interaction energy (E_{sw}), are listed in the table with corresponding structural parameters. The initial configurations for structural relaxation calculations are the same as those in literature [14], with (a) corresponding to (b)[a], (b) corresponding to (a')[a], and (c) corresponding to (c)[a]. ΔOxy is the distance of the oxygen atom away from the positive top site of Na^+. α is the tilt angle of H_2O–NaCl (001), Φ is the angel between the projection of the dipole moment linkage on the (001) surface and the [001] (Na-Cl linkage) direction, and θ is the HOH angle, see Fig. 3.3. Positive energy values represent attraction. The units of energy, length, and angle are eV, Å, and degree (°), respectively

[a] See Fig. 4 and Table 2 in literature [14]

Table 12.4 The surface primitive cells used for calculations of the adsorption of 0.5 ML, 1 ML, 1.5 ML, 1.75 ML and 2 ML of water molecules on the NaCl (001)

Coverage	Primitive cell	N_{H_2O}	E_{ads}	E_{ww}	E_{sw}
0.5 ML (tetramer)	$p(2\sqrt{2} \times 2\sqrt{2})$	4	0.47	0.25	0.22
1 ML (a)	$p(2\sqrt{2} \times 2\sqrt{2})$	8	0.39	0.05	0.34
1 ML (b)	$p(2\sqrt{2} \times 2\sqrt{2})$	8	0.39	0.14	0.25
1 ML (c)	$p(2\sqrt{2} \times 2\sqrt{2})$	8	0.35	0.06	0.29
1 ML (d)	$p(2\sqrt{2} \times 2\sqrt{2})$	8	0.48	0.24	0.24
1 ML (e)	$p(2\sqrt{2} \times 2\sqrt{2})$	8	0.52	0.32	0.20
1 ML (f)	$p(2\sqrt{2} \times 2\sqrt{2})$	8	0.50	0.33	0.17
1.5 ML (a)	$p(2 \times 4)$	12	0.54	0.37	0.17
1.5 ML (b)	$p(2 \times 4)$	12	0.57	0.42	0.15
1.75 ML	$p(2 \times 4)$	14	0.56	0.39	0.17
2 ML	$p(2\sqrt{2} \times 2\sqrt{2})$	16	0.56	0.40	0.16

Number of water molecules contained within the cell (N_{H_2O}), adsorption energy (E_{ads}), water molecule-water molecule interaction energy (E_{ww}), water molecule–substrate interaction energy (E_{sw}). The unit of energy is eV/H_2O

normal direction, with distances from the oxygen atom to the NaCl (001) surface being $Z_{O\text{-Nacl}(001)} \approx 2.6$ Å, 2.9 Å, respectively. Energetically, the water molecules of the $c(2 \times 4)$ adsorption structure are more stable, with a ~ 0.03 eV/H_2O difference in adsorption energy.

The adsorption structure of the 1.75 ML water molecule is constructed on the basis of the 1.5 ML adsorption conformation, with two water molecules added to the $c(2 \times 4)$ unit cell, as shown in Fig. 12.12. After structural optimization, some of the original hydrogen bonds in Fig. 12.11b are broken, while some new hydrogen bonds are formed between the newly added water molecule and its immediate neighbors. The adsorption energy of this structure is 0.56 eV/H_2O, which is very close to the adsorption energy of the structure shown in Fig. 12.11b.

For the adsorption of 2 ML of water molecules, all water molecules occupy the top position of Na^+ or Cl^-, forming a bilayer structure in the normal direction of the plane, as shown in Fig. 12.13. The heights of this bilayer structure are $Z_{O\text{-Nacl}(001)}$ ≈ 2.5 Å and 3.1 Å, which correspond to the water molecules adsorbed at the top

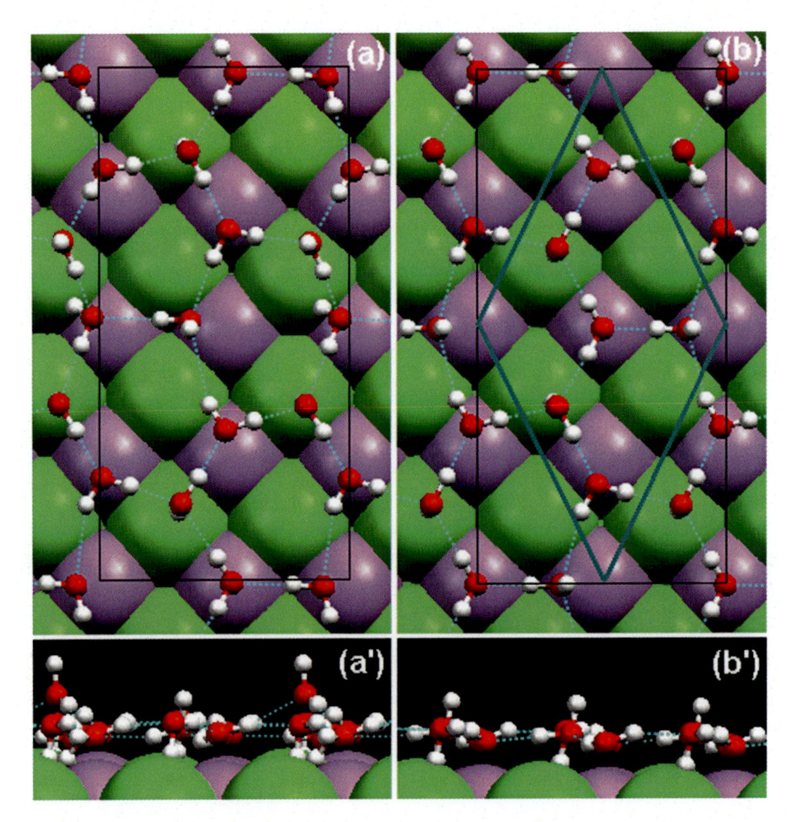

Fig. 12.11 Top and side views of 1.5 ML of water molecules adsorbed on the NaCl (001) surface: (a) and (a') correspond to the water structure with a planar primitive cell of $p(2 \times 4)$; (b) and (b') correspond to the adsorption structure with a planar primitive cell of $c(2 \times 4)$

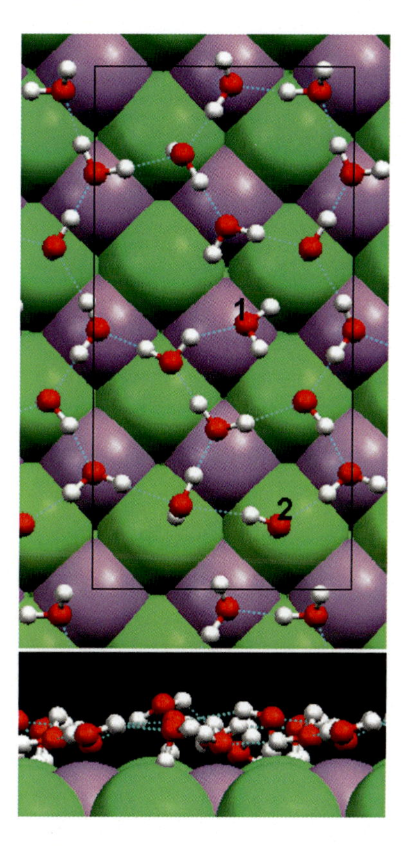

Fig. 12.12 Adsorption of 1.75 ML of water molecules on the NaCl (001) surface. Compared to Fig. 12.11b, two new water molecules are added, indicated by the labels 1 and 2

sites of Na^+ and Cl^-, respectively. This local structure of the adsorbed configuration is very similar to the water molecule tetramer studied earlier: each water molecule forms three hydrogen bonds with its immediate neighbors and a fourth bond with the substrate, which is an O–Na^+ bond when adsorbed at the top site of Na^+ and a H–Cl^-bond when adsorbed at the top site of Cl^-. Thus, the lone pairs of electrons of each water molecule are saturated by hydrogen or O–Na^+ bonds, forming a structure similar to that of the two-dimensional ice. This conformation has an adsorption energy $E_{ads} = 0.56$ eV/H_2O, a water molecule-water molecule interaction energy $E_{ww} =$ 0.40 eV/H_2O, and a water molecule–substrate interaction energy $E_{sw} = 0.16$ eV/H_2O.

Comparing the configuration of the 2 ML model with the (1 × 1) HAS (helium atom scattering) pattern observed experimentally by Bruch et al. [4], it can be seen that the adsorption energies of the two configurations are almost identical (0.56 eV/H_2O vs 0.6 eV/H_2O). Since the positions of Na^+ and Cl^- are equivalent on the NaCl (001) surface, the diffraction pattern obtained should also be of the (1 × 1) structure when the HAS experiment is used to observe the adsorption of 2 ML of water molecules on NaCl (001). This suggests that it is possible that the same diffraction pattern corresponds to different adsorption structures. When interpreting the experimental results, Bruch et al. proposed the adsorption model shown in Fig. 12.10a. However, this model

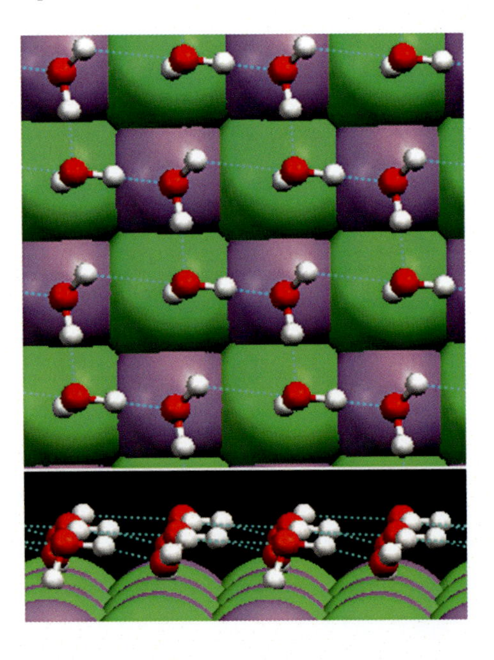

Fig. 12.13 Top (top panel) and side (bottom panel) views of 2 ML of water molecules adsorbed on the NaCl (001) surface. Hydrogen bonds are indicated by dotted lines

gives adsorption energies that are too small compared to the experimental values, while the adsorption structure of 2 ML gives adsorption energies that are very close to the experimental values. So, it is possible that the 2 ML conformation shown in Fig. 12.13 is a more reasonable model for explaining the experimental observations. Furthermore, it can be expected that the adsorption energy will decrease after the coverage of water molecules is greater than 2 ML, because at this coverage all water molecules are saturated with lone pairs of electrons and no more hydrogen bonds are formed. Interestingly, this adsorption configuration with the hydrogen atoms (at the Cl^- top site) and oxygen atoms (at the Na^+ top site) alternately facing downwards is similar to that observed experimentally on the Pt(111) surface previously: the adsorption pattern of the water molecules is almost parallel to the substrate, with both of the hydrogen atoms pointing downwards, only the basic constituent units being different, which is in a quadrilateral shape here but in a hexagonal shape on the Pt(111) surface.

To obtain a general picture of the coverage and energy of water adsorption on the NaCl (001) surface, the water molecule–substrate interaction energy E_{sw}, and the water molecule-water molecule interaction energy E_{ww} versus coverage are shown in Fig. 12.14. For the 1 ML and 1.5 ML cases, the energy parameter for the most stable conformation is taken. At coverage levels below 1.5 ML, E_{sw} decreases while E_{ww} increases as coverage increases. The interaction between water molecules is greatly enhanced by the formation of hydrogen bonds with their immediate neighbors, while the interaction between water molecules and the substrate gradually decreases. The adsorption energy remains almost constant from 1.5 ML to 2 ML in coverage, with only a small fluctuation, see Table 12.4. The increase in coverage of water molecules is followed by the increase in the number of hydrogen bonds, but at the same time, the repulsion between non-hydrogen bonded water molecules also increases. This means that the stability of these adsorption configurations is almost the same for coverages from 1.5 ML to 2 ML. This could be why experimentally some see a $c(4 \times 2)$ conformation (1.5 ML) [5], while others see a (1×1) diffraction pattern [4] (which corresponds to 2 ML here). These two different experimental observations may arise from the different coverages of water molecules adsorbed on the NaCl (001) surface. These two are almost degenerate in energy, but both are stable structures that may transform into each other under certain experimental conditions, such as changes in coverage and temperature, but the potential barrier corresponding to this transformation is not clear and is subject to further study. Nevertheless, the newly proposed model provides a possible atomic mechanism for the experimentally observed shift in the adsorption pattern due to electron irradiation [6]: the incident electrons (~ 90 eV) lead to the desorption of a fraction of the water molecules of the (1×1) HAS pattern and reduce the coverage, which allows the remaining adsorbed water molecules to undergo a structural reorganization from a (1×1)(2 ML) to a $c(4 \times 2)$ conformation (1.5 ML).

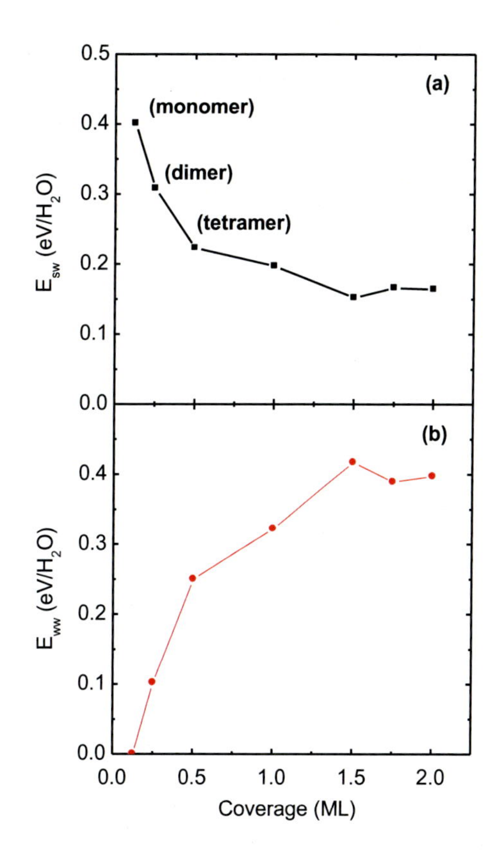

Fig. 12.14 Relationship between water molecule–substrate (layout a)/water molecule-water molecule (layout b) interaction energy and water molecule coverage

From Table 12.4 and Fig. 12.14b, it can be observed that the adsorption energy E_{ads} and the water molecule-water molecule interaction energy E_{ww} show the same trend, which again shows the importance of water molecule-water molecule interactions. When the coverage is greater than 0.5 ML, E_{ww} is always larger than E_{sw}. This suggests that the NaCl (001) surface resembles a hydrophobic plane, a result that helps us understand on a molecular scale why pure NaCl crystals are less prone to water absorption and deliquescence at room temperature and a relative humidity below 75% [7].

12.2 Microscopic Images of Salt Dissolution

The adsorption of water on the surface of salts, dissolution of salts in water and hydration processes are of great importance for surface science, electrochemistry, biophysics and environmental science. Daily life experience tells us that the dissolution of NaCl occurs readily in water, while if heating is used, it takes up to 1074 K to break the Na^+–Cl^- ionic bond [21]. From the thermodynamic point of view, this process is spontaneous at standard condition (1 atm, 298 K):

$$\Delta G_{\mathrm{NaCl}(cryst.)\rightarrow \mathrm{NaCl}(aq.)} = -1.76\,\mathrm{kcal/mol} < 0 \tag{12.8}$$

But the microscopic atomic processes involved are unclear, especially the microscopic mechanism of how ions break the constraint of the crystal and dissolve into water with the help of water molecules. In fact, topics related to salt-water interactions have attracted much attention in the last few years [22–30]. With the help of computer simulations, the dissociation process of individual NaCl molecules in water has been studied [22–24] and the dissolution of Cl^- ions from NaCl nanocrystals into water has been seen [25, 26]. However, a concise and clear microscopic picture of the early dissolution process has not been established.

On the other hand, according to Arrhenius' theory of electrolytes, strong electrolytes are completely dissociated in water, such as NaCl, which should completely decompose into hydrated Na^+ and Cl^- when placed in water. However, this view is inconsistent with recent experimental observations on aqueous solutions of salts. Raman spectroscopy experiments have shown the presence of clusters of salts such as $NaNO_3$ [31], KH_2PO_4, and $(NH_4)H_2PO_4$ [32] in aqueous solutions. In light scattering experiments, Georgalis et al. directly confirmed the presence of submicron-sized clusters of NaCl, $(NH_4)_2SO_4$, and Na-citrate in supersaturated and undersaturated aqueous solutions at room temperature [33]. These experiments show that simple electrolytes, such as NaCl, can aggregate into clusters in aqueous solutions at moderate concentrations. This also makes it necessary to study the dynamic hydration process of salt nanoclusters in water, which is also important for the understanding of the process of salt dissolution. Also, the synthesis of nanoclusters in solution requires the understanding of the dynamic processes at the solid–liquid interface at the level of atoms and molecules. The hydration process of NaCl nanoclusters in liquid water serves as a suitable model for this study.

Since the dissolution process involves the breaking of chemical bonds, this is a small-probability event under the microscopic time scale of the simulation. To properly depict the structure of liquid water, and to ensure that the ratio of the H_2O and NaCl at the macroscopic scale ($N(H_2O):N(NaCl) \sim 9:1$ in a saturated NaCl solution) can cause dissolution of NaCl grains (the smallest crystal structure with (100) plane exposed contains 32 NaCl units), the simulated system must contain at least $32 \times 2 + 32 \times 9 \times 3 = 928$ atoms, and the time scale of simulation is typically above 10^2 ps. Therefore, Yang et al. used classical molecular dynamics simulations to study this problem [34, 35]. The program package used was AMBER 6 [36]. The super primitive cell used for the calculation is shown in Fig. 12.15a. The cubic nanograin in the primitive cell contains 32 Na^+ and 32 Cl^-, with a side length of about 11.3 Å in each of the X, Y, and Z directions, and is surrounded by 625 water molecules in liquid structure with a macroscopic density of about 1 g/cm^3. Thus, the six (100) planes of the nanocrystalline grains are in direct contact with water. The ratio between the water molecules and the NaCl units corresponds to an undersaturated solution of NaCl.

The water molecule-water molecule interactions are depicted by the TIP3P model, and the ion-ion, ion-water interactions are depicted by the PARM94 force field in AMBER 6. The system used to study the hydrated shell of nanocrystalline grains underwent an equilibration process: first equilibrating for 150 ps at 300 K and then slowly heating up to 350 K in 150 ps. The data generated in the equilibration process is excluded from statistics. A temperature of 350 K was chosen here to speed up the dissolution process. During the process of equilibration, the atoms of the nanograins were applied with harmonic bound to prevent dissolution during this process. The dimension of the initial simulated primitive cell was $27.86 \times 27.88 \times 27.50$ Å, and its side dimension did not vary by more than 0.5 Å in an isothermal and isobaric canonical ensemble. Other systems with different initial configurations used to study dissolution events underwent an equilibrium process of at least 250 ps before the molecular dynamics (MD) simulations for data collection. Periodic boundary conditions were used for the primitive cells. The summation of the various interaction potentials between the atoms using the Ewald method with a summation truncation distance of 10 Å. The time step of the simulation is 0.5 fs. The internal OH vibrations of the water molecules are bounded by the SHAKE law. The temperature and pressure of the ensemble were regulated to reach the target temperature (350 K) and pressure (1 bar) using a Berendsen constant temperature and pressure source, respectively. When the simulation time exceeds 1.8 ns, the velocities of all the atoms in the system are reassigned according to Maxwell's distribution, while the configurations of the ions remain unchanged. These configurations are used for subsequent operations.

To verify the validity of the model potential in AMBER 6, the interaction energies of Na^+, Cl^- ions and water molecular clusters $Na^+ (H_2O)_n$ and $Cl^- (H_2O)_n$ ($n = 1,2,3$) were first calculated and compared with previous first-principle calculations, which are in good agreement. The adsorption of water molecules on the NaCl (001) surface was also calculated and compared with the results of VASP calculations, see

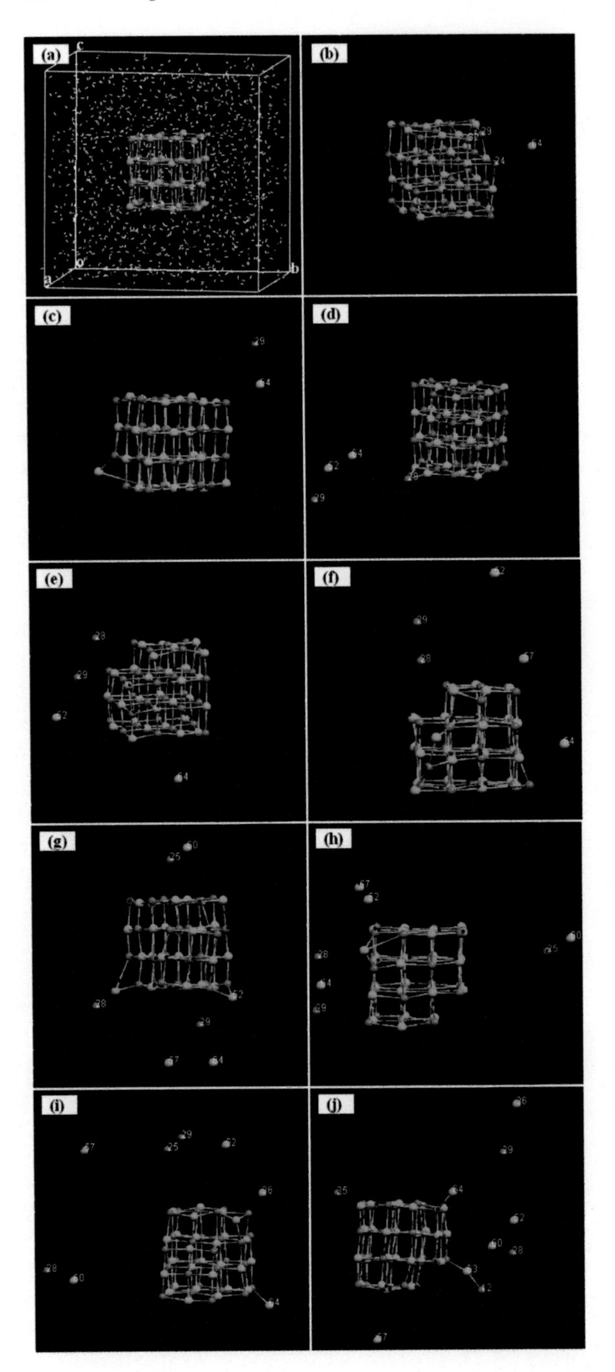

Fig. 12.15 **a** Schematic diagram of the primitive cell used in the simulations. Na^+ and Cl^- are represented by purple (slightly smaller) and green (slightly larger) spheres, respectively, and water molecules are represented by curved sticks. **b–j** Instantaneous configuration diagrams of the MD simulation at moments t = 241.635 ps (**b**), 369.9 ps (**c**), 2013 ps (**d**), 2088 ps (**e**), 2108 ps (**f**), 2450 ps (**g**), 2475 ps (**h**), 3297.5 ps (**i**), 3625 ps (**j**), respectively, showing an ion sequence of Cl^-, Na^+, Cl^-, Na^+ … For the sake of brevity, the hydrogen atoms are not plotted in the instantaenous configuration diagram.

Table 12.5 Comparison of the results of adsorption on the NaCl (001) plane for a single water molecule and 1 ML of water molecules obtained with VASP and AMBER calculations. The HOH planes of single water molecules are identified by "Flat" and "Upright", representing the two typical conformations where the adsorption is mainly through O–Na^+ attraction and H–Cl^- attraction, respectively. For the case of 1 ML adsorption, the HOH planes of water molecules are almost parallel to the NaCl (100) plane

Configuration	VASP			AMBER		
	O–Na (Å)	H–Cl (Å)	E_{ads} (eV/H_2O)	O–Na (Å)	H–Cl (Å)	E_{ads} (eV/H_2O)
Flat on Na^+	2.385	–	−0.401	2.408	–	−0.391
Upright on Cl^-	–	2.232	−0.174	–	2.250	−0.173
1 ML	2.434	–	−0.391	2.426	–	−0.406

Table 12.5. It can be seen that the adsorption configurations and adsorption energies given by both methods are in good agreement. Moreover, the crystal binding energy of NaCl obtained by AMBER calculations is − 188.7 kcal/mol (~ 8.2 eV/ion pair), which is very close to the experimental value (− 185.5 kcal/mol) [37]. This indicates that the model potential used in the calculations is reliable.

12.2.1 Distribution of Water Around Salt Nanograins

The distribution of water molecules at the H_2O/NaCl interface plays an important role for both dissolution and crystallization processes. Figure 12.16 plots the density distribution of water molecules around the nanograins 232 ps before dissolution. The change in the average radial density of water molecules (defined as the quotient of the number N of water molecules within the thin layer of water surrounding the crystal and the volume δV of the thin layer) is plotted with the distance of the water molecules to the NaCl (001) surface as a varying parameter. The density of water molecules is a relative value compared with the bulk water density. Three distinct peaks, labeled A, B, and C, appear in the figure. They are associated with the local structures of the water molecules at the H_2O/NaCl (001) interface. The first peak, at a distance of about 2.4 Å from the NaCl (001) surface, corresponds to the first layer of water covering the surface of the nanograin, which means that the average distance between the water molecules and the NaCl (001) surface is about 2.4 Å. The average coordination number of water molecules of the surface ions of the nanograin is about 2.3 H_2O/NaCl unit, which is the first hydration circle of the nanograin. Similarly, the second and the third peak correspond to the second and the third hydration shell of the nanograin, respectively. The valley of the third peak is located at ~ 5.5 Å from the NaCl (001) plane. This implies the presence of an ordered network of water molecules in the region 5.5 Å below the NaCl (001) plane. This result is supported by the results of a recent experiment at the H_2O–NaCl (001) interface [38]. A previous MD simulation conducted at the H_2O–NaCl (001) interface gave similar results [18], except that the heights of the peaks differed somewhat, and a difference of nearly

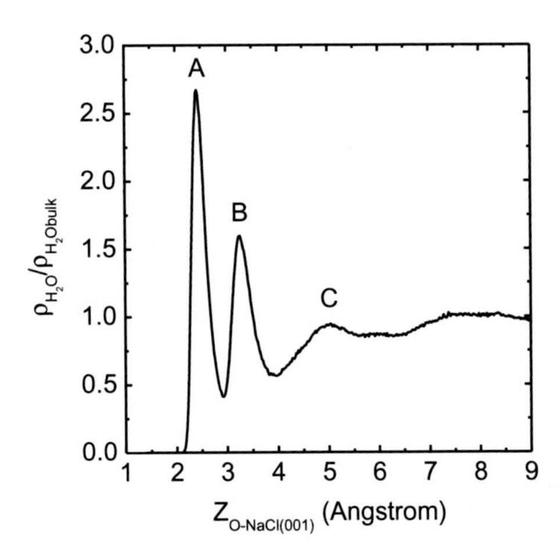

Fig. 12.16 Variation of the radial density distribution of water molecules as a function of the distance from the water molecule to the NaCl (001) surface

1 Å occurred in the position of the third peak. These differences may arise from the fact that the crystals used in the calculations here are finite in size, whereas the X and Y directions of the NaCl (001) plane in the previous MD simulation were extended infinitely with periodic boundaries [18]. In Fig. 6.3, the density of water molecules remains almost constant at ~ 0.87 g/cm^3 for distances $Z_{O\text{–}NaCl(100)}$ = 5.5–6.5 Å, and then reaches the bulk water density at around $Z_{O\text{–}NaCl(100)}$ = 7.5 Å. There is no new peaks appear in the region of bulk phase liquid water with $Z_{O\text{–}NaCl(100)} > 7.5$ Å.

12.2.2 Sequence of Dissolved Ions

Yang Yong et al. then studied the microscopic dissolution process of NaCl nanograins. To speed up the dissolution process, all MD simulations were performed at a temperature of 350 K. The first thing that should be concerned with is the ionic order in which dissolution occurs. A typical dissolution trajectory is given in Fig. 12.15, with the first dissolution event occurring at the moment t ~ 232 ps. For convenience, the Na^+ and Cl^- ions are labeled sequentially with integers from 1 to 64, with the numbers 1–32 labeling all Na^+ in order and the numbers 33–64 labeling all Cl^- in order. The first ion to dissolve in water is the Cl^- in the corner position, labeled Cl^{64} (Fig. 12.15b); the second to dissolve in water is a nearest Na^+ neighbor of Cl^{64}, labeled Na^{29} (Fig. 12.15c). The third to dissolve is Cl^{52} (Fig. 12.15d), and the fourth to dissolve is Na^{28} ($t = 2088$ ps) (Fig. 12.15e). At this point, one edge of the nanograin has been dissolved. The dissolution process then continues with the fifth ion, labelled as a Cl^{57}, dissolving into the water at moment t ~ 2108 ps (Fig. 12.15f). At moment t ~ 2430 ps, Na^{25} and Cl^{60}, two ions previously bonded at edge positions, dissolve into water almost simultaneously as Na^+–Cl^- ion pairs (Fig. 12.15g, h). Thus, the order of the seven ions that undergo dissolution during the previous 2.5 ns simulation can be expressed as Cl^-, Na^+, Cl^-, Na^+, Cl^-... This suggests that nanograins can dissolve into water in both forms, as charged ions and neutral ion pairs, although the latter has few chances to occur during the simulation.

As the dissolution process continues, the first ion to undergo dissolution, Cl^{64}, adsorbs to another location of the remnant of the crystal. Another ion, Cl^{36}, begins to dissolve into water at moment t ~ 3297.5 ps (Fig. 12.15i). About 327.5 ps later, a close neighbor ion of Cl^{36}, Na^{12}, breaks two Na-Cl bonds and shows a tendency to dissolve into water (Fig. 12.15j), and its close neighbor Cl^{63} is also pulled by a Na-Cl bond into the water. In fact, both ions dissolve into water in subsequent, longer simulations.

The dissolution process discussed above shows a tendency to maintain the overall minimum charge of the nanograins: the dissolution of a negative ion is always accompanied by a positive ion, or followed immediately by a positive ion dissolving into the water. This also explains why the dissolution that occurs can be a single ion or a neutral ion pair. It is also seen in the simulations that after all the ions in the corners of a edge have been dissolved, the remaining ions are likely to dissolve into the water as Na^+–Cl^- pairs, followed by ions in other corners and their immediate neighbors. Thus, this microscopic equilibrium can keep until the attraction of the crystal remnant to the ions is negligible relative to the attraction of the water molecules.

In order to gain a deeper understanding of the dissolution process, it is necessary to study its statistical properties. With this aim, a statistical study of the trajectories obtained from MD simulations of nine different initial configurations is made. The statistics show that six dissolution events start from Cl^- and three dissolution events start from Na^+. As for the starting position of the dissolution, eight dissolution events start from the corner positions and one from the edge position. This indicates that during the dissolution of nanograins, dissolution generally starts from the corner position of the crystal, and Cl^- tends to dissolve into water before Na^+.

It is well understood that dissolution starts from the corner position because this is where the number of ionic bonds is the lowest. Then, why does Cl^- have a higher chance of dissolving into water than Na^+, whose hydration heat is greater than that of Cl^- in both small clusters (Table 12.5) and in the bulk phase water? In bulk phase water, the hydration heat of Na^+ is – 94.81 kcal/mol (average coordination number of water molecules is 5.9), whereas the hydration heat of Cl^- is – 93.85 kcal/mol (average coordination number of water molecules is 7.1). Therefore, it seems that Na^+ dissolves more easily than Cl^-. However, this is not the truth. The crux of the problem lies in the potential barriers to be overcome for the dissolution of both ions, and the initial hydration process. For this purpose, Yang, Y. et al. calculated the potential barriers for separating Na^+ and Cl^- in vacuum along two different directions—[111] and [11 $\bar{1}$], respectively (Fig. 12.17). It can be seen that the potential barrier for moving out Na^+ is always about 1 kcal/mol higher than that for Cl^-, regardless of the direction. However, the difference of 1 kcal/mol is small. First-principle calculations with VASP give similar results. Thus, it is the hydration circle of the ion that determines the order of dissolution. As will be seen below, the first hydration circle of Cl^- contains a much larger number of water molecules than that of Na^+ when dissolution has not yet occurred. This makes it easier for Cl^- to dissolve into water than Na^+ when dissolution occurs.

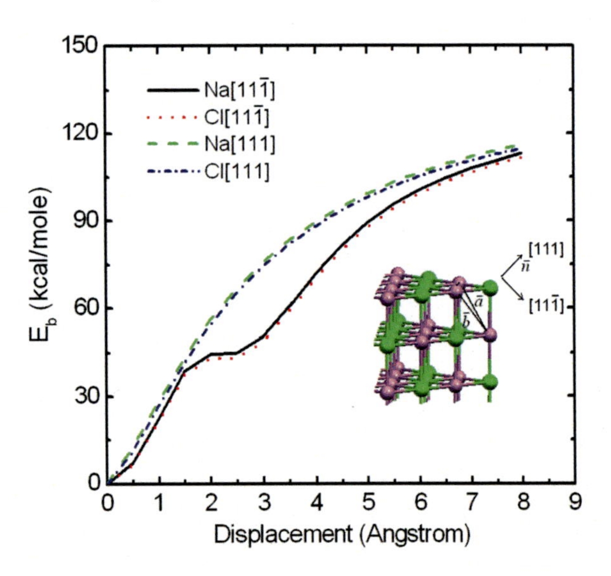

Fig. 12.17 Potential barriers for moving a corner Na^+ (Cl^-) into the vacuum along [11 $\bar{1}$] and [111] respectively. The local coordinate system used in the calculation consists of the three vectors $\vec{a}$, $\vec{b}$, $\vec{n}$

12.2.3 Trajectory Orientation of Dissolution and Force Analysis of Ions

For most dissolution trajectories, the dissolution of either Na^+ or Cl^- at the corner begins by breaking two ionic bonds almost simultaneously and breaking the third ionic bond after a few ps, and then they dissolve into the water, such as the dissolution of Cl^{64} and Na^{29} shown in Fig. 12.18a, c. This allows the initial dissolution trajectory to exhibit direction selectivity, i.e., the ions choose to slip away from the crystal into the water along [11 $\bar{1}$] rather than the [111] direction. Indeed, this can be seen from the potential barriers calculated in Fig. 12.17: when the slip distance is about 2 ~ 3 Å, the ionic bond has been elongated to ~ 5 Å. The potential barrier for either Na^+ or Cl^- to overcome to leave the crystal along the direction [11 $\bar{1}$] is ~ 20 kcal/mol lower than along the [111] direction. This is the main reason for the direction selectivity.

To further clarify the kinetic processes involved here, the two dissolution events in Fig. 12.15 are further discussed below: the Cl^{64} that starts at ~ 232 ps, and the Na^{29} that starts at ~ 324 ps.

Figure 12.19 gives the time evolution of the instantaneous forces acting on the two ions during the dissolution process. The magnitude of the force is determined by Newton's second law, and the rate of change of velocity can be obtained from the difference in velocity between the consecutive two steps. The sign of the acting force is specified as follows: the direction of the force pointing inside the crystal is negative; the direction of the force pointing outside the crystal is positive. Since the nanograins are continuously translating and rotating in water, a local coordinate system bound to the crystal is used to depict the motion of the corner ions and to calculate the direction of the force. The local coordinate system is determined by the small surface of the crystal and its normal, see Fig. 12.17. The magnitude of the

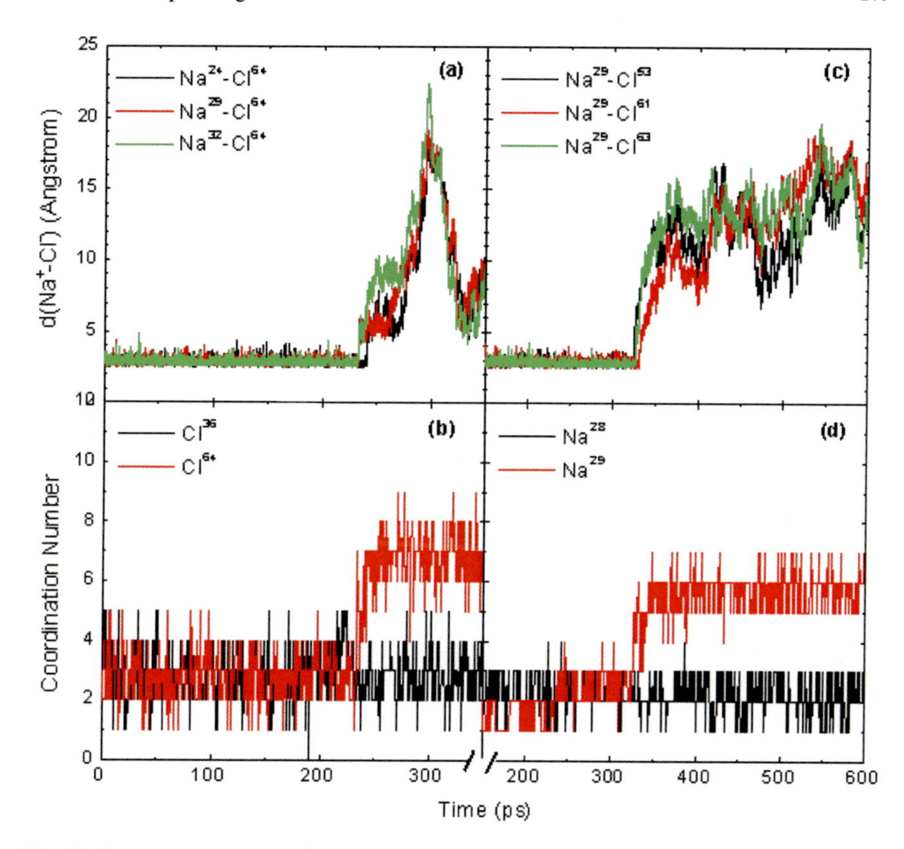

Fig. 12.18 Evolution of the Na^+–Cl^- bond lengths and coordination numbers of water molecules of Cl^{64} (left panel) and Na^{29} (right panel) over time during the dissolution process. The coordination numbers of water molecules of undissolved Cl^{36} and Na^{28} during this time period are also plotted for comparison

force acting on the two ions where dissolution occurs is shown in Fig. 12.19a, c, and the direction of the force is depicted by the angle between the force and the normal of the surface, see Fig. 12.19b, d. The forces acting on the other two corner ions, Cl^{36} and Na^{28}, which did not undergo dissolution during this time period are also plotted for comparison.

As can be seen from Fig. 12.19, the forces acting on the ions during dissolution change very dramatically. The direction of the force can change rapidly from pointing into the crystal to pointing out of the crystal. At the moment when the chemical bond has just broken (marked with an arrow in the figure), the force increases rapidly, pointing out of the crystal and at an angle ~ 90° from the [111] direction. This corresponds to a situation where two ionic bonds are broken simultaneously and the force on the corner ion is directed along the direction $[11\,\overline{1}]$.

The direction of the velocity of the ion undergoing dissolution was further calculated by Yang, Y. et al. [34]. Similarly, the direction of the ion velocity can be defined

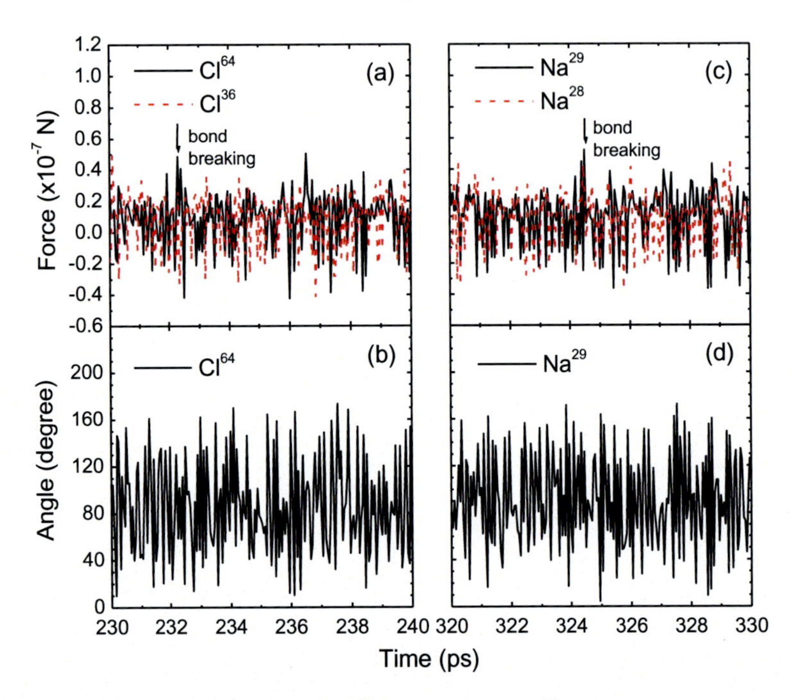

Fig. 12.19 Forces acting on the two ions Cl^{64} (left panel) and Na^{29} (right panel) where dissolution occurs and the angles between the force and the normal of the small surface of the crystal. For comparison, the magnitudes of the forces acting on the two corner ions, Cl^{36} and Na^{28}, that have not yet dissolved during this time period are also plotted in the figure. The arrows indicate the moment when the two ionic bonds begin to break

in terms of the angle between the velocity and the normal $\vec{n}$ of the small surface of the crystal in the local coordinate system (Fig. 6.6):

$$\cos\varphi = \frac{\vec{v} \cdot \vec{n}}{\left|\vec{v}\right| \cdot \left|\vec{n}\right|} \tag{12.9}$$

During the time period of the dissolution, the average value of the velocity angle ϕ of Cl^{64} (230–240 ps) and Na^{29} (320–330 ps) is 91.5° and 85.4° respectively. Both are very close to 90°. The statistics show that the velocity angle ϕ of the ions undergoing dissolution occurs most frequently in the vicinity of 90°. This indicates that at the initial stage of the dissolution of the nanograins, the ions undergoing dissolution are sliding out of the crystal into the water along $[11\bar{1}]$ or other equivalent directions, which is a path with an average effect and the greatest chance of occurrence.

12.2.4 Hydration Shell and Dynamic Properties of Dissolved Ions

As can be seen from Fig. 12.18b, d, when dissolution of an ion occurs, it is always accompanied by a rapid increase in its coordination number of water molecules, while the coordination number of water molecules of ions in other equivalent positions remain unchanged in their original oscillations. This means that there is a substantial fluctuation in the density of water molecules around the dissolved ion. For example, the coordination number of water molecules of Cl^- in the corner position can vary from 0 to 7. This fluctuation in the density of water molecules is the microscopic reason that promotes the dissolution of ions. Studies of pure liquid water at 350 K show that the local density of water molecules is highly discrete, showing a Gaussian distribution centered at the mean value.

The radial distribution functions (RDFs) of the water molecules around the two ions, Cl^{64} and Na^{29}, before and after dissolution occurs are given in Fig. 12.20. Overall, all the radial distribution functions change very significantly before and after dissolution. The first minimum of $g_{Cl\text{-}O}$ determines the radius of the first hydration circle of Cl^-, and this value shrinks from ~ 4.50 Å before dissolution to ~ 3.90 Å after dissolution, indicating that the Cl^- ion hydration circle shrinks after dissolution and that the ion-water interaction is strengthened. It can be seen from $g_{Cl\text{–}O}$ and $g_{Cl\text{–}H}$ that the distribution radius of O atoms around Cl^- changes after dissolution, while the distribution radius of H is almost constant. The integral of $g_{Cl\text{–}O}$ gives an average number ~ 8.3 (7.4 after dissolution) of water molecules in the first hydration circle of Cl^- before dissolution. Note that this value is defined in terms of the geometric distribution of water molecules and is not quite the same as the definition of the coordination number of water molecules, which also takes into account the limitation of the physical range of the H–Cl^- interaction. These two definitions give the same value after the ion is dissolved in aqueous solution. Such a distribution of water molecules makes it easy for water molecules in the shell, that were previously far away from H–Cl^-, to rotate and meet the strong interaction range of H–Cl^-, leading to a rapid increase in the coordination number of water molecules of the corner Cl^-.

In contrast, the radius of the first hydration circle of Na^+ remains almost constant before and after dissolution at ~ 3.25 Å. The integral gives an increase in the average number of water molecules in the first hydration circle of Na^+ from ~ 2.6 before dissolution (which is the same as the average coordination number of water molecules in Fig. 12.18) to ~ 5.7 after dissolution, with the water molecules in the hydration circle being more closely aligned and their interactions enhanced. Moreover, the number of water molecules in the first hydration circle of Cl^- before dissolution is over three times that of Na^+, which can help us understand why corner Cl^- is more likely to dissolve in water than corner Na^+. In addition, the radial distribution functions calculated for the dissolved ions match the available experimental data [39], see Fig. 6.9. The peak positions of all the radial distribution functions match well with the experiments, except that the peaks measured experimentally are a little

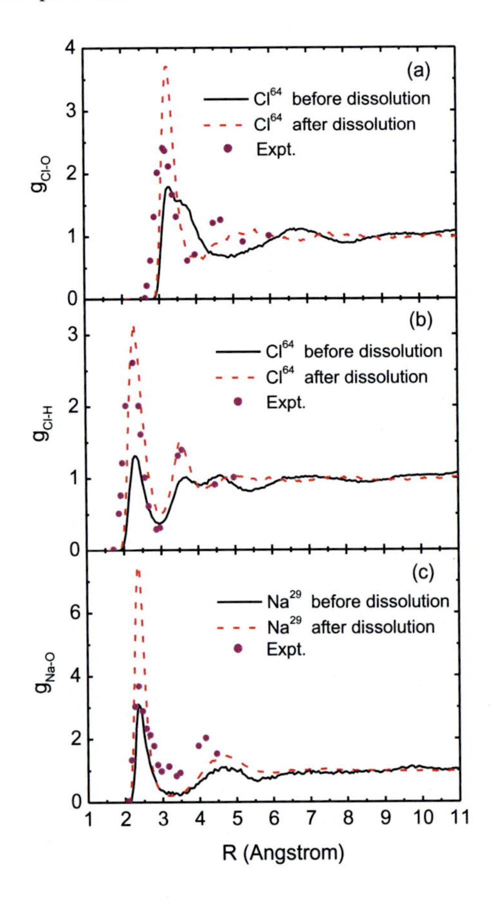

Fig. 12.20 Radial distribution functions $g_{Cl\text{–}O}$, $g_{Cl\text{–}H}$, $g_{Na\text{–}O}$ of water molecules before and after dissolution for two ions (Cl^{64} and Na^{29}) on nanograins. The statistical time periods for Cl^{64} are 0 to 232.05 ps and 232.065 to 300 ps, respectively; for Na^{29} are 300 to 324 ps and 324.015 to 600 ps, respectively. Previous experimental results [39] are presented as discrete data points

wider and lower, which can be attributed to the kinetic spreading and the quantum motion of the H atoms in the water molecules.

12.3 Salt Crystallization Processes at the Solid–liquid Interface

The phenomenon of nucleation and growth in crystals is of undeniable importance for both science and technology. In the last decade or so, a large number of theoretical and experimental studies on the kinetic processes associated with nucleation have been carried out [27, 33, 40–47]. Although many theoretical models based on thermodynamic and rate equations [48] have been proposed, these processes are still poorly understood at the atomic and molecular level [49]. Experimentally, it is now in principle possible to use scanning probes such as STM or AFM to trace atomic nucleation processes on surfaces in real time, but in systems where solutions exist, such as at the gas–liquid interface, inside the solution, or at the solid–liquid interface, direct experimental tracing of atomic processes therein is still a very challenging project. In this regard, computer simulation based on various atomic models can be a powerful research tool. For example, Ohtaki and Fukushima [47], more than a decade ago, and more recently Zahn [27], studied spontaneous nucleation processes in supersaturated solutions of NaCl using classical molecular dynamics methods. In the early nucleation processes, small irregularly shaped clusters are seen in solution [47]. Zahn also found that the vast majority of stable clusters had at their center a Na^+ that was not in direct contact with the water molecules and was directly bonded to six Cl^-, which formed an octahedral conformation [27]. In actual crystal growth, a seed crystal is usually placed in a supersaturated solution and the solute is deposited spontaneously on top of the seed crystal and grows up gradually. Unlike the previous homogeneous nucleation, this heterogeneous nucleation has a much lower potential barrier to overcome and a much smaller critical nucleus size [46]. Therefore, the crystals grown by this method are of much better quality than those grown by homogeneous spontaneous nucleation in supersaturated solutions. However, the microscopic mechanism of this nucleation phenomenon at the solid–liquid interface, which is more important in practice, is not yet known at the atomic level. Oyen and Hentschke have previously studied the kinetic processes at the interface between the NaCl (001) surface and various NaCl solutions with different concentrations, but did not see any crystallization processes [41]. Therefore, one of the aims of the current work is to investigate the early nucleation processes of NaCl at the water-NaCl (001) interface.

Yang, Y. et al. studied a more complex scenario [50]: the early crystallization process of NaCl at the H_2O–NaCl (001) solid–liquid interface. Similar to the study of the dissolution process, the AMBER 6 program package [36] was used to make classical molecular dynamics simulations. The seed crystal used in the simulations consists of a five-layer NaCl (001) crystal plate containing 160 atoms, with both sides of the plate in contact with the aqueous NaCl solution, as shown in Fig. 12.21.

At room temperature (~ 298 K, 1 bar), the concentration of a saturated NaCl solution is 35.96 g/(100 ml H_2O) and the ratio of NaCl units to water molecules is about 1/9. The model system used by Yang et al. [50] contains 100 NaCl units and 600 water molecules. Using the definition of molar fraction, the supersaturation of the solution is

$$\sigma = \frac{c - c_0}{c_0} = \frac{100/(100 + 600) - 1/(1 + 9)}{1/(1 + 9)} \approx 0.43 = 43\% \qquad (12.10)$$

Here, c is the concentration of the supersaturated solution and c_0 is the concentration of the saturated solution.

Water molecule-water molecule interactions are depicted by the TIP3P model, and ion-ion, ion-water molecule interactions are depicted by the PARM94 force field in AMBER 6 (see literature [51] for specific energy and size parameters). A total of eight different trajectories were studied. They had the same number of water molecules and the same number of NaCl ion pairs, while the initial velocities and

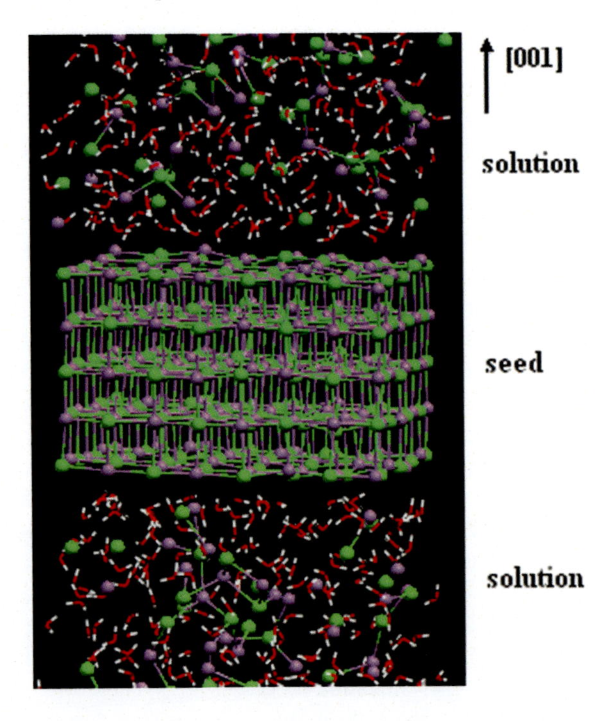

Fig. 12.21 Schematic diagram of the model system used during the MD simulations. The Na^+ and Cl^- ions are represented by slightly smaller purple and slightly larger green balls, respectively; the water molecules are represented by small curved sticks

configurations were different. All these trajectories were thermally equilibrated for at least 300 ps at ~ 300 K before starting to formally collect data. During equilibration, the Na^+ and Cl^- in solution were bound by the harmonic potential to avoid their direct deposition onto the NaCl (001) surface that served as the substrate during thermal equilibration. An isothermal and isobaric canonical ensemble is used, and the dimension of the primitive cell makes small variations around 22.83 × 22.83 × 62.76 Å during the simulation, with the variation of each side length not exceeding 0.5 Å. Under periodic boundary conditions, the summation of energy and force is done using Ewald summation with a real-space truncation distance of 9 Å. The time step of the simulation is 0.5 fs, and the OH vibration of the water molecule is constrained with the SHAKE law. The temperature and pressure are controlled by Berendsen's constant temperature and pressure regulator, guaranteed to be near 300 K and 1 bar.

12.3.1 Critical Size of Crystal Nuclei

In classical nucleation theory [52], the formation of a nucleus in solution is determined by two competing factors: the increase in free energy gained by the particle during the transition from the liquid phase (solution phase) to the solid phase, and the decrease in free energy due to the creation of a new solid–liquid interface during this transition. The first term can be written as $n\Delta\mu$, where n *is the* number of particles in the nucleus and $\Delta\mu = \mu_s - \mu_l$ is the difference between the chemical potentials of the solid and liquid phases. The second term can be written as γA, where γ is the free energy density at the solid–liquid interface, and A is the area of the new interface. Then, the change in Gibbs free energy is:

$$\Delta G(n) = n\Delta\mu + \gamma A \tag{12.11}$$

As the size of the nucleus increases, a critical nucleus size n_c appears, such that $\Delta G(n)$ takes its maximum. $\Delta G(n)$ of the nucleus smaller than this size increases with the number of particles, and the chance of nucleus growth is smaller than the chance of degradation, which is an unstable nucleus; on the contrary, $\Delta G(n)$ of the nucleus larger than this size decreases with the number of particles, and the chance of nucleus growth is larger than the chance of degradation, which is a stable nucleus.

The size of a critical nucleus can be determined using statistical methods. A nucleus with a determined size is unstable if it has a chance of degenerating greater than 0.5, or it is a stable nucleus. 1.2 ns MD simulations were performed for all eight trajectories. Statistics were made for the nuclei containing one, two, and three atoms. Figure 12.22 gives the total number of nuclei produced and degraded during the simulation as a function of time. When making the statistics, a Na^+ (Cl^-) ion is considered adsorbed to NaCl (001) when its distance from the crystal plane is less than 3.25 Å (which is the first minimum of the $g_{Na\text{-}Cl}$ radial distribution function), and it can be considered desorbed from the crystal plane when its distance from the crystal plane is greater than 5.5 Å, since at such a distance the Na^+–Cl^- ionic bond can be considered to be broken. The same criterion is used when counting the number of bonding of each ion adsorbed to the substrate in order to determine the size of the nucleus. When one or more atoms are detached from the nucleus, it is considered that degradation of the nucleus has occurred. In Fig. 12.22, each data point represents the total number of nuclei that are produced or degenerate from the beginning of the simulation to that moment.

For a Na^+ nucleus and a Cl^- nucleus (Fig. 12.22a), the total number of nuclei produced during the 1.2 ps simulation is 138 and 110, respectively, while the degenerate monatomic Na^+ and Cl^- nuclei are 74 and 77, respectively. Thus, the chances of degradation for these two types of monatomic nuclei are $P_{decayNa1} = 74/138 = 0.54 > 0.5$ and $P_{decayCl1} = 77/110 = 0.7 > 0.5$.

For diatomic nuclei (Na^+–Cl^- ion pair, Fig. 12.22b), the total number of nuclei produced during this time period is 90, while the number of nuclei degraded is 40, with a degradation chance $P_{decay2} = 40/90 = 0.44 < 0.5$. This is the critical nucleus size!

For a triatomic nucleus (two Na^+ and one Cl^-, or two Cl^- and one Na^+, Fig. 12.22c), the total number of nuclei produced is 69, the number of nuclei degraded is 23, and the chance of degradation is $P_{decay3} = 23/69 = 0.33 < P_{decay2} < 0.5$. This trend is consistent with classical nucleation theory [52]: as the size of a nucleus exceeds the critical size, the larger the nucleus is, the more stable it is and the smaller the chance of degradation is.

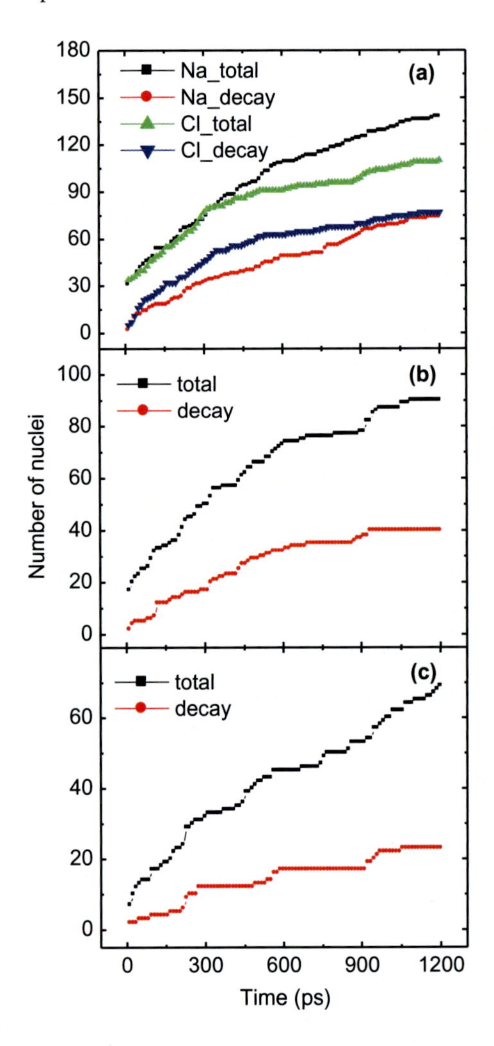

Fig. 12.22 Time evolution of the total number of nuclei produced and degraded during the MD simulations: **a** monoatomic nuclei; **b** diatomic nuclei; **c** triatomic nuclei

It should be noted that the size of the critical nuclei obtained here is closely related to the state of the simulated system, e.g., the temperature and the degree of supersaturation. Simulations done at moderate supersaturation (85 NaCl, 600 H_2O, 24% supersaturation) indicate that the critical nucleus size at 300 K is a nucleus of three atoms (two Na^+ and one Cl^-, or two Cl^- and one Na^+).

12.3.2 Deposition Properties of Solute Ions

A transient diagram of the growth process of a typical diatomic nucleus formed at the H_2O–NaCl (001) interface is given in Fig. 12.23. It grows gradually from a diatomic nucleus (Fig. 12.23a) to a zigzag chain (Fig. 12.23b), then to a three-dimensional island progressively becoming larger (Fig. 12.23c, d), with the two atoms in the diagram denoted by number "1" (Na^+) and "2" (Cl^-).

The statistics of the stable nuclei above the critical nucleus size show that there are more Cl^--centered nuclei than Na^+-centered ones. Larger islands than those in Fig. 12.23d are seen in longer simulations, where both Na^+ and Cl^- have a coordination number of 6 in the island, as in the bulk phase.

Figure 12.24a gives the evolution of the average number of Na^+ and Cl^- ions deposited at the solid–liquid interface as a function of time. It is clear that the deposition rate of Na^+ is much higher than that of Cl^-. At moment t ~ 925 ps, there is a small drop in the deposition rate of Na^+, while the deposition rate of Cl^- increases rapidly, followed by a renewed increase in the amount of deposited Na^+. This interactive fluctuation predicts a new round of Na^+–Cl^- deposition. During the 1.2 ns simulation, the amount of deposited Na^+ is always greater than that of Cl^-. Moreover, this trend is independent of the ratio of the very small amounts of Na^+ and Cl^-

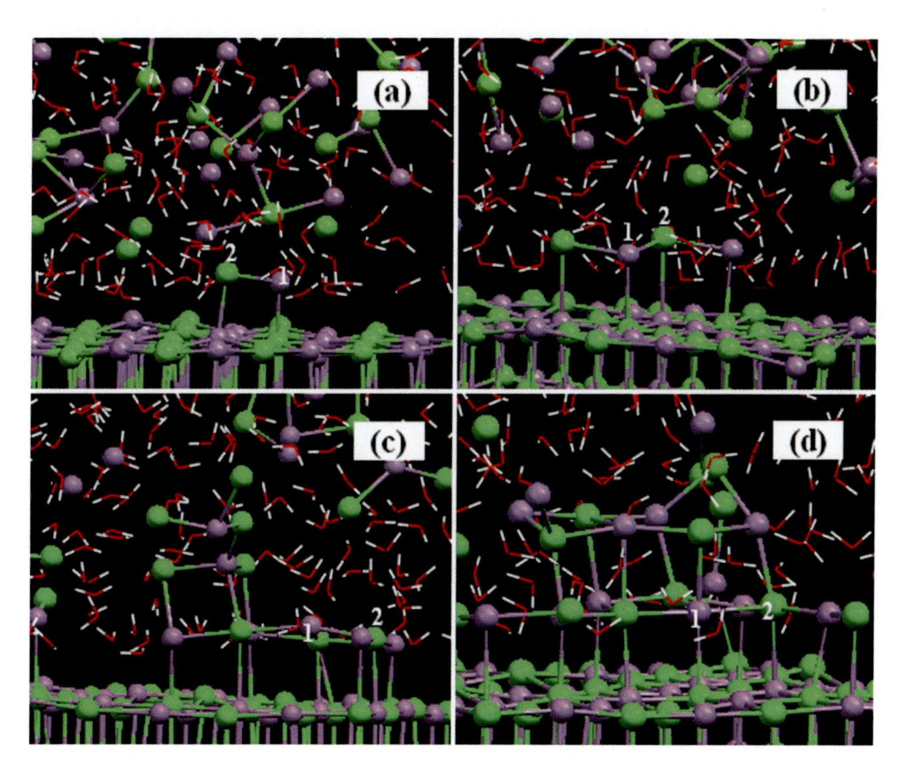

Fig. 12.23 Evolution of a diatomic nucleus over time (Na^+ is labeled with "1" and Cl^- with "2"). As before, Na^+ and Cl^- are represented by smaller purple balls and larger green balls, respectively, and water molecules are represented by curved sticks. The instantaneous pictures correspond to moments **a** t = 0.3 ps (300 K), **b** 19.59 ps (300 K), **c** 1200 ps (300 K), and **d** 2550 ps (320 K), respectively

that may be adsorbed on the substrate during equilibrium. This means that the H_2O–NaCl (001) interface is positively charged during the early nucleation. This positive charge accumulation at the solid–liquid interface will attract negatively charged Cl^- from the solution and get them deposited on the substrate. Therefore, Yang et al. proposed that in addition to the chemical potential difference between the solid–liquid phases in the classical nucleation theory [52], there is a new driving force for nucleation at the solid–liquid interface due to the Coulombic imbalance.

To better see this difference, Yang Yong et al. also made MD simulations of the epitaxial growth of NaCl for the same system at 300 K under vacuum conditions, see Fig. 12.24b. It can be seen that the amount of Na^+ and Cl^- deposited is almost the same, and therefore the substrate is electrically neutral during the epitaxial growth. This clearly shows the important influence of water molecules on the process of early nucleation at the solid–liquid interface.

As can be seen in Figs. 12.22 and 12.24, the nucleation rate as well as the deposition rate in the first 300 ps are greater than those in any later time period. This is due to the decrease in the degree of supersaturation. In a supersaturated solution, the nucleation ratee as well as the deposition rate are proportional to the degree of supersaturation [53]. At moment t ~ 300 ps, the average deposition amount is ~ 7.5 for Na^+ and ~

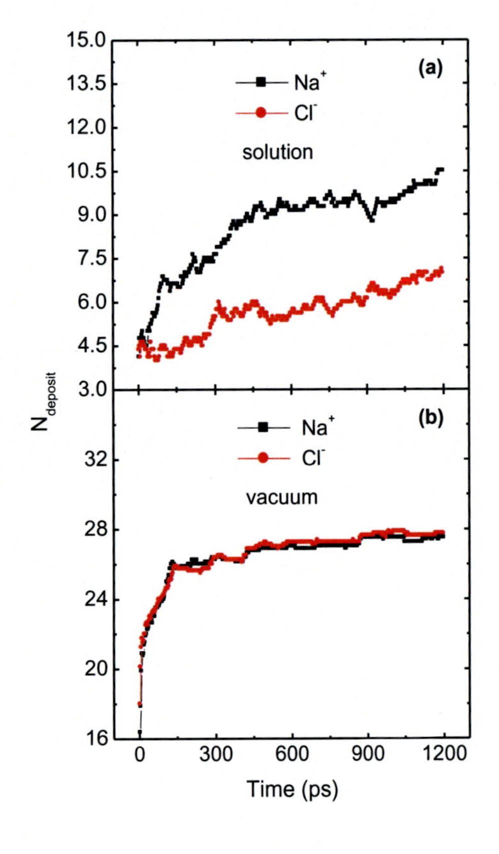

Fig. 12.24 Evolution of the amount of Na^+ and Cl^- deposited on the NaCl (001) surface in aqueous solution (**a**) and in vacuum (**b**) as a function of time during the MD simulation. Each data point represents an average of eight different trajectories

6 for Cl^-, and the supersaturation of the solution correspondingly decreases from 43% to ~ 30%.

12.3.3 Stable Water Network at the Interface

The different deposition properties of Na^+ and Cl^- at the water-NaCl (001) interface arise from the presence of a relatively stable network of water molecules at the solid–liquid interface, see Fig. 12.25. When Na^+ in solution reaches the NaCl (001) surface, the vast majority of them will displace the water molecules adsorbed at the top site of Cl^-, due to the Coulombic attraction between Na^+ and Cl^-. For the same reason, when Cl^- in solution are adsorbed to the NaCl (001) surface, they will take the place of the water molecules adsorbed at the top site of Na^+. However, the adsorption of water molecules to the top site of Cl^- is achieved mainly through H–Cl^- hydrogen bonding, whereas the adsorption of water molecules to the top site of Na^+ is achieved mainly through O–Na^+ attraction, which is much stronger than H–Cl^- hydrogen bonding. When the adsorption of water molecules on the NaCl (001) surface is calculated using first principles, it is found that the adsorption energy of a single water molecule at the top site of Cl^- (0.174 eV) differs by a factor of ~ 2.3 from that of the top site of Na^+ (0.401 eV). From Fig. 12.25 it can be seen that many more water molecules stay at the top site of Na^+ than at the top site of Cl^-. The average residence time of water molecules near the top site of Na^+ on the substrate surface is 8.95 ps, while the average residence time near the top site of Cl^- on the substrate surface is 4.12 ps. These results indicate that water molecules adsorbed at the top site of Na^+ are more stable than those adsorbed at the top site of Cl^-. For this reason, Na^+ in a supersaturated solution has a better chance to approach and stay on the NaCl (001) surface than Cl^-, showing a greater deposition rate (Fig. 12.24a). Further analysis shows that the stable network of water molecules at the solid–liquid interface exists in the region 5.5 Å below the NaCl (001) surface, which is a quasi-bilayer structure normal to the NaCl (001) surface (Fig. 12.25a, b). The diffusion coefficient of a single water molecule in a supersaturated solution is $4.77 \times 10^{-9} m^2 s^{-1}$ (the TIP3P model has a diffusion coefficient of $\sim 5.5 \times 10^{-9} m^2 s^{-1}$ [54] for water molecules in pure water), while in the network of water molecules at the solid–liquid interface, this value is $3.02 \times 10^{-9} m^2 s^{-1}$. The average number of hydrogen bonds (HB) around water molecules in solution is given in Fig. 12.25c as ~ 1.4 HB/H_2O, while the average number of hydrogen bonds (HB) at the interfacial network of water molecules is ~ 1.2 HB/H_2O. This is due to the fact that more water molecules are bonded to the substrate at the interface.

As comparison, in an ideal ice structure, the average number of hydrogen bonds per water molecule in the bulk and interfacial phases is 2 and 1.75, respectively. Moreover, surface X-ray diffraction experiments at the water-NaCl (001) interface confirmed the existence of a stable and ordered network of water molecules [38].

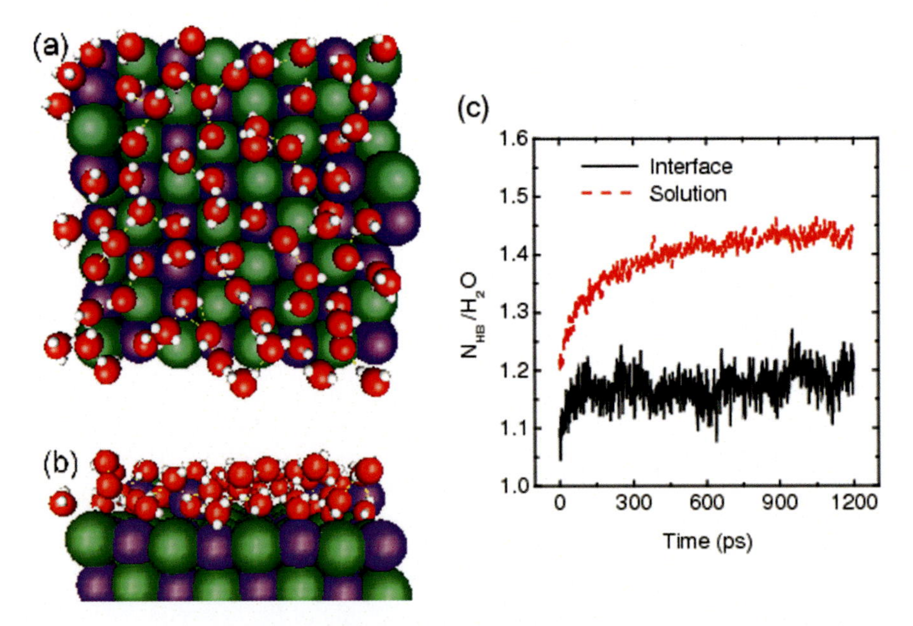

Fig. 12.25 Top (**a**) and side (**b**) views of the network of water molecules at the solid–liquid interface, with the hydrogen bonds between the water molecules represented by dotted lines. This instantaneous diagram corresponds to one side of NaCl (001) at $t = 577.5$ ps. **c** Evolution of the average number of hydrogen bonds between water molecules at the interface and inside the solution as a function of time. The continuous increase in the number of hydrogen bonds in solution is due to a continuous nucleation process; while the increase/decrease in hydrogen bonds at the interface corresponds to the process of ion desorption/deposition on the NaCl (001) plane

12.3.4 Temperature Dependence

Since the temperature dependence of the solubility of NaCl in water is not obvious, the supersaturation of the solution will change very little when the temperature of the solution is increased moderately. For example, at ~ 273 K, the solubility of NaCl is 35.65 g/100 ml H_2O, while at ~ 373 K, its solubility becomes 38.99 g/100 ml H_2O [21], which correspond to the salt-water molar ratios of N(NaCl):N(H_2O) = 1:9.1 and 1:8.3, respectively. However, when the temperature is increased, both solute ions and water molecules show a higher diffusion capacity, and the influence of water molecules at the interface will thus be reduced, while at the same time the solute deposition rate will increase. To confirm this, after 1.2 ns of simulation, the solution temperature was gradually increased from 300 to 320 K. The deposition rates of both Na^+ and Cl^- were greater than that at 300 K. Gradually, at the solid–liquid interface, the NaCl nuclei show a three-dimensional island growth pattern.

To investigate the effect of water, Yang Yong et al. also used molecular dynamics to study the epitaxial growth of NaCl in vacuum for the same solute and substrate at the same temperature (300 K). The same three-dimensional island growth pattern was observed. According to thermodynamics, two-dimensional growth should be the most stable form of homogeneous epitaxy. It can be seen that kinetic factors play a major role here. At room temperature, ions have a limited ability to diffuse on the substrate. The high supersaturation allows for a large deposition rate of atoms, which makes the chances small for an ion to cross the ES potential barrier and find the most energetically stable position; instead, the deposited ions tend to stay at the point with the local minimum of energy they first contacted, which leads to three-dimensional growth. MD simulations of solutions with moderate supersaturation (σ = 24%) found that while 3D island growth is still visible, the size of the islands has become smaller and less pronounced. Therefore, it can be expected that at higher solution temperatures and lower supersaturation, a two-dimensional island growth pattern will be observed.

In summary, classical molecular dynamics methods simulating the early nucleation growth of NaCl at the solid–liquid interface of aqueous solution-NaCl (001) at 43% supersaturation at 300 K indicate that the critical nucleation size is one Na^+–Cl^- ion pair at this condition. Na^+ and Cl^- in solution exhibit different deposition rates on the NaCl (001) substrate due to the presence of a relatively stable network of water molecules at the solid–liquid interface. The different deposition characteristics of positive and negative ions cause the substrate to be positively charged, and thus a nucleation mechanism that is distinct from the supersaturation of the solution exists at the solid–liquid interface—an imbalance in the Coulombic interaction between the solution and the substrate acts as an additional driving force attracting more solute Cl^-ions to the substrate. At room temperature and high supersaturation, the nuclei grow in a three-dimensional growth pattern, which is a result of the dominance of kinetic factors. These results still provide valuable information for understanding the crystallization process at the solid–liquid interface, which is instructive for how to improve the quality of crystal growth in solution.

References

1. W. Shockley, Electronic energy bands in sodium chloride. Phys. Rev. **50**, 754 (1936)
2. M.Y. Kiriukhin, K.D. Collins, Dynamic hydration numbers for biologically important ions. Biophys. Chem. **99**, 155 (2002)
3. P. Jungwirth, D.J. Tobias, Specific ion effects at the air/water interface. Chem. Rev. **106**, 1259 (2006)
4. L.W. Bruch, A. Glebov, J.P. Toennies, H. Weiss, A helium atom scattering study of water adsorption on the NaCl(100) single crystal surface. J. Chem. Phys. **103**, 5109 (1995)
5. S. Fölsch, A. Stock, M. Henzler, Two-dimensional water condensation on the NaCl(001) surface. Surf. Sci. **264**, 65 (1992)
6. J.P. Toennies, F. Traeger, J. Vogt, H. Weiss, Low-energy electron induced restructuring of water monolayers on NaCl(100). J. Chem. Phys. **120**, 11347 (2004)
7. S.J. Peters, G.E. Ewing, Water on salt: An infrared study of adsorbed H_2O on NaCl(100) under ambient conditions. J. Phys. Chem. B **101**, 10880 (1997)
8. M. Foster, G.E. Ewing, An infrared spectroscopic study of water thin films on NaCl (100). Surf. Sci. **427**, 102 (1999)

9. A. Verdaguer, G.M. Sacha, M. Luna, D.F. Ogletree, M. Salmeron, Initial stages of water adsorption on NaCl (100) studied by scanning polarization force microscopy. J. Chem. Phys. **123**, 124703 (2005)
10. J. Guo, X.Z. Meng, J. Chen, J.B. Peng, J.M. Sheng, X.Z. Li, L.M. Xu, J.R. Shi, E.G. Wang, Y. Jiang, Real-space imaging of interfacial water with submolecular resolution. Nat. Mater. **13**, 184 (2014)
11. B. Wassermann, S. Mirbt, J. Reif, J.C. Zink, E. Matthias, Clustered water adsorption on the NaCl(100) surface. J. Chem. Phys. **98**, 10049 (1993)
12. K. Jug, G. Geudtner, Quantum chemical study of water adsorption at the NaCl(100) surface. Surf. Sci. **371**, 95 (1997)
13. A. Allouche, Water adsorption on NaCl(100): a quantum ab-initio cluster calculation. Surf. Sci. **406**, 279 (1998)
14. H. Meyer, P. Entel, J. Hafner, Physisorption of water on salt surfaces. Surf. Sci. **488**, 177 (2001)
15. J.M. Park, J.-H. Cho, K.S. Kim, Atomic structure and energetics of adsorbed water on the NaCl(001) surface. Phys. Rev. B **69**, 233403 (2004)
16. D.P. Taylor, W.P. Hess, M.I. McCarthy, Structure and energetics of the water/NaCl(100) interface. J. Phys. Chem. B **101**, 7455 (1997)
17. O. Engkvist, A.J. Stone, Adsorption of water on NaCl(001). I. Intermolecular potentials and low temperature structures. J. Chem. Phys. **110**, 12089 (1999)
18. E. Stöckelmann, R. Hentschke, A molecular-dynamics simulation study of water on NaCl(100) using a polarizable water model. J. Chem. Phys. **110**, 12097 (1999)
19. Y. Yang, S. Meng, E.G. Wang, Water adsorption on NaCl (001) surface: a density functional theory study. Phys. Rev. B **74**, 245409 (2006)
20. J.E. Nickels, M.A. Fineman, W.E. Wallace, X-ray diffraction studies of sodium chloride-sodium bromide solid solutions. J. Phys. Chem. **53**, 625 (1949)
21. D.R. Lide (ed.), *CRC Handbook of Chemistry and Physics*, 77th edn (CRC Press, Boca Raton, Florida, USA, 1996)
22. P.L. Geissler, C. Dellago, D. Chandler, Kinetic pathways of ion pair dissociation in water. J. Phys. Chem. B **103**, 3706 (1999)
23. J. Martí, F.S. Csajka, The aqueous solvation of sodium chloride: a Monte Carlo transition path sampling study. J. Chem. Phys. **113**, 1154 (2000)
24. J. Martí, F.S. Csajka, D. Chandler, Stochastic transition pathways in the aqueous sodium chloride dissociation process. Chem. Phys. Lett. **328**, 169 (2000)
25. H. Ohtaki, N. Fukushima, E. Hayakawa, I. Okada, Dissolution process of sodium chloride crystal in water. Pure & Appl. Chem. **60**, 1321 (1988)
26. H. Ohtaki, N. Fukushima, Dissolution of an NaCl crystal with the (111) and (-1-1-1) faces". Pure & Appl. Chem. **61**, 179 (1989)
27. D. Zahn, Atomistic mechanism of NaCl nucleation from an aqueous solution. Phys. Rev. Lett. **92**, 040801 (2004)
28. E. Oyen, R. Hentschke, Molecular dynamics simulation of aqueous sodium chloride solution at the NaCl(001) interface with a polarizable water model. Langmuir **18**, 547 (2002)
29. P. Jungwirth, How many waters are necessary to dissolve a rock salt molecule? J. Phys. Chem. A **104**, 145 (2000)
30. E.M. Knipping, M.J. Lakin, K.L. Foster, P. Jungwirth, D.J. Tobias, R.B. Gerber, D. Dabdub, B.J. Finlayson-Pitts, Experiments and simulations of ion-enhanced interfacial chemistry on aqueous NaCl aerosols. Science **288**, 301 (2000)
31. I.T. Rusli, G.L. Schrader, M.A. Larson, Raman spectroscopic study of $NaNO_3$ solution system—solute clustering in supersaturated solutions. J. Cryst. Growth **97**, 345 (1989)
32. M.K. Cerreta, K.A. Berglund, The structure of aqueous solutions of some dihydrogen orthophosphates by laser Raman spectroscopy. J. Cryst. Growth **84**, 577 (1987)
33. Y. Georgalis, A.M. Kierzek, W. Saenger, Cluster formation in aqueous electrolyte solutions observed by dynamic light scattering. J. Phys. Chem. B **104**, 3405 (2000)
34. Y. Yang, S. Meng, L.F. Xu, E.G. Wang, S.W. Gao, Dissolution dynamics of NaCl nanocrystal in liquid water. Phys. Rev. E. **72**, 012602 (2005)

35. Y. Yang, S. Meng, E.G. Wang, A molecular dynamics study of hydration and dissolution of NaCl nanocrystal in liquid water. J. Phys.: Cond. Matter **18**, 10165 (2006)
36. D.A. Case, D.A. Pearlman, J.W. Caldwell, T.E. Cheatham III., W.S. Ross, C. Simmerling, T. Darden, K.M. Merz, R.V. Stanton, A. Cheng, J.J. Vincent, M. Crowley, V. Tsui, R. Radmer, Y. Duan, J. Pitera, I. Massova, G.L. Seibel, U.C. Singh, P. Weiner, P.A. Kollman, *AMBER 6* (University of California, San Francisco, 1999)
37. B.E. Conway, *Ionic Hydration in Chemistry and Biophysic* (Amsterdam; New York: Elsevier Scientific Pub. Co.: distributors for the U.S. and Canada, Elsevier/North-Holland, 1981)
38. J. Arsic, D.M. Kaminski, N. Radenovic, P. Poodt, W.S. Graswinckel, H.M. Cuppen, E. Vlieg, Thickness-dependent ordering of water layers at the NaCl(100) surface. J. Chem. Phys. **120**, 9720 (2004)
39. D.H. Powell, G.W. Neilson, J.E. Enderby. The structure of Cl^- in aqueous solution: an experimental determination of $g_{ClH}(r)$ and $g_{ClO}(r)$. J. Phys. Condens. Matter **5**, 5723 (1993)
40. J. Anwar, P.K. Boateng, Computer simulation of crystallization from solution. J. Am. Chem. Soc. **120**, 9600 (1998)
41. M. Asta, F. Spaepen, J.F. van der Veen, Solid-liquid interfaces: molecular structure, thermodynamics, and crystallization. MRS Bull. **29**, 920 (2004)
42. T. Koishi, K. Yasuoka, T. Ebisuzaki, Large scale molecular dynamics simulation of nucleation in supercooled NaCl. J. Chem. Phys. **119**, 11298 (2003)
43. S. Auer, D. Frenkel, Line tension controls wall-induced crystal nucleation in hard-sphere colloids. Phys. Rev. Lett. **91**, 015703 (2003)
44. S. Auer, D. Frenkel, Prediction of absolute crystal-nucleation rate in hard-sphere colloids. Nature **409**, 1020 (2001)
45. S. Auer, D. Frenkel, Suppression of crystal nucleation in polydisperse colloids due to increase of the surface free energy. Nature **413**, 711 (2001)
46. A. Cacciuto, S. Auer, D. Frenkel, Onset of heterogeneous crystal nucleation in colloidal suspensions. Nature **428**, 404 (2004)
47. H. Ohtaki, N. Fukushima, Nucleation processes of NaCl and CsF crystals from aqueous solutions studied by molecular dynamics simulations. Pure. Appl. Chem. **63**, 1743 (1991)
48. J.A. Venables, Rate equation approaches to thin film nucleation kinetics. Philos. Mag. **27**, 697 (1973)
49. J. Maddox, Colloidal crystals model real world. Nature **378**, 231 (1995)
50. Y. Yang, S. Meng, Atomistic nature of NaCl nucleation at the solid-liquid interface. J. Chem. Phys. **126**, 044708 (2007)
51. M. Patra, M. Karttunen. Systematic comparison of force fields for microscopic simulations of NaCl in aqueous solutions: Diffusion, free energy of hydration, and structural properties. J. Comput. Chem. **25**, 678 (2004) (The size and energy parameters of Na^+ and Cl^- are the same in PARM94 and PARM99 force field)
52. K.F. Kelton, Crystal nucleation in liquids and glasses. Solid State Phys. Adv. Res. Appl. **45**, 75 (1991)
53. J. Garside, A. Mersmann, J. Nyvlt, Measurement of crystal growth and nucleation rates, 2nd edn (IChemE, UK, 2002)
54. P. Mark, L. Nilsson, Structure and dynamics of the TIP3P, SPC, and SPC/E water models at 298 K. J. Phys. Chem. A **105**, 9954 (2001)

Chapter 13
Ice Surface and Its Ordering

13.1 Structure of Ice

The most common bulk-phase ice in nature is the hexagonal ice phase Ih, which is widespread in nature and is extremely important in atmospheric and geological studies such as cloud formation, ozone catalysis, and rock weathering. The oxygen atoms in ice Ih are regularly arranged on a hexagonal lattice, i.e., four nearest-neighbor oxygen atoms exist around each oxygen atom, and these four nearest-neighbor oxygen atoms form a quasi-regular tetrahedral arrangement that is compressed along the [0001] direction. Compared with the angle of 109.47° for a perfect regular tetrahedron, the angle between the oxygen atoms forming the tetrahedron within the ice Ih decreases to 109.33° along the [0001] direction and increases to 109.61° perpendicular to the [0001] direction (Fig. 13.1a). Unlike the regular arrangement of oxygen atoms, the hydrogen atoms in ice Ih are in a somewhat disordered arrangement, obeying the constraints of the Bernal-Fowler-Pauling ice rules [1, 2]: (1) every two hydrogen atoms and one oxygen atom are covalently bonded; and (2) there is one and only one hydrogen atom between two oxygen atoms. The ice rules suggest that there is no regular pointing of hydrogen atoms within the bulk ice Ih. It has been found that the disordered hydrogen atoms in ice Ih undergo a phase transition from disorder to order and form ice XI when cooled to 72 K [3]. As shown in Fig. 13.1b, the hydrogen atoms in ice XI are arranged in an ordered manner, giving ice XI a large polarity along the [0001] direction.

13.2 Surface Ordering of Ice

The surface of ice plays an important role in catalysis of various physicochemical reactions in which ice Ih is involved. Compared to what is known about the structure of the bulk phase of ice Ih, the ice surface is poorly studied. An important aspect of the surface properties of ice is whether the direction of the hanging OH bonds on the

S. Meng and E. Wang, *Water*,
https://doi.org/10.1007/978-981-99-1541-5_13

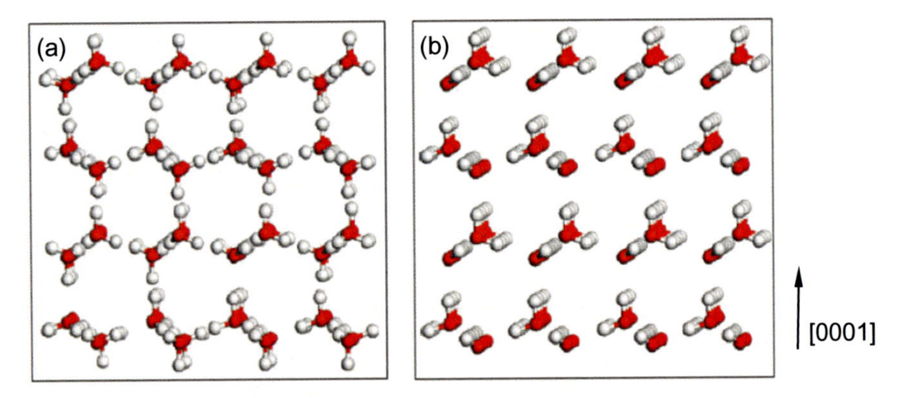

Fig. 13.1 Structures of bulk ice XI and Ih. **a** Ice XI, with hydrogen atoms arranged in the [0001] direction in an orderly fashion. **b** Ice Ih, with the internal hydrogen atoms arranged disordered

ice surface is ordered, i.e., whether the surface hydrogen atom distribution is ordered. Recent neutron scattering experiments have suggested that some of the experimentally small diffraction peaks are related to the arrangement of the hanging OH bonds on the ice surface [4]. Since the experimental signal of the hanging OH bonds on the ice surface is very weak, further structural information cannot be obtained, let alone answers to questions like whether the hydrogen atoms on the ice surface are ordered or whether there is an order–disorder phase transition as the case of hydrogen atoms in bulk. This dilemma was first investigated by the group of Enge Wang, Pan Ding et al. defined a new order parameter to describe the ordered state of the hanging OH bonds on the ice surface [5, 6]:

$$C_{\mathrm{OH}} = \frac{1}{N_{\mathrm{OH}}} \sum_{i=1}^{N_{\mathrm{OH}}} c_i \tag{13.1}$$

where N_{OH} is the total number of the hanging (i.e., not hydrogen-bonded) OH bonds of the surface system and c_i is the number of nearest-neighbor hanging OH bonds around the i-th hanging OH. In this way the ordered state of hydrogen atoms on the ice surface can be fully quantitatively described: $C_{OH} = 2$ corresponds to the antiferroelectric-like arrangement, $C_{OH} = 3$ corresponds to the most disordered surface, and $C_{OH} = 6$ corresponds to the ferroelectric-like (XI ice) or all upward-facing OH arrangement. Several typical situations of distribution and arrangement of the hanging OH bonds on ice surfaces are shown in Fig. 13.2. This is the first time that one can propose a quantitative way to describe the ordered state of an ice surface and this method is proved to be quite effective. For example, it was found that the surface energy of the ice surface and C_{OH} are simply linearly correlated (Figs. 13.2 and 13.3), which can be used to conveniently infer the stability of the formed ice surface. Analysis of this correlation showed that the surface energy of ice is determined by the electrostatic repulsion between the hanging OH bonds. Namely:

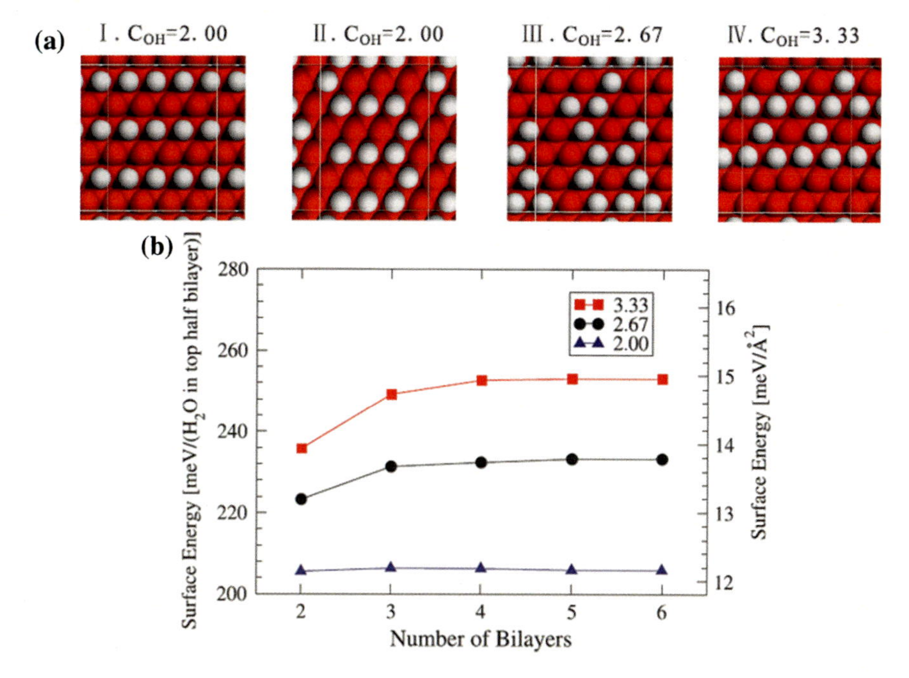

Fig. 13.2 **a** Several typical distributions of hanging OH bonds on ice surface. **b** Evolution relationship of surface energy versus ice thickness for ice layers with different surface orders

$$\gamma_{Ih} \approx \gamma_{C_{\mathrm{OH}}=2} + \frac{q^2}{4d_{\mathrm{HH}}\sigma}(C_{\mathrm{OH}} - 2) \tag{13.2}$$

13.3 Structural Phase Transition of Hydrogen Orientation on Ice Surface

Unlike the disorder-order phase transition of hydrogen atoms in bulk-phase ice at 72 K, the hydrogen atom arrangement on the ice surface shows different properties with thermodynamic fluctuations. Pan Ding et al. used the Monte Carlo method to study the orderliness of the hydrogen atom arrangement on the ice surface at different temperatures [5] (Table 13.1). They simulated an ice structure including six layers, where Layer 1 is the outermost layer and Layer 6 is the innermost layer. To avoid the effect of melting at the ice surface, they fixed the position of the oxygen atoms and considered only the energy changes caused by the different pointing of the hydrogen atoms, i.e., the hanging OH bonds. They found that below the melting point of the ice surface, the order parameter of the ice Ih surface (Layer1) is always close to 0, which means that the OH bond arrangement is very stable and no order–disorder phase transition of hanging OH bond arrangement occurs before the ice Ih surface

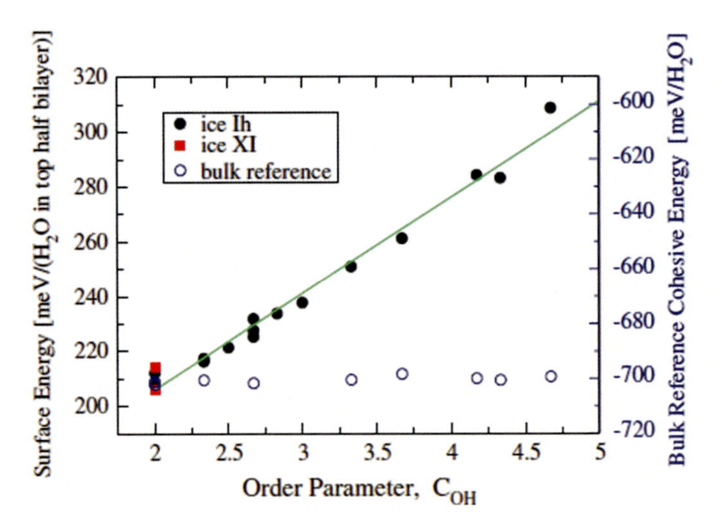

Fig. 13.3 Ice surface energy versus surface order parameter C_{OH}

Table 13.1 Monte Carlo simulation of order parameters as a function of temperature

T (K)	Layer1	Layer2	Layer3
40	2.00 (0.000)	2.42 (0.041)	2.44 (0.007)
100	2.00 (0.000)	2.48 (0.020)	2.58 (0.026)
200	2.13 (0.003)	2.70 (0.022)	2.75 (0.028)
500	2.33 (0.011)	2.71 (0.020)	2.74 (0.023)

melts. Both Layer2 and Layer3 in Table 13.1 have relatively large order parameters, reflecting the disordered hydrogen atom arrangement inside ice Ih.

13.4 Vacancies on Ice Surface

Based on their study of the orderliness of the ice surface, Enge Wang's group and their British collaborators further investigated the vacancy structure of the ice surface and its chemical activity [7]. Using first-principle calculations, they made an unexpected finding that the energy required to form a vacancy of a water molecule on the ice surface (vacancy formation energy) can be as small as 0.1–0.2 eV (Fig. 13.4). This is considerably less than the energy required to form a vacancy of a water molecule in bulk-phase ice, which is 0.74 eV. This suggests that there may be a large quantity of vacancies on the ice surface, and the presence of them necessarily affects the chemical activity of the surface ice. This changes the previously common understanding that ice activity has little to do with surface vacancies, and explains why ice surfaces are so active in atmospheric chemistry. Their calculations further reveal that the formation energy of ice vacancies on the surface varies over a wide range, from 0.2 to 0.9 eV. This mainly depends on the orderliness of the surface and the conformational relaxation of the surface molecules. For example, the distribution of dipole moments of water molecules on the ice surface also varies dramatically compared to the fixed dipole moments of water molecules in bulk ice, which is in agreement with the trend of vacancy energy variation (Fig. 13.5).

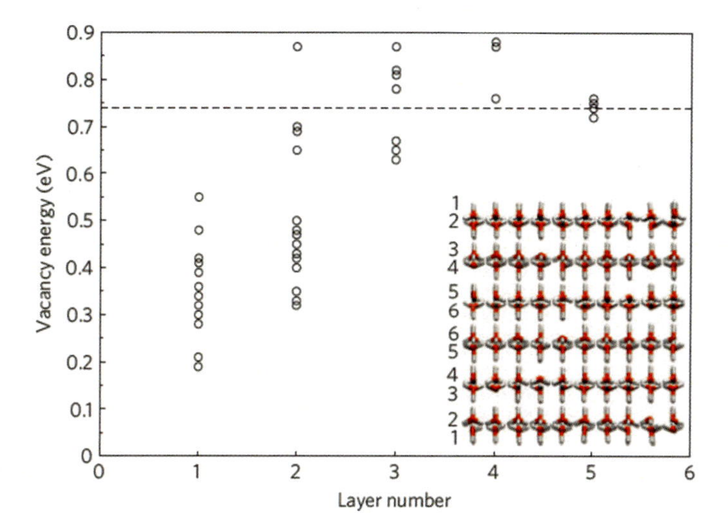

Fig. 13.4 Distribution of vacancy formation energy of water molecules over different ice depths. The inset represents the structure of the ice layers and the labeling of each layer

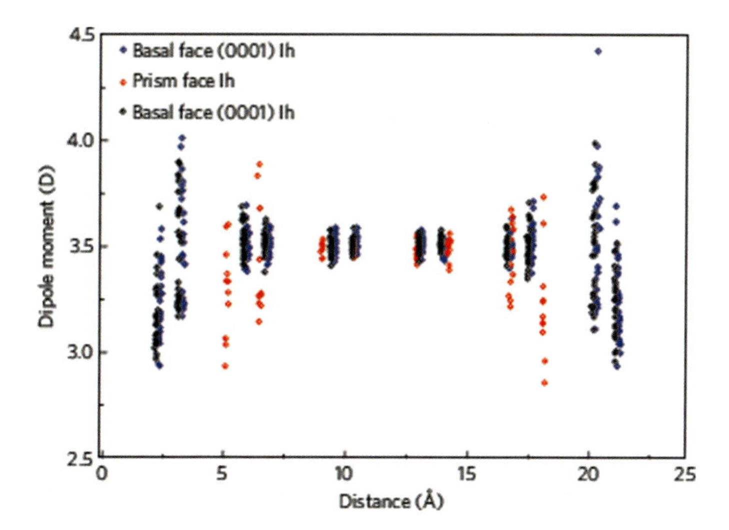

Fig. 13.5 Distribution of dipole moments due to water vacancies in the ice at different depths. The top and side surfaces of ice Ih (0001) (using two different OH distributions) are considered

13.5 Adsorption on Ice Surface

The chemical properties of the ice surface have recently been further studied directly by Enge Wang and his team. They used the adsorption of polar small molecules (e.g., H_2O, H_2S, etc.) on the ice surface as an example to study the effect of different structures of the ice surface on its reaction properties and adsorption characteristics [8]. They found that there are two types of adsorption sites for water molecules on the ice surface. One is that the water molecule is adsorbed on the vacant site via two to three hydrogen bonds (type A1), while the other is that the water molecule is adsorbed on the top site and forms a hydrogen bond with a water molecule on the ice surface (type A2) (see Fig. 13.6). These two types differ in their adsorption energies, with the adsorption energy of type A1 being slightly higher than that of type A2.

However, regardless of the adsorption configuration, the adsorption energy of polar water molecules on the ice surface varies largely and is directly related to the proton order parameter of the ice surface: the most ordered ice surface has the strongest ability to adsorb molecules, while the surface with the smallest order parameter has essentially the weakest adsorption ability, as shown in Fig. 13.7.

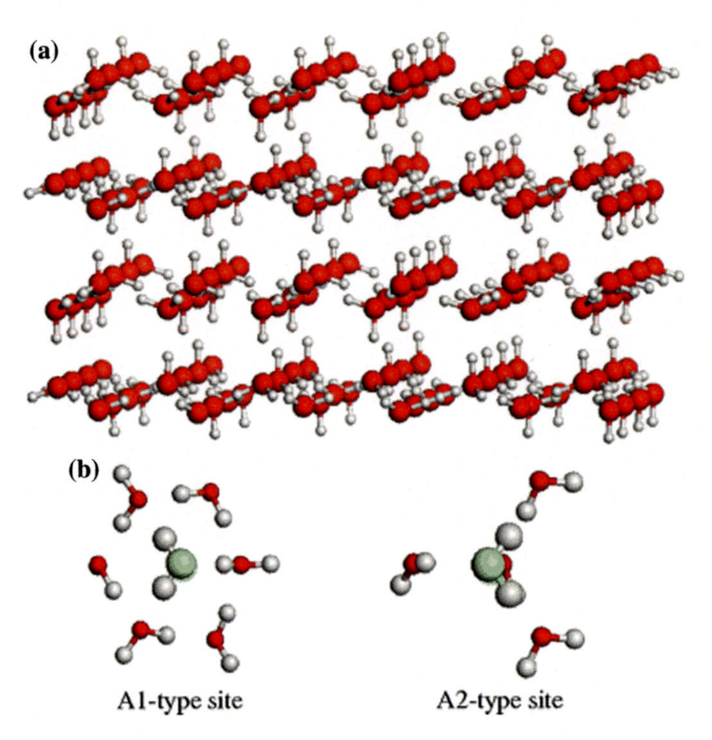

Fig. 13.6 Adsorption of polar molecules on the ice surface. **a** Ice layer structure; **b** two typical adsorption sites A1 and A2

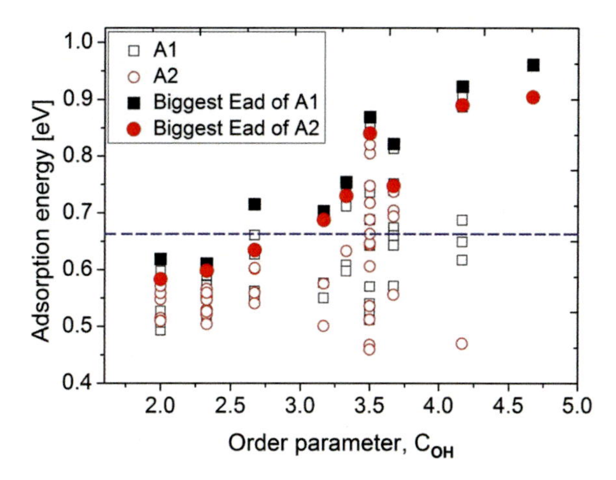

Fig. 13.7 Dependence of the adsorption energy of water molecules on the ice surface on the surface order parameter C_{OH}

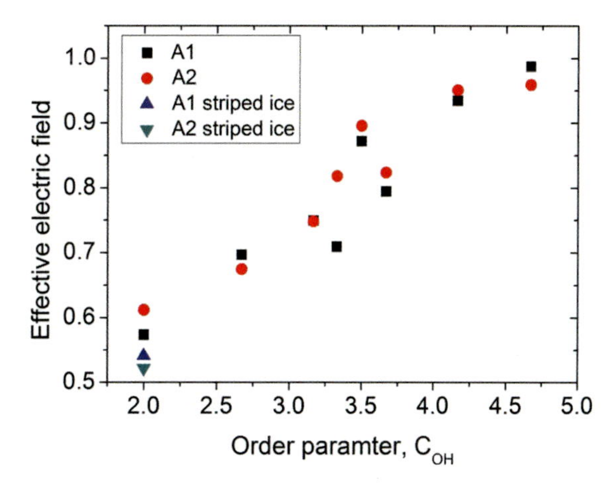

Fig. 13.8 The effective local electric field strength at the ice surface as a function of the surface order parameter C_{OH}

Ultimately, they attribute this effect to changes in the electric field distribution resulting from the ordered distribution of surface OH (see Fig. 13.8): the ordered surface produces such an extremely large surface electric field that it naturally has the strongest adsorption of polar molecules. This holds for polar molecules such as H_2O and H_2S, although the adsorption energy of H_2S molecules on the ice surface is 0.2–0.4 eV smaller than that of H_2O. This finding has extremely important implications for ice growth kinetics and cloud chemistry reaction.

References

1. L. Pauling, The structure and entropy of ice and of other crystals with some randomness of atomic arrangement. J. Am. Chem. Soc. **57**, 2680 (1935)

2. J.D. Bernal, R.H. Fowler, A theory of water and ionic solution, with particular reference to hydrogen and hydroxyl ions. J. Chem. Phys. **1**, 515 (1933)
3. S. Kawada, Dielectric dispersion and phase transition of KOH doped ice. J. Phys. Soc. Jap. **32**, 1442 (1972)
4. V. Buch, H. Groenzin, I. Li, M.J. Shultz, E. Tosatti, Proton order in the ice crystal surface. Proc. Nat. Acad. Sci. USA **105**, 5969 (2008)
5. D. Pan, L.M. Liu, G.A. Tribello, B. Slater, A. Michaelides, E.G. Wang, Surface energy and surface proton order of ice Ih. Phys. Rev. Lett. **101**, 155703 (2008)
6. D. Pan, L.M. Liu, G.A. Tribello, B. Slater, A. Michaelides, E.G. Wang, Surface energy and surface proton order of the ice Ih basal and prism surfaces. J. Phys. Condens. Matter. **22**, 074209 (2010)
7. M. Watkins, D. Pan, E.G. Wang, A. Michaelides, J. Vande Vondele, B. Slater. Large variation of vacancy formation energies in the surface of crystalline ice. Nat. Mater. **10**, 794 (2011)
8. Z.R. Sun, D. Pan, L.M. Xu, E.G. Wang, Role of proton ordering in adsorption preference of polar molecule on ice surface. Proc. Nat. Acad. Sci. USA **109**, 13177 (2012)

Chapter 14
Quantum Behaviors of H in Water

14.1 Quantum Behaviors in Bulk Water

Hydrogen is the least massive chemical element in nature, with a mass of only 1836 times that of the electron, and thus often exhibits quantum behavior in condensed systems. The classical model of hydrogen as a mass in these cases is not quite appropriate and the quantum nature of hydrogen needs to be considered. Hydrogen is the main constituent element of water and thus the quantum nature of hydrogen can often be demonstrated when dealing with water interactions.

One of the most obvious examples is the transport of H in water. According to the theory of chemical equilibrium, the pH of pure water is 7, i.e. the concentration of H^+ in water is 10^{-7} mol/L (the same concentration of OH^- ions), i.e. there is about 1 H^+ for every 500 million water molecules. The H^+ in water exists as the hydrate H_3O^+. This H^+ is not fixed to a particular water molecule but is transported through the water at a certain rate along the hydrogen bonding direction. On the surface, it appears as if H_3O^+ is being transported, and this is the Grotthuss mechanism of H^+ ion transport, which is essential for aqueous solution chemistry and biological processes. Due to the small mass of H^+, it exhibits strong quantum effects when transferred between different water molecules.

The bonding and motion of H_3O^+ in water simulated with path integral molecular dynamics are shown in Fig. 14.1. Each atom, and in particular the H atom, is represented by a density distribution of multiple spheres. Each bead corresponds to a path in the path integral. As seen in the figure, the extra H^+ is largely shared by the two surrounding O atoms and may be transferred in the subsequent process. The process is essentially thermodynamic, but the quantum effect of hydrogen plays an important role (Fig. 14.2). For example, in the classical image, the transfer potential of H in the water cluster of this configuration at the time of transfer is 24 meV, much smaller than the zero-point vibrational energy of H [1]. If the quantum effect of H is considered, there is no potential barrier. Besides, the quantum effect increases the radius of gyration of H^+ (i.e., the range in which H^+ is shared) from 0.9 Å in the classical image to 1.3 Å, making H^+ more nonlocal.

S. Meng and E. Wang, *Water*,
https://doi.org/10.1007/978-981-99-1541-5_14

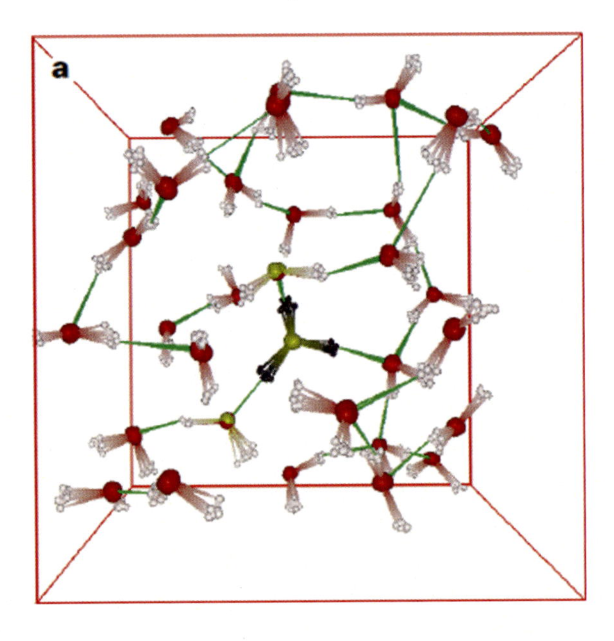

Fig. 14.1 A bulk water configuration in a path-integral molecular dynamics simulation [1]. Boxes indicate the size of the simulated primitive cell

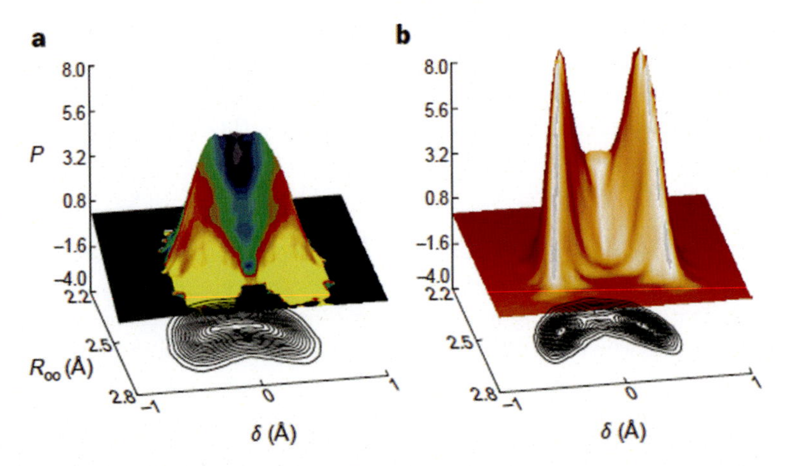

Fig. 14.2 Distribution of H in water during transport at 300 K [1]. r_{OO} is the distance between the two oxygen atoms connected by this H atom; δ denotes the distance difference between this H atom and these two O atoms. **a** Quantum image; **b** classical image

The transport of OH^- in water is very similar, but also different [2]. Path-integrated molecular dynamics simulations show that OH^- is surrounded by four nearest-neighbor water molecules, unlike H^+ transport which forms an $H_5O_2^+$ transition structure, OH^- transport forms $H_3O_2^-$ structure (Fig. 14.3). Considering the quantum effect of H, the OH^- transfer potential decreases from 55 meV in the classical image to 15 meV (Fig. 14.4). Also, since there is a double potential well in the $H_3O_2^-$ structure even when quantum effects are taken into account, this suggests that H is

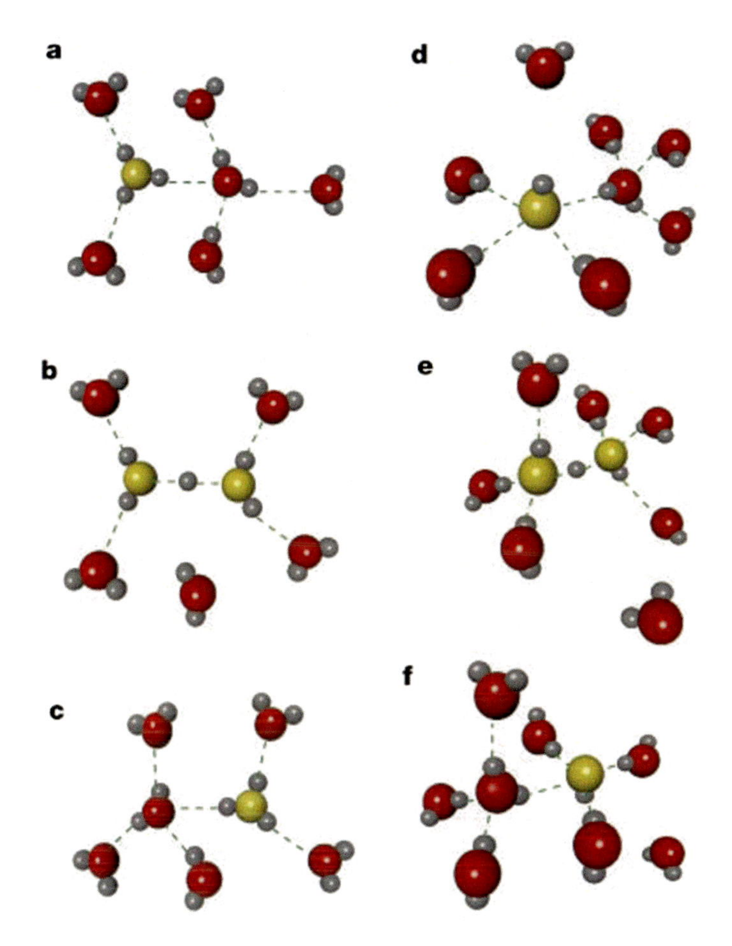

Fig. 14.3 **a–c** Common pathways for H^+ transport in water; **d–f** common pathways for OH^- transport [2]

only briefly shared in $H_3O_2^-$ and is an unstable transition state. This is different from the quantum state formed by $H_5O_2^+$ where H is more stably shared, indicating that the OH transport process cannot be simply treated as the transport of "H vacancies".

14.2 Quantum Behaviors in Confined Water

Water under confined conditions exhibits physicochemical properties that are different from those of the bulk state. It has been found that confined water chains can form within carbon nanotubes and that rapid proton transport can be achieved through such confined water chain channels [3]. Chen et al. compared in detail the

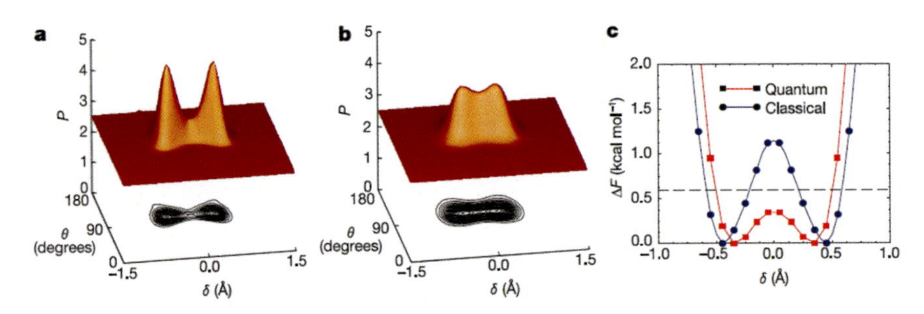

Fig. 14.4 The distribution of H during OH^- transfer [2]. The angle θ is the angle between OH^- and the two O atoms to which the transferred H is attached; δ denotes the difference in distance between that H atom and those two O atoms. **a** Classical image; **b** quantum image; **c** classical and quantum reaction potentials

proton transport in bulk water and confined water chains in carbon nanotubes [4] and found that the quantum behavior of protons is more pronounced in confined water chains than in the bulk state. The quantum effect in bulk water makes the transported protons shared by two oxygens forming $H_5O_2^+$ structure, which is considered as the stable state of protons existing in bulk water. In contrast, in the confined water chain, $H_5O_2^+$ and even $H_7O_3^+$ are only transient states in the steady-state of the quantized protons. The protons exhibit strong nonlocal properties and their distribution can cover all the five water molecules on the water chain all the time. This non-local proton distribution is evident in confined water (Fig. 14.5).

14.3 Quantization Diffusion of Water Dimers on Surface

The adsorption of small clusters formed from one to several water molecules on metal surfaces has also attracted a lot of attention. It is generally believed that individual water molecules are stable on metal surfaces only at very low temperatures (< 20 K) and they readily diffuse (potential barrier: 0.1–0.3 eV) to form clusters. This fast diffusion process was directly observed by Mitsui et al. using STM on Pd(111) [5]. At 40 K, individual water molecules diffuse rapidly and collide to form water molecular dimers, trimers, etc. Surprisingly the formed water dimers will diffuse on Pd(111) at a rate four orders of magnitude faster than that of single molecules [5]. In 2002, Sheng et al. calculated the adsorption structure of water molecule dimers on metal surfaces [6, 7]: the water molecules are all in the top position; one molecule in the dimer is lower and has a configuration close to that of the adsorbed monomolecule, and the other is higher and accepts a hydrogen bond to the former. Potential barrier calculations indicate that if a simple diffusion model of water molecules jumping from the top atomic position to the next top position is used, the dimer and monomer of water have similar potential barriers ~ 150 meV, which is not sufficient to explain the observed phenomena [7]. To explain this phenomenon,

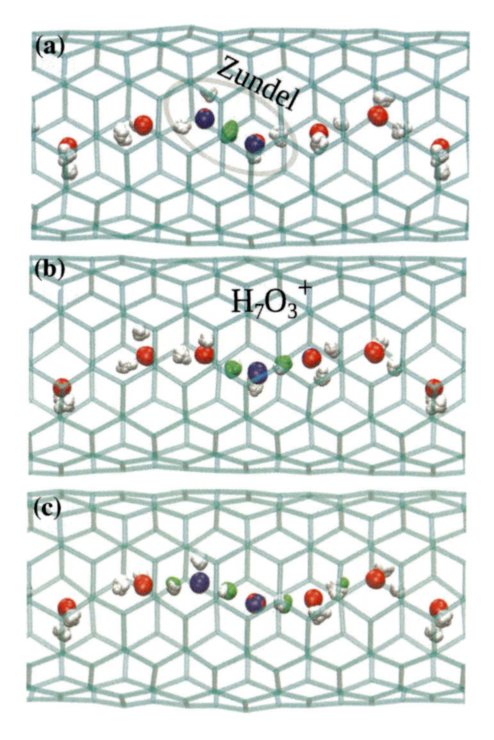

Fig. 14.5 The distribution of H as it is transported on the confined water chain of a carbon nanotube [4]. The configuration of Zundel ($H_5O_2^+$) is shown in (**a**); **b** shows the configuration of $H_7O_3^+$; **c** the non-local distribution of protons over multiple hydrogen bonds

Ranea et al. proposed that the quantum behavior of H plays a key influence (Fig. 14.6) [8]: dimer diffusion is achieved through three steps: upper molecule rotation—upper and lower molecule exchange—new upper molecule rotation: the potential barrier for the rotation process is small, about 5 meV. While for the second process upper and lower molecule exchange is mainly involved in the position change of H. If we consider the quantum effect of H, tunneling through this exchange potential barrier (220 meV), then dimer diffusion can proceed very quickly, being several orders of magnitude faster than thermal diffusion at low temperatures. This is an excellent example of explaining that the dynamics of surface water adsorption have to take into account the quantum effect of H.

14.4 Quantum Properties of Surface Hydrogen Bonds

For surface water layers, the quantum effect of hydrogen also plays a key role in their structural conversion and decomposition.

As previously mentioned, there is a hydrogen pointing up bilayer structure and a hydrogen pointing down bilayer structure on the surface of Pt(111). The transition between them can be achieved by the rotation of water molecules in the upper

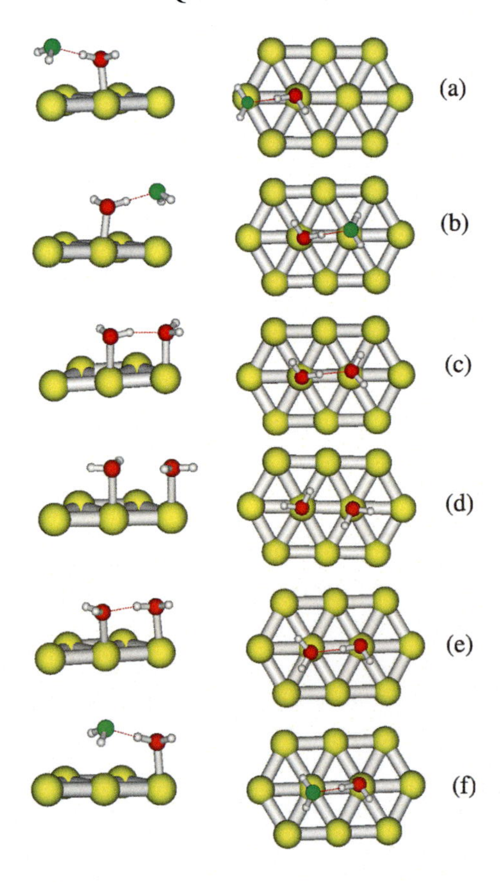

Fig. 14.6 Quantum behavior of water diffusion on the surface of Pd(111) [8]. (**a**) Water dimmer absorbed on Pd(111). The diffusion of the water dimer is achieved through three steps: upper molecular rotation—upper and lower molecular exchange—new upper molecular rotation. The second of these steps (from **b**–**f**) mainly involves the motion of H, which is achieved at low temperatures mainly through quantum tunneling

half of the bilayer. By the NEB method in first-principles calculations, we calculated the transition potential barrier and fitted it with a quadratic function to obtain a potential energy profile as in Fig. 14.7. potential wells A and B represent the hydrogen-pointing-up and hydrogen-pointing-down bilayer structures, respectively, with an energy difference of about 24 meV. The transition from potential well A to potential well B requires overcoming a classical migration potential barrier of about Ec = 76 meV.

We consider the correction of this transition potential by the zero-point energy of hydrogen motion. Assuming that the mass-heavy O atom is stationary while the water molecule undergoes quantum rotation, then the solution of the Schrödinger equation corresponding to the rotation of water in this potential defines some low-energy bound state. Figure 14.7 also shows the eigenenergies and eigenwave functions corresponding to the rotation of water molecules in this double potential well. We find that the lowest energy bound states are mainly distributed in the potential well B, as ground states (not shown). And the next lowest energy bound state has an eigenenergy of $E_{A0} = 33$ meV. This state is mainly distributed in the potential well A. The other bound states with slightly higher energy are located no lower than $E_{A1} =$

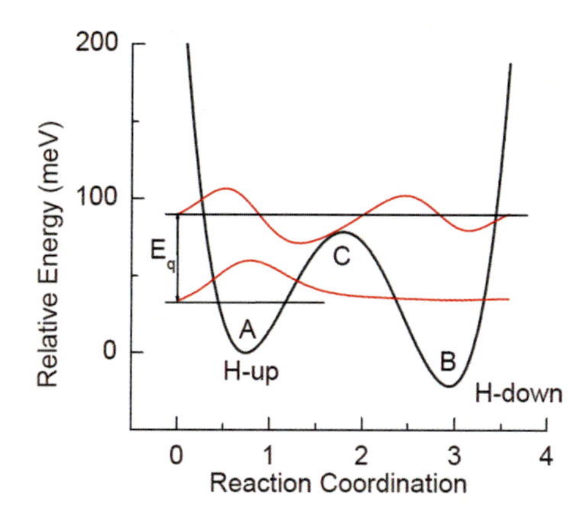

Fig. 14.7 Fitted potential energy curves for the transition from the hydrogen pointing up to the hydrogen pointing down bilayer structure on the Pt(111) surface, and the corresponding two eigenvalues and eigenwave functions

90 meV, which is already above the classical migration potential Ec and whose wave function extends into the entire double potential well. Thus, considering the quantum rotation in water, the quantum potential barrier corresponding to the transition from hydrogen pointing up to hydrogen pointing down structure is $Eq = E_{A1} - E_{A0} = 90 - 33 = 57$ meV, which is 19 meV lower than the classical migration potential barrier [9]. Replacing the water molecule with a D_2O molecule would increase its quantum potential barrier by 4 meV.

At low temperatures, the process of proton tunneling through the potential barrier is very important and even plays a dominant role. The tunneling probability obtained using the WKB approximation is:

$$\Gamma = \mathrm{e}^{-\frac{2}{\hbar}\int\sqrt{2m[V(x)-E]}\mathrm{d}x} \cong \mathrm{e}^{-\alpha\sqrt{E_q}} \tag{14.1}$$

where the value of α is taken to be $24\,\mathrm{eV}^{-1/2}$, corresponding to the mass of two H atoms and two potential wells separated by 1.4 Å. The probability of quantum tunneling and thermal motion across the potential barrier are, respectively.

$$\mathrm{D_q} = We^{-\alpha\sqrt{E_q}} \text{ corresponding to the quantum tunneling process} \tag{14.2}$$

$$\mathrm{D_c} = We^{-E_q/k_BT} \text{ corresponds to the thermal motion process} \tag{14.3}$$

where W is the attempted frequency of H motion, $\mathrm{W} = \frac{\mathrm{k_B}T}{h}\left(e^{\frac{\hbar\omega}{k_BT}} - 1\right)$. A reasonable choice of parameters at temperatures below 115 K leads to the conclusion that the quantum tunneling rate dominates the transition between hydrogen pointing up and hydrogen pointing down structures, i.e., at this point Dq > Dc. At this temperature, the tunneling rate is $2 \times 10^{11}\ \mathrm{s}^{-1}$.

Using the same approach and considering the quantum motion of the hydrogen atoms, we can also conclude that quantum effects lower the decomposition potential barriers of the double-layer D_2O and H_2O structures on the Ru(0001) substrate by

30 meV and 100 meV, respectively. The classical decomposition potential barrier previously obtained using the NEB method is 0.62 eV. Compared with the decomposition potential barrier of 0.53 eV taking into account the quantum effects, the decomposition potential barrier of the H_2O layer is lower than that of the water desorption, while the decomposition potential barrier of the D_2O layer is still higher than that of the water desorption potential. This indicates that H_2O decomposes first before desorption at elevated temperatures; while D_2O desorption first and cannot decompose. The different decomposition potentials barriers of the H_2O and D_2O layers also explain exactly why a fraction of the H_2O would decompose in the low-temperature experiments. The fact that no D_2O would decompose indicates the importance of quantum effects in the surface water layer.

In addition, through systematic path integral molecular dynamics simulations, Xinzheng Li et al. summarized in detail the effect of quantum properties of hydrogen in aqueous layers on metal surfaces [10]. They found that in the mixed system of hydroxyl and water, the potential barrier for H transfer is greatly dissipated with the H quantum effect taken into account. Moreover, the difference between part of OH chemical and hydrogen bonds in the system becomes either small (Pt(111), Ru(0001)) or even disappears (Ni(111)) on different metal surfaces. As a result, H is completely shared by the two nearest O atoms and it is no longer possible to distinguish which is the OH bond and which is the hydrogen bond. A comparison between the O–H distribution and O–O distribution in the classical picture and the quantum picture, is shown in Fig. 14.8.

Based on these results, they proposed the correction of the quantum effect on the hydrogen bond [11]. For strong hydrogen bonds, the quantum effect of H tends to pull the heavy atom (e.g. O–O) spacing shorter; while for weaker hydrogen bonds, the correction of the quantum effect pulls the heavy atom spacing longer. This is caused by the fact that quantization makes the distribution of H more nonlocal. If the ratio of the vibrational frequency of the OH that forms a hydrogen bond to the vibrational frequency of the free unbonded O and H is used as a scale for the strength of the hydrogen bond (0.5—very strong; 1—very weak), the cut-off point for such strong and weak hydrogen bonds is around 0.7 (Fig. 14.9). This hydrogen bond strength scale has a one-to-one correspondence with the energy of the hydrogen bond: the weaker the hydrogen bond, the closer its scale is to 1; conversely, the stronger the hydrogen bond, the smaller its scale value. The above law holds universally for hydrogen bonds in water, HF, and organic molecules, and also for hydrogen bonds in gas, liquid, and solids, so it is a universal law. Due to the larger mass of D, its quantum effect is weaker and lies between the H quantum behavior and the classical behavior, which is also very important for understanding the isotope calibration experimental structure and isotope effects.

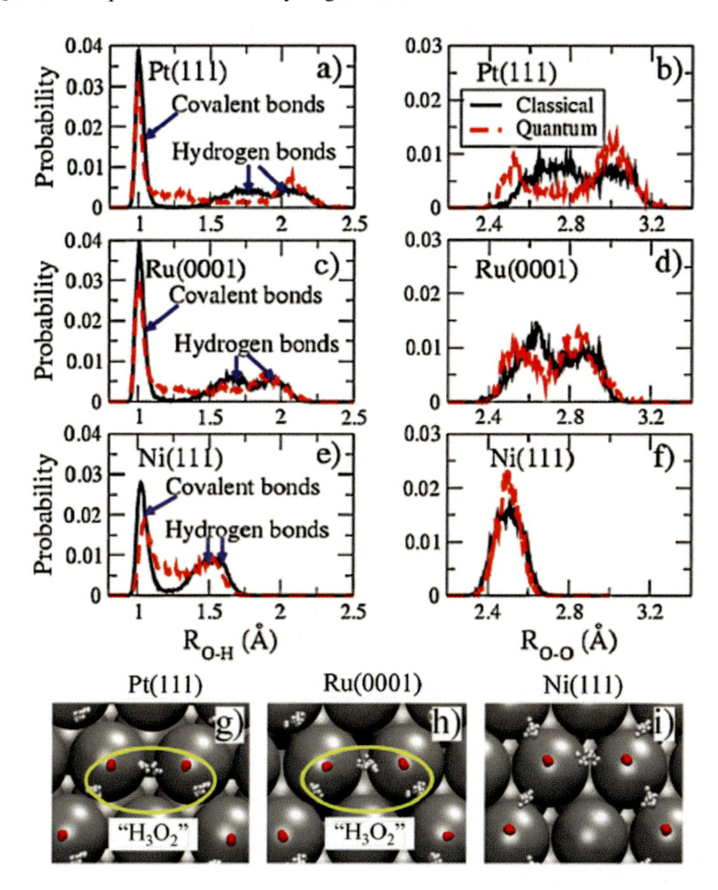

Fig. 14.8 Probability of distribution of O–H spacing and O–O spacing on different metal surfaces [10]. The results obtained from both classical and quantum images are displayed in the figure. The following figure shows the structure of water obtained from path integral molecular dynamics simulations

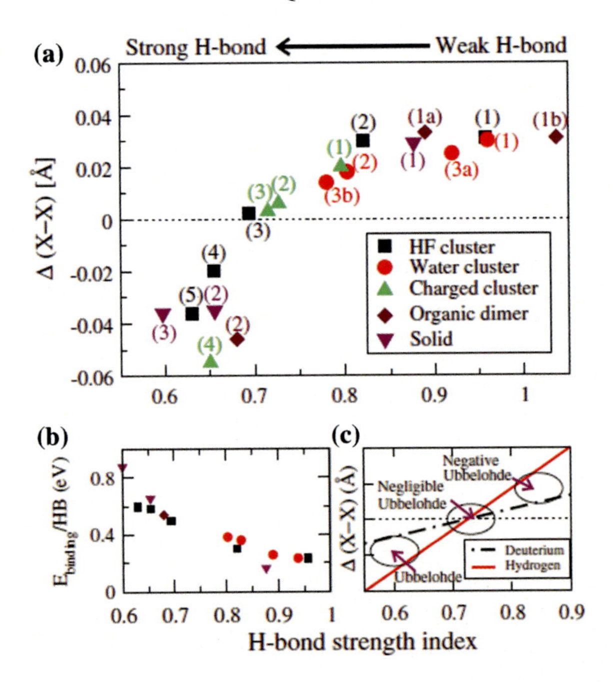

Fig. 14.9 **a** Correction of quantum effects on the distance (X-X) between heavy elements forming hydrogen bonds [11]: Δ(X-X) > 0, i.e. the quantum X-X distance is larger than the classical value, and Δ(X-X) < 0, i.e. the quantum X-X distance is smaller than the classical value. **b** Relationship between the energy of the hydrogen bond and the hydrogen bond strength scale. **c** Comparison between the quantum effects of H and D

References

1. D. Marx, M.E. Tuckerman, J. Hutter, M. Parrinello, The nature of the hydrated excess proton in water. Nature **397**, 601 (1999)
2. M.E. Tuckerman, D. Marx, M. Parrinello, The nature and transport mechanism of hydrated hydroxide ions in aqueous solution. Nature **417**, 925 (2002)
3. C. Dellago, M.M. Naor, G. Hummer, Proton transport through water-filled carbon nanotubes. Phys. Rev. Lett. **90**, 105902 (2003)
4. J. Chen, X.Z. Li, Q. Zhang, A. Michaelides, E.G. Wang, Nature of proton transport in a water-filled carbon nanotube and in liquid water. Phys. Chem. Chem. Phys. **15**, 6344 (2013)
5. T. Mitsui, M.K. Rose, E. Fomin, D.F. Ogletree, M. Salmeron, Water diffusion and clustering on Pd(111). Science **297**, 1850 (2002)
6. S. Meng, L.F. Xu, E.G. Wang, S.W. Gao, Vibrational recognition of hydrogen-bonded water networks on a metal surface. Phys. Rev. Lett. **89**, 176104 (2002)
7. S. Meng, E.G. Wang, S.W. Gao, Water adsorption on metal surfaces: a general picture from density functional theory studies. Phys. Rev. B **69**, 195404 (2004)
8. V. A. Raneal, A. Michaelides, R. Ramírez, P. L. de Andres, J. A. Vergés, and D. A. King, Water dimer diffusion on Pd{111} assisted by an H-bond donor-acceptor tunneling exchange. Phys. Rev. Lett. **92**, 136104 (2004)
9. Z.J. Ding, Y. Jiao, S. Meng, Quantum simulation of molecular interaction and dynamics at surfaces. Front. Phys. **6**, 294 (2011)
10. X.Z. Li, M.I.J. Probert, A. Alavi, A. Michaelides, Quantum nature of the proton in water-hydroxyl overlayers on metal surfaces. Phys. Rev. Lett. **104**, 066102 (2010)
11. X.Z. Li, B. Walker, A. Michaelides, Quantum nature of the hydrogen bond. Proc. Nat. Acad. Sci. USA **108**, 6369 (2011)

Chapter 15
Phase Transitions of Water Under Surface Confinements

15.1 Phase Transitions of Water in Nanotubes

Under the restriction of the surface potential field, the laws and conditions of water phase transition change. An interesting example is the phase transition of water in one-dimensional carbon nanotubes. Through classical molecular dynamics simulations, Xiaocheng Zeng et al. at the University of Nebraska-Lincoln found that inside carbon nanotubes with different diameters, water molecules join together to form peculiar tetragonal, pentagonal, hexagonal, and heptagonal tubular ices (see Fig. 15.1). It is obvious that these ice structures are very different from the natural bulk-phase ice Ih with hexagonal symmetry and snowflakes, etc. These nanotubes of ice are able to undergo a solid-to-liquid phase transition when induced by temperature increase or axial pressure, one manifestation of which is a change in the density distribution of water molecules from tubular to a more continuous distribution from the center of the carbon tube to the wall (Fig. 15.2). Even more surprising is the presence of a phase transition critical point in the finer tubes, above which there is no difference between ice and liquid water, and a continuous phase transition between them. This is evident from the gradual disappearance of the stepwise abrupt changes in both potential energy and volume at the point of pressure-induced solid–liquid phase transition as the temperature increases to 330–360 K (see Fig. 15.3). This suggests that water under confined conditions has properties completely different from those of bulk water [1].

15.2 Phase Transitions of Water Monolayers

Water also exhibits unusual phase transition characteristics in the confinement of two parallel flat plates. A flat plate with a hydrophobic surface ensures that there is only weak van der Waals forces between the surface and water. The distance between the two layers of plates is gradually reduced. Through molecular dynamics

S. Meng and E. Wang, *Water*,
https://doi.org/10.1007/978-981-99-1541-5_15

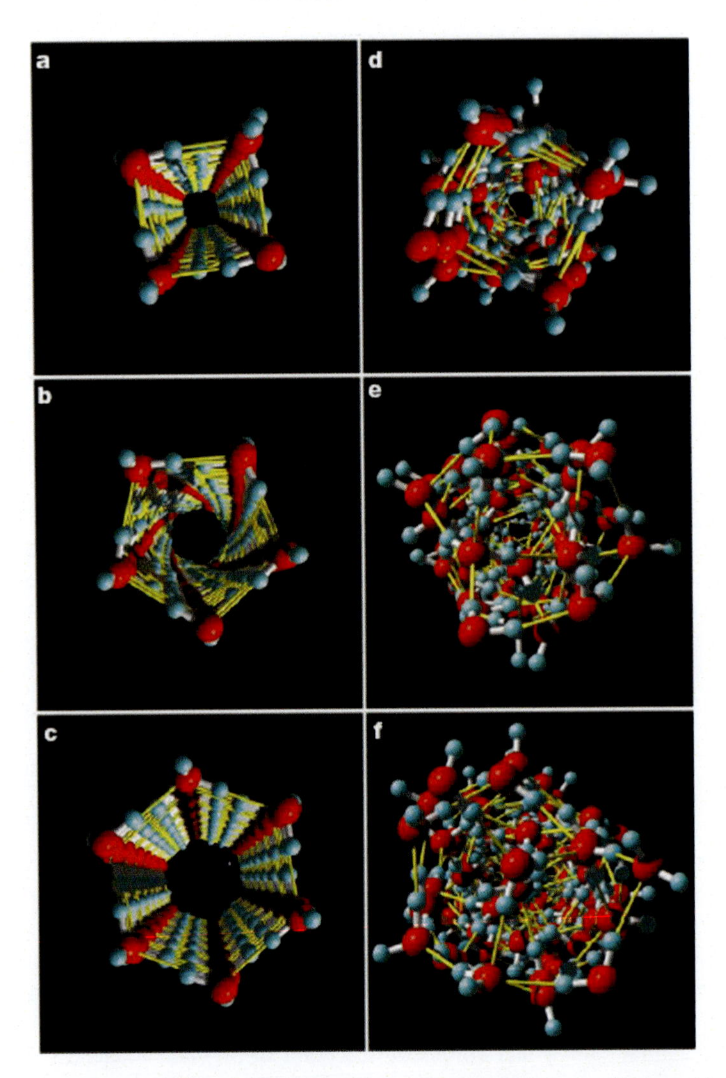

Fig. 15.1 Water exhibits different structures in the confinement of carbon nanotubes with different tube diameters [1]. **a–c** are the ice structures in (14,14), (15,15), and (16,16) carbon nanotubes (not shown), respectively; **d–f** are the corresponding disordered structures after melting of the ice

simulations, Zangi and Mark [2] found that liquid water suddenly condenses when the plate spacing is at 0.51–0.55 nm. This is manifested in the diffusion coefficient in the horizontal direction, which drops by about three orders of magnitude, from 10^{-5} to 10^{-8} cm^2/s (Fig. 15.4). This indicates that the water undergoes a liquid to solid phase transition. The single layer of ice that forms between the two plates at this point has a rhombic quadrilateral structure, as shown in Fig. 15.5. This is a special ice phase of water that differs from the normal hexagonal ice Ih.

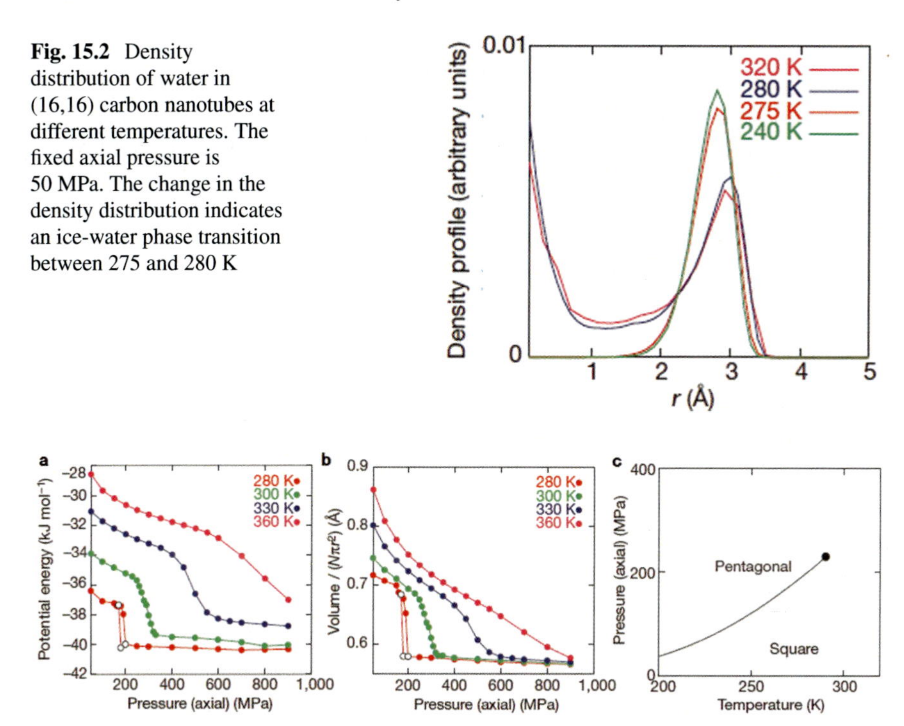

Fig. 15.2 Density distribution of water in (16,16) carbon nanotubes at different temperatures. The fixed axial pressure is 50 MPa. The change in the density distribution indicates an ice-water phase transition between 275 and 280 K

Fig. 15.3 Variation of potential energy (**a**) and volume (**b**) of water in nanotubes with axial pressure at different temperatures. The sudden changes in potential energy and volume disappear in the phase transition at higher temperatures, indicating that the phase transition changes from a first-order phase transition to a continuous phase transition. **c** The phase diagram corresponding to this phase transition. The critical point of the phase transition is at higher temperatures and pressures

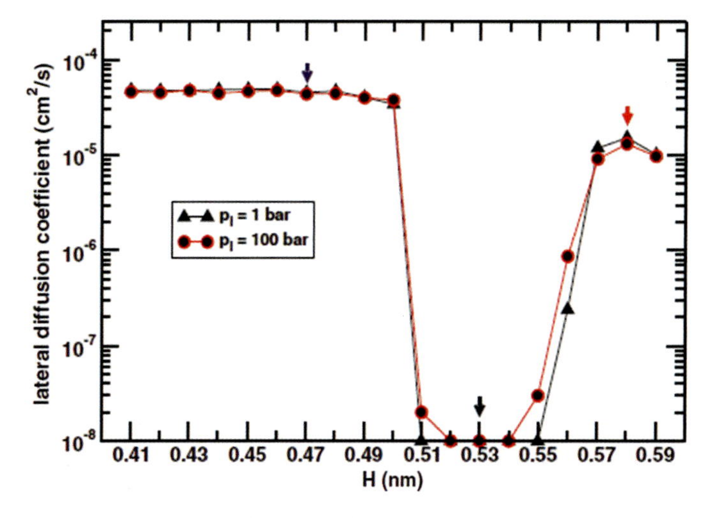

Fig. 15.4 Variation of the lateral diffusion coefficient of the water layer between sparse horizontal plates with the layer spacing [2]

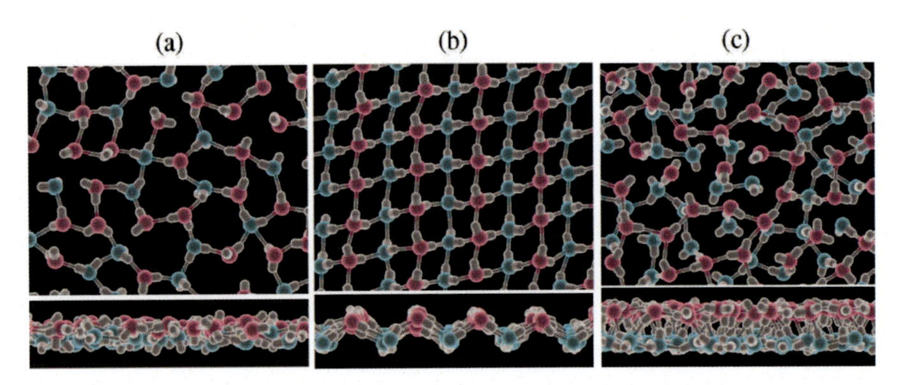

Fig. 15.5 Structure of the water layer before and after the phase transition [2]. **a–c** correspond to the cases in the above figures where the distance between the plate and H is 0.47, 0.53, 0.58 nm, respectively

Further compression of the layer spacing results in a sharp increase in the diffusion coefficient of water molecules back to 5×10^{-5} cm^2/s, indicating that the confined water layer changes from a rhombic ice phase to a liquid water phase again. Structural analysis shows that the rhombic structure of water becomes a single layer of irregularly arranged liquid water layer.

The phase transition of water under the plates confinement was further investigated. Using molecular dynamics simulations, Han et al. [3] found that at low densities ($\rho < \rho_c = 1.33$ g/cm^3) ice forms a hexagonal structure, and a phase transition from ice to water occurs when warming the ice layer, accompanied by an abrupt change in system potential energy of about 3–6 kJ/mol. This suggests that the ice layer undergoes a first-order phase transition. However, at higher water densities ($\rho > \rho_c = 1.33$ g/cm^3) ice forms a tetragonal rhombic structure and the system potential energy increases gradually during the ice to water phase transition and no abrupt change occurs (Fig. 15.6). This indicates that the ice to liquid water phase transition is a continuous phase transition in a dense confined water layer, and there is no significant abrupt change in the structure before and after the phase transition. That is, as in the one-dimensional confined system in nanotubes, in the two-dimensional confined system there is a critical point of phase transition beyond which the structural distinction between solid ice and liquid water disappears. In bulk water, the critical point exists only for the phase transition from the liquid state to the gaseous state; there is no critical point for the phase transition between the solid state and the liquid state. However, this critical phenomenon in solid–liquid phase transitions is also present in confined systems, which is very interesting! The phase diagram for two-dimensional confined water, summarized from the simulated structure, is shown in Fig. 15.7.

It can be expected that such strange phase transition critical phenomena in confined water systems can occur at the surface even if there is only one layer of surface confinement or if there is a nanoscale rough structure on the surface, provided that

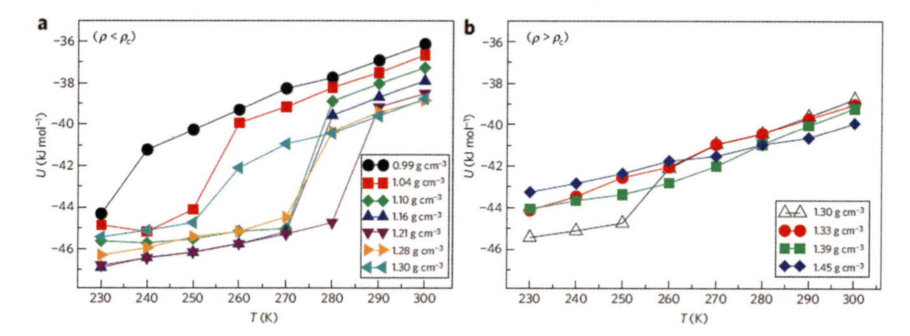

Fig. 15.6 Variation of potential energy of water with temperature for different densities in the flat plate limit [3]. The left panel shows the case where the density is less than the critical density (i.e., the first-order phase transition system); the right panel shows the case where the water density is greater than the critical density (the continuous phase transition system)

Fig. 15.7 Phase diagram of water under the confinement of two flat layers [3]

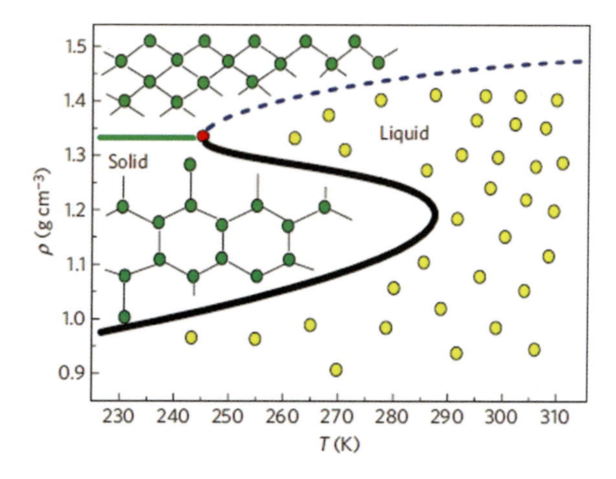

the confinement effect is relatively significant. In addition to solid–liquid phase transitions, it is possible that for other phase transitions, such as liquid–liquid phase transitions in supercooled water, there may exist phase transition laws that differ from those of the bulk phase. This awaits further experimental verification and continuous exploration.

15.3 Pressure-Induced Phase Transition of Water

As an example, we discuss in detail the process of the ice-water phase transition occurring in two metallic flat plates under pressure. In particular, we try to understand the process of phase transition occurring under confined conditions from the point of view of electronic action and the first-principles.

The confinement and compression of water lead to two different effects: (1) the reduction of geometric space leads to shorter hydrogen bonds between water molecules; (2) and it leads to the breaking and resynthesis of hydrogen bonds between water molecules. Both effects depend on the atomic structure of the confined water and ultimately on the interactions of electrons between these atoms. The latter is related to the macroscopic boundary conditions that arise after the introduction of confinement and pressure. Therefore, a self-consistent description of the interatomic interactions and geometric confinement is important for understanding confined water and its response to pressure. Previous calculations are based on classical molecular dynamics simulations, which use model potentials for water-water and water-surface interactions. The water-surface interaction is usually described by some artificially determined approximate model potentials, such as a hard wall model, or an effective potential that incorporates a mirror force. These model potential approaches can give a general characterization of confined water, but cannot explain in detail the electron interactions at interfaces under dynamic constraints and pressures, and cannot provide insight into pairwise electron distribution and charge transfer processes. To date, describing confinement and pressure by ab initio methods has been a great challenge for computer simulations, mainly due to the complex atomic and chemical structures at confined interfaces and the high computational consumption of ab initio methods.

Based on the description of atomic interactions by density functional theory, we investigated the melting process of confined water caused by pressure using ab initio computational molecular dynamics methods [4]. A very thin ice layer (Ice Ih) was placed between metal plate surfaces (Pt(111)) and the thickness of the film was gradually reduced to simulate the melting process of ice caused by pressure. We found that at about 0.5 GPa, the ice starts to melt, and accordingly, the volume shrinks by 6.6%, consistent with the volume change during the phase transition of bulk ice. The continuing increase of pressure leads to a linear increase in the temperature of the liquid water. The liquid phase exhibits diffusion of water molecules and frequent hydrogen bond breakage and recombination. The non-equilibrium kinetic energy distribution in ice and liquid water due to confinement and pressure regains its equilibrium state after a relaxation time on the order of picoseconds. Electron redistribution at the interface due to confinement and pressure is observed. Our results demonstrate the feasibility and necessity of first-principles calculations to describe confined water and its associated phenomena.

A four-layer Pt(111) surface is taken and four bilayers $2\sqrt{3} \times 2\sqrt{3}R30°$ of thin ice films containing 32 H_2O molecules are placed in the middle. One Pt-H_2O contact interface (bottom layer) matches the best adsorption configuration of the bilayer ice, while the other interface (top layer) does not match the ice structure. This allows us to study both the water at the interface and the water in the confined domain. We simulate the process of applying pressure to the ice film by varying the length of the primitive cell in the z-direction. The distance is compressed from the equilibrium value $z = 23.41$ Å ($\Delta z = 0$) to $z = 21.91$ Å ($\Delta z = -1.5$ Å). The distance above and below the thin ice decreases equivalently. The Pt atoms remain in a fixed position in the bulk material during compression (lattice constant of 3.99 Å) due to the small position relaxation (2%) of the Pt atoms on a free Pt(111) surface or a surface at several GPa pressures. The calculations are performed with the ultra-soft pseudopotential (USPP), and the exchange–correlation energy of PW91. The energy cutoff is taken to be 300 eV for a single k-point (1/2, 1/2, 0). The simulation is performed with a time step of 0.5 fs.

Figure 15.8 shows the relationship between the average kinetic energy (i.e., temperature) of atoms in a water film as a function of their thickness (Δz). In the MD simulation, these kinetic energies are averaged over all configurations in the 0.5–1 ps channel. It is clear that this curve is split into two regions: (1) A smooth region from $\Delta z = 0$ to $\Delta z = -0.875$ Å. (2) A nearly linear region from $\Delta z = -0.875$ Å to $\Delta z = -1.5$ Å. This smooth region indicates that the internal potential energy of the ice does not change much during the initial stages of ice film compression, because the potential energy of the ice block is very flat near the equilibrium volume [5]. This is also the origin of the latent heat in ice-water or other solid–liquid phase transition. So this smooth region corresponds to solid ice. Constantly increasing the pressure, when $-\Delta z > 0.875$ Å, the kinetic energy of the water film increases rapidly, along with the potential energy, showing the process of conversion of external work into internal energy of the water film. Thus, the rapid increase in kinetic energy marks the formation of liquid water. The sudden increase in kinetic energy at $\Delta z = -0.875$ Å shows that this is a first-order phase transition of a solid–liquid state. The change in the thickness of the water film at the critical point corresponds to a volume change of 6.6% and agrees well with the experimental result [6] that the volume changes by 6.4% during the bulk ice phase transition. The compression leads to the shortening of hydrogen bonds and the transfer of energy, which is responsible for the ice melting in Fig. 15.8. In the MD simulation, by averaging the forces of surface Pt atoms, we can find the pressure added to the Pt surface, which is also the value of the pressure added to the water film. By estimation, a compression of $\Delta z = -1$ Å corresponds approximately to a pressure of 0.5 GPa.

Using the obtained pressure P, temperature T, and volume V, we have plotted the graphical curve PV/nRT as a function of film thickness (solid line) in Fig. 15.9. It can be seen that this quantity increases almost linearly with the film thickness. The larger oscillations, because of the resulting oscillatory pressure, arise from the finite MD simulation time, and finite system size. The dashed line is obtained from the resolved equation of state, i.e., the equation of state for liquid water in the literature [7] (19). Despite some oscillations, the equation of state obtained from simulations in the whole phase transition region still agrees well with the analytic equation. In particular, it is important to note that for the ice film that is not compressed ($\Delta z =$ 0 Å), the two results agree, i.e., $PV/nRT \approx 0$. It is interesting to note that this analytical curve shows a weak inflection point near the phase transition point $\Delta z = -0.875$ Å. The agreement of the two curves proves the correctness of our pressure estimation method and that the equation of state for the bulk state essentially describes the confined water system here correctly.

In molecular dynamics simulations, the trajectories of the atoms can show the phase transition process of the film. Figure 15.10 shows typical trajectories of O (left panel) and H (right panel) in the water molecule before (black line) and after

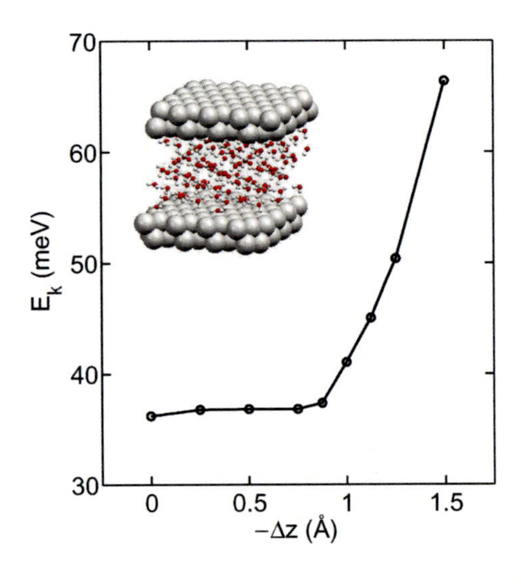

Fig. 15.8 The average kinetic energy of atoms as a function of water film thickness Δz. The kinetic energies correspond to temperatures of 280, 284, 285, 285, 289, 317, 348, 375, and 513 K. The inset shows the model used for the simulations

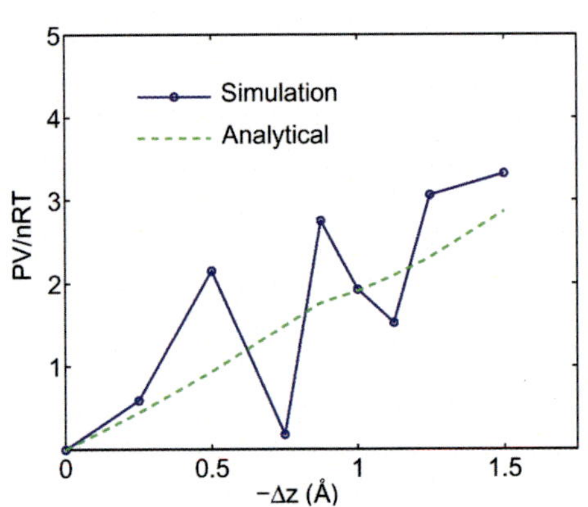

Fig. 15.9 The relationship between the physical quantity *PV/nRT* obtained from the simulation and the variation of the water film thickness Δz. The dashed line is the result obtained from the resolved equation of state

(gray line) the phase transition. They correspond to the systems with Δz = − 0.75 Å and Δz = − 1 Å, respectively, and are three-dimensional spatial trajectories of the atoms projected in the xy plane within 1.8 ps. After the phase transition occurs, the trajectories in the water molecules deviate to a large extent from their original position and become delocalized. Liquid water's size is approximately 2–3 times larger than ice in a single dimension of phase space. The diffusion coefficient D of H_2O can be obtained from the slope of the oxygen atom's mean square deviation with time. Before the ice melts (Δz = − 0.5 Å), D < 0.008 $Å^2$/ps; when the ice melts (Δz = − 1 Å) D reaches 0.5 $Å^2$/ps. The latter is in general agreement with the diffusion coefficient of liquid water at 300 K (0.24 $Å^2$/ps) [8]. In addition, the correlation of the H and O atomic trajectories in Fig. 15.10 shows that the water molecules in liquid water are rotating and vibrating frequently. The MD trajectories indicate that the motion of the water molecules in the bottom layer is more delocalized than that in the middle layer, which also indicates the layering effects of the confined water at the interface. The same phenomenon has been recently observed in the simulation of the Ag-water interface using ab initio MD [9]. Other properties of liquid water, such as the frequent breaking of hydrogen bonds and the formation process of new hydrogen bonds, have also been observed. The process in which one hydrogen bond is temporarily broken and rapidly recreated (black line) and the process in which another hydrogen bond is completely broken (gray line) are depicted in Fig. 15.11. Once again, the atomic trajectories and molecular diffusion demonstrate that the ice film undergoes a phase transition.

To quantitatively show the changes that occur in the water film structure under pressure induction, we obtained the pair correlation function (PCF) distributions of the atoms by averaging over the MD channels and plotted them in Fig. 15.12. The three plots each reflect the OO (top panel), OH (middle panel), and HH (bottom panel) pair correlation distribution functions when $\Delta z = 0, -1$, and -1.5 Å. For bulk water, the PCF is known, but the PCF for confined water films remains unclear. For comparison with the data for bulk water, we chose the middle two layers of the water bilayer for the analysis. The solid black line in Fig. 15.12 ($\Delta z = 0$) is very close to the PCF distribution of bulk ice, with a sharp peak corresponding to

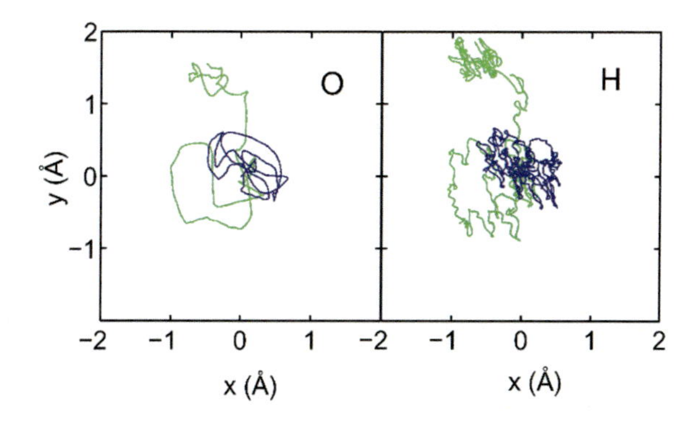

Fig. 15.10 Projection of the trajectories of the O and H atoms in the xy plane before (black line) and after (gray line) the phase transition in 1.8 ps

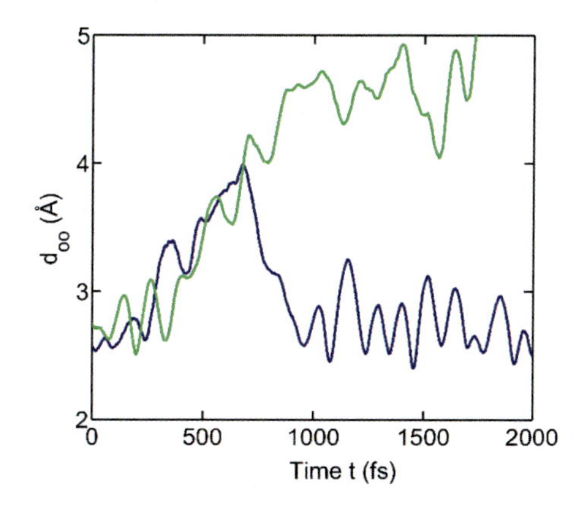

Fig. 15.11 Dynamics of hydrogen bond breaking and regeneration at $\Delta z = -1$ Å

the average bond length of Ice Ih structures. From $\Delta z = 0$ to -1.5 Å, all peaks of the PCF gradually broaden and shift, which also indicates changes in hydrogen bonding and the delocalization of the atomic motion orbitals in Fig. 15.10. The apparent change in the PCF is shown in the second peak of $g_{OO}(r)$ ($r = 4.8$ Å), which corresponds to the tetrahedral coordination characteristic of solid ice (Ice Ih) in water structure. This peak gradually shifts down during the phase transition and eventually disappears almost completely when $\Delta z = -1.5$ Å, i.e., when it becomes liquid water. At this point, the distribution of $g_{OO}(r)$ after the first peak is rather flat. Such a uniform distribution over long distances reflects the unique properties of hydrogen bonding dynamics in liquid water, where directional hydrogen bonding and coordination are broken. Apart from some very small peak shifts, the full character of the PCF curve in Fig. 15.12, that is, the positions and widths of the peaks, agrees well with those obtained experimentally with neutron scattering [10, 11] and X-ray diffraction [12]. The closest OO distance in the top graph is 2.7 Å, which is 6% smaller than the experimental value of 2.87 Å. This small difference stems from the effect of restriction and the shorter hydrogen bond lengths obtained by the USPP-PW91 method we used.

The effect of non-adiabatic compression on the water layers and the change of electronic states during the phase transition are discussed next. They are interconnected because the sudden compression of the ice layer generates a large force on the outermost layer of the water film in the direction perpendicular to the interface. This creates a non-equilibrium distribution of kinetic energy in the innermost and

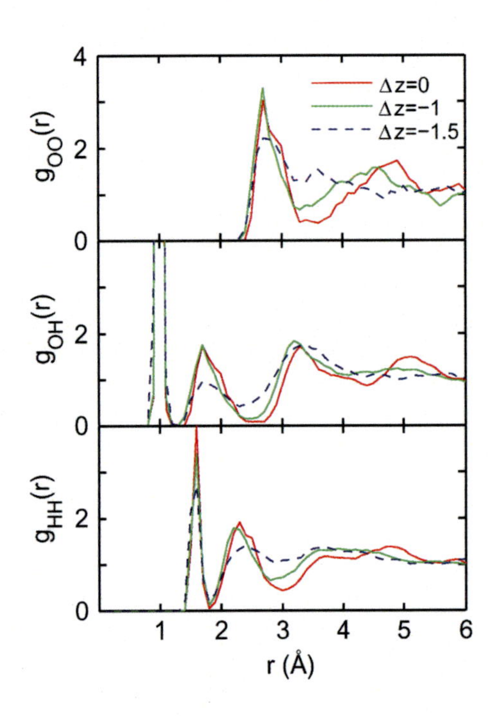

Fig. 15.12 Pair correlation function (PCF) distributions for OO, OH, and HH. The black, gray, and dashed lines correspond to the cases at $\Delta z = 0, -1$, and -1.5 Å, respectively

outermost layers of the film and requires some relaxation time to regain equilibrium. We are concerned with how the confined and compressed water film recovers its equilibrium distribution. This is partly a technical problem because it tells us whether the system is in nonequilibrium or equilibrium after a given period of MD simulation time. It is also a physical problem because it helps us to determine the non-adiabatic nature of the compression process. In principle, when the compression time is less than the relaxation time of the film, we should use a non-adiabatic and non-equilibrium dynamic system to deal with the compressive phase transition.

Figure 15.13 depicts the stratified kinetic energy distribution resulting from compressing the ice film 1 Å from its original equilibrium state. The time-dependent distribution of the kinetic energy fraction (E_{kxy}) parallel to the xy-plane and the vertical fraction (E_{kz}) along the z-axis is plotted in the time period of 3 ps. Initially, there is a sharp increase in kinetic energy in the vertical direction of the bottom (first) layer, due to the compression creating a large initial force in the vertical direction (for some other cases, this increase is seen in both the bottom and top layers). The kinetic energy of the other layers is affected relatively little. Afterward, this increase in kinetic energy in the z-direction is transferred to the parallel direction. After 2–3 ps, the kinetic energy of each layer returns to its equilibrium distribution, reaching 24 meV and 12 meV for E_{kxy} and E_{kz}, respectively. A similar study has been done for other states during the phase transition. Although the details differ slightly, the essence is the same: the non-adiabatic compression leads to a non-equilibrium distribution of kinetic energy that takes 3–5 ps to return to the equilibrium state. Very interestingly, the same phenomenon appears in previous classical MD simulations of pressure-induced melting of small clusters of ice [13], where gradually melting of ice into water from the outer to the inner layers was observed. The relaxation time of 3 ps is much shorter than the time scale of the mechanical response to any compression, so we conclude that the compression leading to the phase transition of the ice film is essentially an equilibrium process.

Finally, we discuss the effect of confinement and compression on the electronic structure of confined water. This can only be obtained from ab initio calculations. Figure 15.14 depicts the redistribution of the one-dimensional (1D) charge density along the z-axis. It is obtained by averaging the difference between the charge density of the whole system (Pt and water film) and the charge densities of Pt and water film alone in the same geometry in the parallel direction. The three curves describe the variation of the electron density of the ice film in equilibrium when confined ($\Delta z = 0$ Å); when stressed ($\Delta z = -1.0$ Å); and the variation of the electron density of liquid water obtained from MD simulations after a phase transition. The curves describing confined ice show the bonding features at the ice-Pt interface (left region) and the electron polarization effects arising at the interface with disproportionate contact (right region). In the region at the Pt–O interface, electrons are transferred from O to Pt, reflecting the $d_z{}^2$–p_z bonding feature. The interface on the right does not have any bonding features but shows electron polarization. Compression of the ice film 1 Å results in a shift of the peak position and leads to a large increase in the deformation charge density at the bonding interface. In contrast, the intermediate water layer remains almost unchanged. The redistribution of charge at the interface

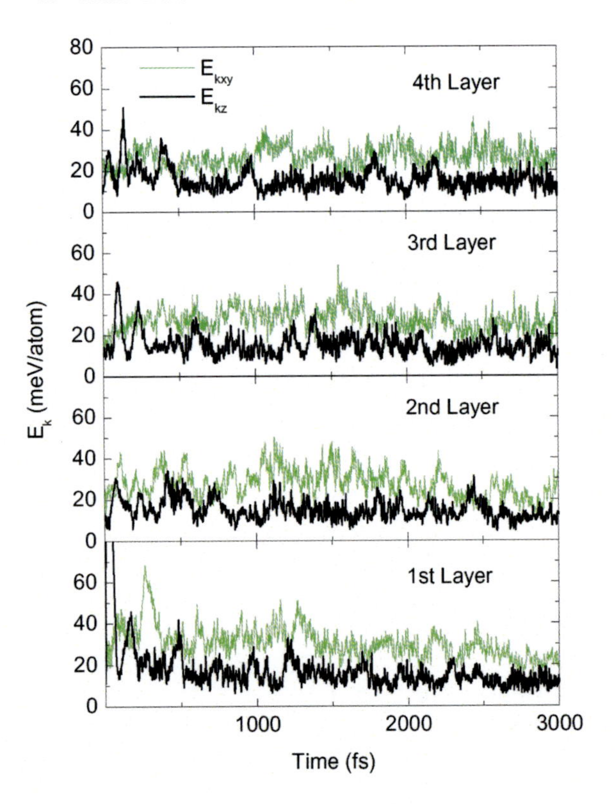

Fig. 15.13 Non-equilibrium kinetic energy distribution as a function of time due to compression of ice film 1 Å

is the driving force leading to the ice-water phase transition. When liquid water is generated, the charge distribution becomes smoother, especially in the intermediate water layer. This smoothing of charge distribution is not caused by the delocalization of atomic positions in liquid water. It mainly reflects the purely electronic state change in liquid water. Interestingly, the characteristics of the electronic structure at the interface do not change as the water moves from confined ice to compressed ice, and finally to a liquid water layer. This reflects the fact that the geometric structure at the interface does not change significantly during the phase transition. In other words, the confined water behaves like "supercooled water", which is consistent with the structural analysis results of previous theoretical [9] and experimental studies [14]. The redistribution of electrons in Fig. 15.14 shows the role of electronic structure and charge transfer in the confined, compressed, phase transition process of the water film. This information about the electron distribution cannot be obtained by classical molecular dynamics simulations, illustrating the importance of first-principles computational methods when dealing with dynamic processes in confined aqueous layers.

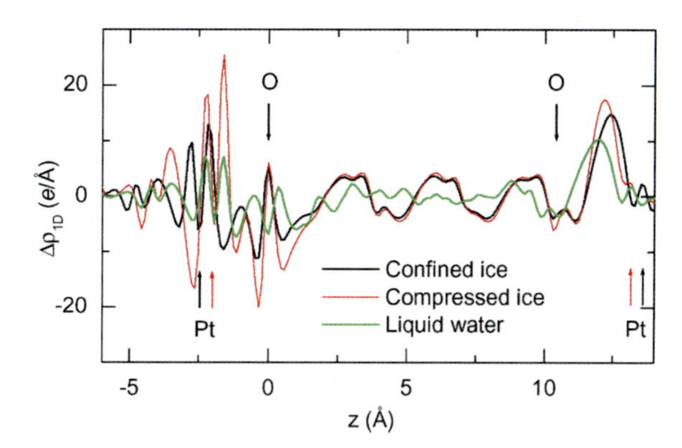

Fig. 15.14 One-dimensional differential charge density distributions in the confined ice film ($\Delta z = 0$, thick black line), the compressed ice film ($\Delta z = -1$ Å, thin black line), and the liquid water film obtained after melting (gray line). Arrows indicate Pt, O atomic positions

References

1. K. Koga, G.T. Gao, H. Tanaka, X.C. Zeng, Formation of ordered ice nanotubes inside carbin nanotubes. Nature **412**, 802 (2001)
2. R. Zangi, A.E. Mark, Monolayer ice. Phys. Rev. Lett. **91**, 025502 (2003)
3. S.H. Han, M.Y. Choi, P. Kumar, H.E. Stanley, Phase transitions in confined water nanofilms. Nat. Phys. **6**, 685 (2010)
4. S. Meng, E.G. Wang, S.W. Gao. Pressure-induced phase transition of confined water from ab initio molecular dynamics simulation. J. Phys. Cond. Matt. **16**, 8851 (2004)
5. D.R. Hamann, H_2O hydrogen bonding in density-functional theory. Phys. Rev. B **55**, R10157 (1997)
6. V.R. Brill, A. Tippe, Gitterparameter von eis i bei tiefen temperature. Acta Crystallogr. **23**, 343 (1967)
7. C.A. Jeffery, P.H. Austin, A new analytic equation of state for liquid water. J. Chem. Phys. **110**, 484 (1999)
8. K. Laasonen, M. Sprik, M. Parrinello, R. Car, Ab-initio liquid water. J. Chem. Phys. **99**, 9080 (1993)
9. S. Izvekov, G.A. Voth, Ab initio molecular dynamics simulation of the Ag(111)-water interface. J. Chem. Phys. **115**, 7196 (2001)
10. A.K. Soper, Orientational correlation-function for molecular liquids—the case of liquid water. J. Chem. Phys. **101**, 6888 (1994)
11. A.K. Soper, The radial distribution functions of water and ice from 220 to 673 K and at pressures up to 400 MPa. Chem. Phys. **258**, 121 (2000)
12. A.H. Narten, H.A. Levy, Liquid water: molecular correlation functions from X-ray diffraction. J. Chem. Phys. **55**, 2263 (1971)
13. T.A. Weber, F.H. Stillinger, Pressure melting of ice. J. Chem. Phys. **80**, 438 (1984)
14. M.F. Reedijk, J. Arsic, F.F.A. Hollander, S.A. de Vries, E. Vlieg, Liquid order at the interface of KDP crystals with water: evidence for icelike layers. Phys. Rev. Lett. **90**, 066103 (2003)

Chapter 16
Summary and Prospect

"H_2O: O determines its structure, H determines its dynamics."

In this world, water is abundant, important, and wondrous, it is also one of the current interdisciplinary research hotspots of physics, chemistry, biology, ecology, and geography. This book attempts to emphasize that an accurate description of molecular interactions is the key to understanding and unraveling the peculiar properties of water. On the one hand, this is obvious that the macroscopic phenomena of water depend on its microscopic mechanisms, and on the other hand, there are many behaviors of confined water at the molecular scale, that are different from the bulk state, which can only be explained rationally by studying the interactions between molecules. In addition, the interaction between water and the outside usually occurs mediated to the surface. The focus of this book is therefore on theoretical and experimental advances in the study of interactions between water molecules and between water and surfaces from a basic science perspective, with a particular emphasis on the use of first-principles calculations to understand the interactions between water and different surfaces at the atomic and even electronic level.

The book begins with a brief introduction to some peculiar properties of water and the need for fundamental research at the molecular scale, pointing out that an accurate understanding and precise description of the hydrogen bonding interactions between water molecules is the key to revealing a range of abnormal properties of water (more than a dozen phases, thermal shrinkage and cold expansion, hot water cooling fast, great heat capacity, critical points, etc.) and the mechanisms by which water interacts with various external surfaces.

The molecular structure of water is composed of two H atoms and one O atom in a V-shape, the bond length and bond angle are 0.9572 Å and 104.52° respectively; the electronic structure of water is a tetrahedral configuration composed of sp^3 hybridization, with two unbonded lone pairs of electrons, they are easy to interact with the outside world to form hydrogen bonds or other chemical bonds. These properties of the water molecule determine the form and pattern of its interaction with the outside world.

S. Meng and E. Wang, *Water*,
https://doi.org/10.1007/978-981-99-1541-5_16

Two water molecules combine to form a stable dimer structure. The two water molecules are connected by hydrogen bonds with an OO spacing of 2.98 Å. The OH bond forming the hydrogen bond is elongated by 0.01 Å and the energy of action is about 250 meV. The water dimer structure is a representative model for the study of hydrogen bonds and has also made a non-negligible contribution to the absorption of microwave radiation in the atmosphere. Density functional theory based on quantum mechanics can predict the structure and energy of water molecules and dimers essentially correctly; the newly developed charge transfer model also improves the accuracy of the traditional classical model to describe water dimers. Water molecules further condense to form cyclic trimers, tetramers, pentamers, and hexamers, as well as a variety of sub-stable structures, whose atomic structures have been accurately determined by means of vibrational rotation tunneling spectroscopy. Larger water clusters adopt a three-dimensional structure and gradually transition to the bulk phase, the Ih hexagonal ice structure, forming the ice, frost, and snowflakes common in nature.

Theoretical descriptions and calculations of water are currently performed at three levels: the continuum media model (which ignores details of molecular interactions and is used in traditional fluid dynamics and some thermodynamic studies), the classical model, and first-principles accurate calculations. The latter two are the cornerstones for conducting theoretical studies in the basic science of water. In practice, classical models allow molecular dynamics simulations of systems of thousands to millions of water molecules up to the microsecond scale, and thus are used to understand fundamental problems such as liquid nature, confined phase transitions, critical phenomena, surface infiltration, etc., but the numerical results depend on the choice of models, each with several to tens of empirical parameters. No classical model can give a comprehensive description of water properties consistent with experimental results. First-principles calculations enable a quantum mechanical understanding of water and its effects, but the system size, simulation time, and accuracy are limited by current computational resources and methods. Various commonly used exchange–correlation functionals in density functional theory can describe hydrogen bond interactions basically, but with varying accuracy, needing further development, and comparison with quantum chemistry and quantum Monte Carlo methods. There is currently considerable progress in van der Waals force calculations and rapid developments in nuclear quantum effects and excited state dynamic properties of hydrogen in water. Large-scale accurate calculations are the focus of the next theoretical method development.

Because of the "fragile" of the water interaction system, traditional experimental methods are often not convenient for studying water. Precise experimental studies of water and surface interactions require high vacuum, low-temperature experimental environments, and surface energy spectroscopy analysis methods including photoelectron spectroscopy, electron energy loss spectroscopy and low energy electron diffraction can be used to measure the structure of surface water and its electronic structure. Recent developments in scanning probe techniques, particularly scanning tunneling microscopy and atomic force microscopy, have been successfully used to study the details of the interaction between water and surfaces such as metals, semiconductors, and thin insulators at the molecular scale. Using the angstrom-scale spatial resolution of scanning probes, and the femtosecond-scale temporal resolution of ultrafast lasers, the combination of the two may enable simultaneous ultra-high spatial and temporal resolution in surface water studies, revealing mechanisms such as water photocatalytic decomposition, hydrogen bond recombination, and proton transfer. Further directions for water surface science are comparisons with real systems at higher pressures and temperatures. For the study of bulk and surface water at room temperature and pressure, nonlinear optical techniques (e.g., sum-frequency vibrational spectroscopy), synchrotron spectroscopy and imaging, and neutron scattering methods that are particularly sensitive to hydrogen elements have unique advantages and are among the few experimental methods that can give the structure and electronic and vibrational properties of water. Using these methods, "room temperature ice" on mica surfaces and the ordering of water molecule orientation at the surface have been discovered. Due to the extreme importance and sensitivity of water research, effective experimental methods are still lacking and new experimental techniques for basic research need to be vigorously developed.

There have been significant advances in the field of experimental research on surface water in the last decade, with a series of important results. Using scanning tunneling microscopy, it has been possible to observe not only individual water molecules on a surface, but also the structure of individual water molecules as they diffuse and condense into clusters up to ice directly. And recent studies have also achieved orbital imaging of individual water molecules, i.e., the observation of the internal degrees of freedom of water molecules—molecular orbitals. As a result, the direction of hydrogen bonds in water clusters and two-dimensional ice layers can also be discerned using this method. Water molecules diffuse rapidly on the surface, forming small clusters of multimers with six-membered rings as the basic structure, and they can furthermore form petal-like and band-like structures that lie between the zero-dimensional clusters and the two-dimensional ice layer, as well as various one-dimensional chain-like structures. High-resolution imaging of the two-dimensional ice layer reveals not only a large number of six-membered rings, but also five-membered ring to seven-membered ring hydrogen bond structures in the surface infiltrated water layer, and even a large number of Bjerrum hydrogen bond defects in the mixed layer of water and OH roots, which completely breaks the rules of surface ice construction envisioned in the 1980s.

In order to obtain a basic theoretical understanding of the molecular-scale picture of water-surface interactions, the interaction of water and Pt(111) surfaces is discussed in this book as an example of a typical system. Density functional calculations show that water molecules are adsorbed at the top position, basically lying flat, and the angle θ between the molecular plane and the surface is 13°; the barrier of H_2O molecules turnover is 140–190 meV, while the azimuthal angle φ around the axis perpendicular to the surface can rotate almost without obstruction. The water molecule diffuses via the bridge position, corresponding to a potential barrier of about 0.17 eV. The lone pair of electrons of O in water, especially the $3a_1$, $1b_1$ orbitals, are coupled with the d electron band (mainly d_{xz} and d_{z^2}) on the Pt(111) surface and cause a transfer of about 0.02 electrons from the water molecule to the surface. As a result of the charge transfer, the OH bond length of the water molecule elongates by 0.006 Å and the HOH bond angle expands by 0.8°.

If more water molecules are adsorbed on the Pt(111) surface, the water forms a small cluster structure similar to the free case. Each water molecule is adsorbed to the top site and lies as flat as possible to allow the strongest interaction between metal and water. The folded hexagonal ring arrangement of $(H_2O)_6$ is the most stable, in agreement with experimental observations. The hydrogen bond in the bimolecular $(H_2O)_2$ is strongest with an energy of about 450 meV, indicating that the hydrogen bond is greatly strengthened due to adsorption. On the <110>/{100} step represented by the Pt(322) surface, only the sawtooth one-dimensional (1D) H_2O chains are stable, with weak hydrogen bonds; however, the absorption of single molecules on the step is strong.

At higher coverings, water forms bilayer and multilayer structures. Three water layer structures have been found on Pt(111): bilayer ice $\sqrt{3} \times \sqrt{3}R30°$ ($\sqrt{3}$) very similar to the bulk ice (Ice Ih), and more complex periodic structures $\sqrt{39} \times \sqrt{39}R16.1°$ ($\sqrt{39}$) and $\sqrt{37} \times \sqrt{37}R25.3°$ ($\sqrt{37}$). First-principles total energy calculations and molecular dynamics simulations reveal that H-up and H-down have similar $\sqrt{3}$ bilayer structures, with almost energy degenerate and a potential barrier of 76 meV between them, so that in practice both states are possible. The semi-decomposed structure proposed by Feibelman is not dominant on the Pt(111) surface. As the coverage increases, the distance between the bottom H_2O molecule of the bottommost bilayer and the surface decreases. In structures with 1 to 6 bilayers, the distances are 2.69, 2.63, 2.56, 2.49, 2.52, 2.47 Å. With increasing coverage, at 3 bilayers, the $\sqrt{3}$ phase becomes more stable than the other two phases. This trend is consistent with the experimental results.

By extension, this book also discusses the adsorption of water on the densely packed surfaces of transition metals and noble metals such as Rh, Pd, and Au, and open metal surfaces such as Cu(110), as well as the interactions between water and the surfaces. The single-molecule adsorption conformation and the structure of bilayer ice are essentially similar to the situation on Pt(111). We found a general pattern of water adsorption on the surfaces of such materials. From Ru, Rh, Pd, Pt, to Au, as the metal d electronic state occupies and the atomic radius increases, or as the chemical reactivity of the metal element decreases, the water molecule adsorbs at increasingly higher distances (from 2.28 to 2.67 Å), with smaller increases in bond lengths and bond angles, smaller adsorption energies (from 409 to 105 meV), and the thickness of the ice bilayer becomes smaller, but the upper molecules remain essentially constant in height (at 3.40 Å in the H-up structure and 3.20 Å in the H-down structure). These phenomena originate from the interaction of water with the d-band electrons on these surfaces. The aqueous layer decomposition adsorption and molecular adsorption energies on the open Cu(110) surface are very close to each other and are dividing lines of two water adsorption states. In summary, we can conclude that on metal surfaces, the water structure forms chemical bonds through the lone pair electrons of O and the substrate electrons (especially the surface state d electrons); this water-surface interaction is rather localized, concentrating mainly on the underlying water molecules that are directly bonded to the surface.

Further, this book discusses the interactions of water with some representative insulators. On simple metal oxides such as MgO, TiO_2, and ZnO surfaces, water readily decomposes with increasing coverage or temperature, or is in a transition dynamic of decomposition and non-decomposition. Light can facilitate the water decomposition process, but micromechanical studies are just beginning, which is important for practical applications such as energy conversion and pollution removal. Silicon dioxide is a model for common glass, gravel, and other surfaces. Silicon dioxide interacts strongly with water through surface OH groups, with single-molecule adsorption energies around 700 meV, and forms several hydrogen bonds with the surface, but does not decompose. The water molecules on silicon dioxide have a strong orientation effect, resulting in a "mosaic ice" structure with a special tetragonal-octagonal hydrogen bond network. The interaction between water and the model carbon surface—graphene is highly debated, and the measured infiltration angle varies between 30°–90°. Accurate calculations on this system not only reveal the basic interaction of water with the carbon system, but also accurately calibrate various first-principles theoretical calculations. The final quantum Monte Carlo approach gives the strength of the water-graphene interaction at around 80 ± 15 meV.

The adsorption of water on the surface usually results in enhanced intermolecular hydrogen bonds due to the charge transfer between water and the surface, and the automatic regulation of its structure by the small clusters at the surface and the water layer in the adsorption environment. Hydrogen bonds in the adsorbed state bilayers are much stronger than in the free bilayers. In particular, there are two hydrogen bonds of different strengths in the bilayer structure of the molecular state: the top water molecule contributes one hydrogen atom, which forms a strong hydrogen bond with the neighboring water molecule, while the bottom water molecule contributes two hydrogen atoms, which form two weaker hydrogen bonds with the neighboring water molecule. These hydrogen bonds correspond to OH vibrational energies that differ by nearly 40 meV, around 424 and 384 meV, respectively. This finding has been confirmed by relevant experiments. Moreover, the disorder of H pointing at the surface has little effect on the hydrogen bond strength.

From the molecular scale, macroscopic surface hydrophilic and hydrophobic phenomena are determined by surface-water interactions, as well as by competition of hydrogen bond interactions between water and water. Further studied from the perspective of electronic interactions, they are ultimately determined by localized charge transfer and longer-range charge polarization interactions. By analyzing the adsorption energy and hydrogen bond energy of the adsorbed water structure, we give a microscopic quantity to describe the hydrophilicity of the surface: $\omega = E_{HB}/E_{ads}$, which is the ratio of the hydrogen bond energy to the adsorption energy (of a single water molecule). The smaller ω is, the more hydrophilic the surface is. $\omega = 1$ is a rough cut-off between the hydrophilic and hydrophobic regions. Using this approach, we obtain the order of surface hydrophilicity as Ru > Rh > Pd > Pt > Au, which is the same order as their d-band electron filling. In particular, it is worth pointing out that Pt(111) is a hydrophilic surface and Au(111) is a hydrophobic surface, which is consistent with the experimental results.

To compensate for the lack of experimental understanding of molecular structure and action details, this book proposes a vibrational identification approach. By comparing the vibrational spectra measured in experiments (HREELS, IRAS, and SFG, etc.) with those obtained from theoretical calculations of each known molecular structure, one can identify the molecular configuration and interaction dynamics of the actual structures. This provides a bridge between experimental and theoretical work. In particular, OH stretching vibrations are very sensitive to hydrogen bond information in hydrogen bond structures such as water on the surface, and the identification of OH vibrations can be a simple and universal method for identifying the structure of hydrogen bond networks.

Applying this approach, we found two hydrogen bonding modes in the water bilayer on the Pt(111) surface and, by comparison with experimental data, concluded that the D_2O water layer on Ru(0001) does not decompose half of the molecules. The experimental vibrational spectrum and the molecular state of the bilayer structure are in the best agreement. The reason is that although the adsorption energy of the semi-decomposed structure is much lower, they are already detached from the surface before the water molecules have reached the decomposed state. The water layer structure on Ru(0001) may decompose only when the intensity of the probe photon beam or electron beam is high. Both non-decomposed and half-decomposed water layer structures were observed in the experiments.

Applying the knowledge already gained about the interaction of water and solid surfaces, water and water at surfaces, this book further gives some dynamic processes of water at surfaces. For example, the process by which K and Na ions form a two-dimensional hydration ring at the graphite surface. Unlike the three-dimensional structure in free clusters and bulk solutions, this structure arises from the 2D confinement of the surface and the charge transfer between ions and the surface. For K and Na, the first hydration loop layer has three and four water molecules, respectively. As the number of water molecules increases, the water gradually transforms from a 2D structure to a 3D hydration layer. The K ions remain on the surface and form a hemispherical 3D hydration circle around them, while the Na ions gradually move away from the surface until they are completely dissolved in the water. Their difference is determined by the competition of hydration and ion-surface interactions. The stronger interaction of the K-surface and the Na-water molecules results in different dissolution dynamics.

The molecular process of salt dissolution and nucleation is also interesting. Microscopically, the salt surface is quite inert to the water layer, with dissolution occurring only at surface steps and corners, and in the order $Cl^- \rightarrow Na^+ \rightarrow Cl^- \rightarrow Na^+ \rightarrow \ldots$ starting with Cl^- ions. Nucleation proceeds in the opposite way, with Na^+ deposited more readily on the surface (in greater numbers), but with more Cl^--centered nuclei (Cl^- deposited first). At room temperature and moderate supersaturation, the critical nucleation size is 3 atoms. In the early stages of nucleation, the number of Na^+ ions is much more than that of Cl^- (ratio $\approx$ 9:5), suggesting that, in addition to chemical potential differences, the nonequilibrium Coulomb attraction of charged nuclei may also be an important driving force for nucleation.

The surface of ice is critical for environmental issues such as cloud formation and ozone catalysis. Unexpectedly, unlike the random distribution of OH orientation in the bulk phase, the OH orientation at the ice surface is very ordered and does not undergo an order-to-disorder transition at the ice melting temperature. The ordered nature of the ice surface greatly affects the formation of water molecular vacancies at the ice surface, resulting in a large range and overall small distribution of surface vacancy formation energy; it also greatly promotes the adsorption of polar molecules such as H_2O and H_2S, which has important implications for environmental chemistry studies.

Finally, this book discusses two challenging issues in the field of water research: nuclear quantum effects and phase transitions. Due to the small mass of the H atom, which is only 1836 times that of the electron, nuclear quantum effects can be more easily demonstrated in the formation of structures and dynamical changes in water molecules. Theoretical results show that molecular dynamics simulations that take into account nuclear quantum effects give more accurate structural distributions of liquid water, and H_3O^+ and OH^- transport behavior. Under constrained conditions such as water chains in nanotubes, the H quantum effect is more pronounced, leading to a non-local distribution of H^+ on five water molecules. On metal surfaces, experimental and theoretical studies reveal that H quantum effects lead to a more decomposable H_2O layer than the D_2O aqueous layer. Quantum effects on molecular rotation lead to easier interconversion of the two ice bilayer structures, H-up and H-down; and the diffusion of water dimers on the Pd(111) surface is 10,000 times faster than that of single molecules at low temperatures. First-principles molecular dynamics simulations reveal that nuclear quantum effects make strong hydrogen bonds stronger and weak hydrogen bonds weaker.

Phase transition is a difficult problem in water research. On one hand, the phase transition simulations are theoretically very computationally intensive and require accurate interaction models, and on the other hand, certain extreme conditions are extremely difficult to achieve experimentally (e.g., the "no man's land" of liquid water at 150–230 K). Under the spatial constraints of nanotubes, not only novel phase structures such as tetragonal and pentagonal icicles were discovered, but also a possible critical point of solid–liquid phase transition in this system—a point at which there is no difference between solid and liquid. The restriction of the two-layer flat plate also leads to new structures such as single-layer ice, double-layer ice, and transitions between them. We have also studied the phase transition process and the mechanism of electron interaction in the melting of a thin layer of ice confined between two metal plates into liquid water at a certain pressure (0.5 GPa). The volume of the ice-thin layer is compressed by 6.6% during the phase transition, which is consistent with that of bulk ice. The state equation of the thin layer water, the structure of water, molecular diffusion, and the dynamics of hydrogen bond breaking and resynthesis near the point of phase transition are obtained. The compression effect is found to first cause an increase in the kinetic energy of the water layer at the interface in the vertical direction, and then the increased kinetic energy is transferred to the parallel direction and the inner water layer, and re-equilibrates within 3–5 ps. The compression causes a redistribution of electrons between the surface and water.

As the ice melts into liquid water, the electrons at the interface remain in an ordered distribution similar to that in ice, while those in the inner water layer are uniformly distributed, losing the polarization distribution characteristic of the ice layer.

To recapitulate these results, this book mainly summarizes the progress of theoretical and experimental studies of interactions and dynamics between water molecules, between water and surfaces on the molecular scale, especially images obtained from the perspective of quantum mechanical first principles. These studies are mainly concerned with water molecules, small clusters, and bulk phase water, as well as their adsorption, infiltration, ordered-disordered transitions, co-adsorption, and other important phenomena on surfaces such as transition metals, oxides, graphite, and table salt. A general conclusion is obtained that looking at water from the molecular level, O determines its structure and H determines its dynamic. Since for a water molecule, the O atom is 16 times more massive than H (8 times more massive than D), the position of the O atom is essentially the center of gravity of the water molecule. The H atom is tightly bound to O through OH covalent bonds. And because the H atom is lighter, when free it often rotates around O without a defined position, such as the H in a single H_2O molecule on Pt(111). And when forming a hydrogen bond, the H is between two adjacent O atoms. More importantly, the arrangement of O in bulk ice is periodically ordered, while the position of H is often disordered. Knowing the position of the O atom gives you a rough idea of the position of the water molecule and the structure of the entire water system.

Hydrogen bonds formed by H atoms are the source of almost all of the peculiar properties of water. Water molecules are bonded to each other by hydrogen bonds, and the O atom with which H forms a hydrogen bond determines the orientation of the water molecule. The number and strength of the hydrogen bonds formed determine the stability of the entire water structure. The competition between hydrogen bond and water-external interactions produces all sorts of interesting phenomena such as surface hydrophilicity and hydrophobicity, melting of ice, and dissolution of salt. Moreover, the vibration of H relative to O, that is, OH vibration, is very sensitive to the hydrogen bond between water molecules, making our molecular-scale identification of the water structure effective and easy. Also, the disorder of H orientation in the bulk phase water structure, the order of H orientation on the ice surface, the decomposition of water molecules by breaking OH bonds, and the transport of H_3O^+ and OH^- on the surface and in the nanotubes are mediated by the motion of H atoms. So we say: H determines the dynamic of water molecules and water structure.

Looking forward to future research in basic water science, we believe that the molecular picture of water-surface interactions will continue to be a very compelling and colorful area, with many unanswered questions. This requires more in-depth and detailed theoretical and experimental work, but also provides excellent opportunities for basic research and scientific and technological applications. First, many of the current suspenseful questions are about the position and dynamics of H in water, such as whether Pt(111), Rh(111), etc. surfaces are H-up or H-down double layers? What is the ratio and conversion of the two? What is the molecular structure of the undecomposed water layer on Ru(0001)? Where is the H located after decomposition? How does it resynthesize water in the desorption experiment? Is the rapid diffusion of the double water molecule on Pd(111) caused by the quantum motion of H? How do the disordered distribution of H in ice and the ordered orientation of the surface affect the structure and phase transition? What peculiar properties does confined water exhibit? In what cases the quantum nature of H must be taken into account? etc. We need to further develop theoretical methods, such as simple calculations on the quantum dynamics of H and even the whole water molecule; and also to improve the existing experimental methods (many current experimental means of probing structures, such as low-energy electron diffraction and X-ray diffraction, are insensitive to H) to be used to directly study the position and dynamic processes of H in water.

Second, the adsorption of water on the surfaces of noble metals and the co-adsorption with alkali metal atoms and O atoms is relatively well understood, but the adsorption of water on the surfaces of stronger or weaker reactive materials is more complex and is still poorly studied. The next step of research should include the adsorption, co-adsorption, and reaction dynamics of water on the surfaces of complex oxides, semiconductors, salts, etc.

Third, due to the development of nanoscience, the study of interactions between water and nanoscale materials (e.g., small salt particles, self-assembled quantum dots on surfaces, surface alloys, two-dimensional atomic crystals, etc.) will be important and interesting.

Fourth, the study of surface/interface interactions between water and biological systems will be important for understanding life processes at the molecular level and will be one of the hotspots for future surface science research.

Fifth, the system of water and surfaces in a solution environment, at room temperature and pressure, needs more research with effective methods; how it behaves differently from the system in a high vacuum experimental environment needs more attention.

Finally, the properties of the interaction between water and surfaces in non-ground states, such as electron excited states, charged water clusters, and other systems will be a new hot spot in the field of water-surface research in the future. All these works are yet to be realized by further close collaboration between experiment and theory.

Printed in the United States
by Baker & Taylor Publisher Services